List of tables

KV-556-503

4. Trends in household expenditure

Content

Carshalton College

00032223

Learning Centre
Carshalton College
Nightingale Road
Carshalton, SM5 2EJ
Tel: 020 8544 4344
This book is due for return on or before the last date shown below.

R 314.2

1: Overview

2: Housing expenditure

3: Equivalised income

4: Trends in household expenditure

5: Impact of the recession on household expenditure

Appendix A:

Household expenditure tables can be accessed by clicking on the table names in the appendix

Appendix B: Methodology

5. Impact of the recession on household expenditure

Appendix A

Household expenditure tables can be accessed by clicking on the table names in the appendix

Appendix B: Methodology

List of figures

Overview

Housing expenditure

Equivalised income

Trends in household expenditure over time

Impact of the recession on household expenditure

Symbols and conventions used in this report

[ ]   Figures should be used with extra caution because they are based on fewer than 20 reporting households.

..   The data is suppressed if the unweighted sample counts are less than 10 reporting households.

-            No figures are available because there are no reporting households.

Rounding: Individual figures have been rounded independently. The sum of component items does not therefore necessarily add to the totals shown.

Averages: These are averages (means) for all households included in the column or row, and unless specified, are not restricted to those households reporting expenditure on a particular item or income of a particular type.

Period covered: Calendar year 2010 (1 January 2010 to 31 December 2010).

List of contributors

Editor:	Giles Horsfield
Authors:	Alice Jefford
	Emma Jarvis
	Laura Keyse
	Louise Skilton
	Paul Hossack
	Suzanne Fry
Living Costs and Food Survey team:	Karen Watkins
	Linda Williams
	Scott Symons
	Sian Wilson
	Tracy Lane
	Chris Payne
	Rebecca Ayres
	Field Team and Interviewers
	Coders and Editors
Reviewers:	Andrew Barnard
	Ed Dunn
	Gareth Clancy
	Richard Jones
	Felix Ritchie

Acknowledgements

A large scale survey is a collaborative effort and the authors wish to thank the interviewers and other ONS staff who contributed to the study. The survey would not be possible without the co-operation of the respondents who gave up their time to be interviewed and keep a diary of their spending. Their help is gratefully acknowledged.

Introduction

This report presents the latest information from the Living Costs and Food Survey for the 2010 calendar year (January to December). The Expenditure and Food Survey (EFS) was renamed as the Living Costs and Food Survey (LCF) in 2008 when it became a module of the Integrated Household Survey (IHS).

The current LCF is the result of the amalgamation of the Family Expenditure and National Food Surveys (FES and NFS). Both surveys were well established and important sources of information for government and the wider community, charting changes and patterns in Britain's spending and food consumption since the 1950s. The Office for National Statistics (ONS) has overall project management and financial responsibility for the LCF while the Department for Environment, Food and Rural Affairs (DEFRA) sponsors the specialist food data.

The survey continues to be primarily used to provide information for the Retail Prices Index; National Accounts estimates of household expenditure; the analysis of the effect of taxes and benefits, and trends in nutrition. However, the results are multi purpose, providing an invaluable supply of economic and social data.

The 2010 survey

In 2010 5,116 households in Great Britain took part in the LCF survey. The response rate was 50 per cent in Great Britain and 59 per cent in Northern Ireland. The fieldwork was undertaken by the Office for National Statistics and the Northern Ireland Statistics and Research Agency.
Further details about the conduct of the survey are given in Appendix B.

The format of the Family Spending publication changed in 2003/04 so that the tables of key results which were found in the main body of the report are now in Appendix A. This year's report includes an overview chapter outlining key findings, two detailed chapters focusing upon expenditure on housing and the impact of equivalising income when calculating results, a fourth chapter looking at trends in household expenditure over time and finally a chapter on the impact of the recession on household expenditure.

Data quality and definitions

The results shown in this report are of the data collected by the LCF, following a process of validation and adjustment for non-response using weights that control for a number of factors. These issues are discussed in the section on reliability in Appendix B.

Figures in the report are subject to sampling variability. Standard errors for detailed expenditure items are presented in relative terms in Table A1 and are described in Appendix B. Figures shown for particular groups of households (for example income groups or household composition groups), regions or other sub-sets of the sample are subject to larger sampling variability, and are more sensitive to possible extreme values than are figures for the sample as a whole.

The definitions used in the report are set out in Appendix B, and changes made since 1991 are also described. Note particularly that housing benefit and council tax rebate (rates rebate in Northern Ireland), unlike other social security benefits, are not included in income but are shown as a reduction in housing costs.

Income and Expenditure Balancing

The LCF is designed primarily as a survey of household expenditure on goods and services. It also gathers information about the income of household members, and is an important and detailed source of income data. However, the survey is not designed to produce a balance sheet of income and expenditure either for individual households or groups of households. For further information on the balancing of income and expenditure figures, see 'Description and response rate of the survey', page 223.

Related data sources

Details of household consumption expenditure within the context of the UK National Accounts are produced as part of Consumer Trends (www.ons.gov.uk/ons/publications/all-releases.html?definition=tcm%3A77-23619). This publication includes all expenditure by members of UK resident households. National Accounts figures draw on a number of sources including the LCF: figures shown in this report are therefore not directly comparable to National Accounts data. National Accounts data may be more appropriate for deriving long term trends on expenditure.

More detailed income information is available from the Family Resources Survey (FRS), conducted for the Department for Work and Pensions. Further information about food consumption, and in particular details of food quantities, is available from the Department for Environment, Food and Rural Affairs, who are continuing to produce their own report of the survey (www.defra.gov.uk/statistics/foodfarm/food/familyfood/).

In Northern Ireland, a companion survey to the GB LCF is conducted by the Central Survey Unit of the Northern Ireland Statistics and Research Agency (NISRA). Households in Northern Ireland are over-sampled so that separate analysis can be carried out, however these cases are given less weight when UK data are analysed.

Additional tabulations

This report gives a broad overview of the results of the survey, and provides more detailed information about some aspects of expenditure. However, many users of LCF data have very specific data requirements that may not appear in the desired form in this report. The ONS can provide more detailed analysis of the tables in this report, and can also provide additional tabulations to meet specific requests. A charge will be made to cover the cost of providing additional information.

The tables in Family Spending 2010 are available as Excel spreadsheets.

Anonymised microdata from the Living Costs and Food Survey (LCF), the Expenditure and Food Survey (EFS) and the Family Expenditure Survey (FES) are available from the United Kingdom Data Archive. Details on access arrangements and associated costs can be found at www.data-archive.ac.uk or by telephoning 01206 872143.

Overview

This chapter presents the key findings of the 2010 Living Costs and Food Survey (LCF), formerly the Expenditure and Food Survey. The chapter provides an overview of household income and expenditure, characterised by different household types and regions, as well as a summary of the ownership of a limited range of durable goods.

All of the tables (except Table 1.1) referred to in this chapter can be found in Appendix A of the report (page 125).

Household expenditure

Table 1.1 shows total weekly household expenditure in the United Kingdom (UK) by the 12 Classification Of Individual COnsumption by Purpose (COICOP)[1] categories. In 2010 average weekly household expenditure in the UK was £473.60, £18.60 more than in 2009 when it was £455.00, reversing the change seen in the previous year. As in previous years, spending was highest on transport at £64.90 per week. This was £6.50 more than in 2009, a large increase of 11 per cent.

Housing, fuel and power (£60.40) and recreation and culture (£58.10) were the categories with the next highest expenditure. Housing, fuel and power saw an increase of £3.10 making it the second highest category, now higher than recreation and culture which saw only a small increase of 20p. The average weekly expenditure on food and non-alcoholic drinks in 2010 was £53.20 per week.

Table 1.1   **Expenditure by COICOP category and total household expenditure, 2010**
United Kingdom

COICOP category	£ per week
Transport 64.	90
Housing, fuel and power	60.40
Recreation and culture	58.10
Food and non-alcoholic drinks	53.20
Restaurants and hotels	39.20
Miscellaneous goods and services	35.90
Household goods and services	31.40
Clothing and footwear	23.40
Communication	13.00
Alcoholic drinks, tobacco and narcotics	11.80
Education	10.00
Health	5.00
Total COICOP expenditure	406.30
Other expenditure items	67.30
Total expenditure	473.60

Totals may not add up due to independent rounding of component categories.

Of the £64.90 spent on transport each week, approximately half (49 per cent) was spent on the operation of personal transport (£33.30), see Table A1. This was an increase of 14 per cent on the previous year, in line with the overall increase observed in spending on transport. As in 2009 petrol, diesel and other motor oils (£21.60 per week) was the largest expenditure in the operation of personal transport category, an increase from £19.20 in 2009. Households spent on average £12.10 per week on transport services, including rail, tube and bus fares, compared to £9.60 in 2009, however the purchase of vehicles, remained the same at £19.50 per week in 2010.

Approximately a third (31 per cent) of spending on recreation and culture (£17.80 per week) was spent on recreational and cultural services: sports admissions, leisure class fees and equipment hire accounted for £5.30 per week; cinema, theatre and museums etc (£2.40 per week); TV, video, satellite rental, cable subscriptions and TV licenses (£6.00 per week); and gambling payments (£2.70 per week).

Average weekly spend on package holidays fell from £13.20 per week in 2009 to £12.60 in 2010. Of this, £11.60 was spent on holidays abroad, £0.70 less than in 2009. Spending on audio-visual, photographic and information processing equipment (£7.20 per week) remained relatively constant, while spending on other recreational items and equipment, gardens and pets increased to £11.40 per week. The average weekly spend on TV, video and computers increased to £5.20. (Table A1).

Of the £53.20 average weekly spend on food and non-alcoholic drinks , £7.10 was spent on fresh fruit and vegetables; £3.10 on fruit and £4.00 on vegetables; £11.60 was spent on meat, the highest proportion (48 per cent) of which was spent on other meat and meat preparations (£5.60 per week); £5.00 was spent on bread, rice and cereals; £3.20 was spent on buns, cakes, biscuits, etc; and £4.30 was spent on non-alcoholic drinks (Table A1). 81 per cent, £43.10 per week of food and non-alcoholic drinks were purchased from large supermarket chains (Table A2), an increase of £5.40 on the previous year.

Household expenditure by income

Household incomes have been ranked in ascending order and divided into decile groups in order to examine expenditure patterns between different income groups. Households with the smallest income lie in the first decile group and those with the largest income lie in the top decile group. Average weekly household expenditure in 2010 ranged from £185.60 in the lowest of the 10 income decile groups to £1,018.50 in the highest (Figure 1.1, Table A4); expenditure in this highest decile was £26.40 higher than in 2009, reversing the drop seen in the previous year.

Figure 1.1   **Household expenditure by gross income decile group, 2010**
United Kingdom

£ per week

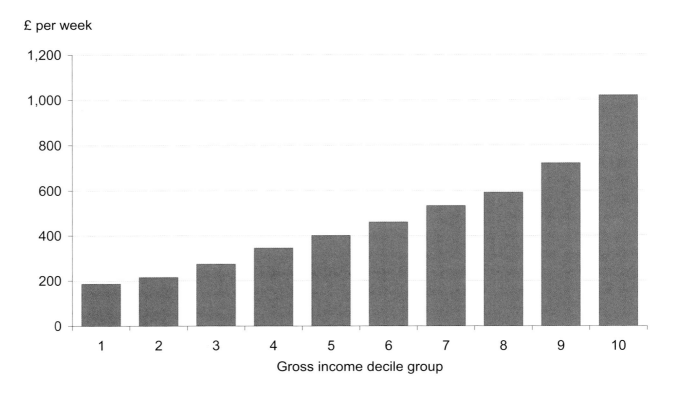

Gross income decile group

Households in the lowest income decile group spent a larger proportion of their total average weekly expenditure on housing, fuel and power (23 per cent), and food and non-alcoholic drinks (15 per cent), than those in the highest income decile group (8 per cent in both expenditure categories). However, households in the highest income decile group spent a greater proportion on transport (15 per cent) and recreation and culture (13 per cent) than those in the lowest income decile group (9 and 10 per cent respectively) (Table A5).

Household expenditure by age

Average weekly expenditure varied with the age of the household reference person (HRP). As in 2009, households whose HRP was aged 30 to 49 years had the highest average expenditure (£573.10 per week) while those with an HRP aged 75 years and over had the lowest average household expenditure (£240.40 per week). It should be noted that households with an HRP aged 30 to 49 years contained an average of 3 people, whereas households with an HRP aged 75 years and over contained an average of 1.4 people (Table A9).

Spending on housing, fuel and power in households whose HRP was aged less than 30 years fell from £97.30 in 2009 to £91.70, a drop of 6 per cent (Table A9). This compares with 20 per cent of total household expenditure for households with a HRP aged less than 30, whereas households with an HRP aged 75 years or over spent 18 per cent of their total household expenditure on housing, fuel and power (Table A10).

The proportion of expenditure spent on food and non-alcoholic drinks increased with age, from 9 per cent among households with an HRP aged less than 30 years to 16 per cent among households with an HRP aged 75 years and over.

The pattern of spending on restaurants and hotels, as a proportion of total expenditure, was relatively constant among age groups, with the percentage of total expenditure ranging from 9 per cent among households with an HRP aged less than 65, to 6 per cent among households with an HRP aged 75 years and over (Table A10). When the amount spent is considered, household expenditure on restaurants and hotels was greatest in households with an HRP aged between 30 and 49 (£49.10 per week), but much lower in households with HRP over 75 (£14.10 per week). This compares with an average expenditure across all ages of £39.20 (Table A11).

Expenditure on recreation and culture, as a proportion of total spending, increased from 9 per cent among households with an HRP aged less than 30 years to a maximum of 15 per cent among households with an HRP aged 65 to 74 years (Table A10).

Figure 1.2 **Expenditure on selected items as a proportion of total spending by age of the HRP, 2010**
United Kingdom

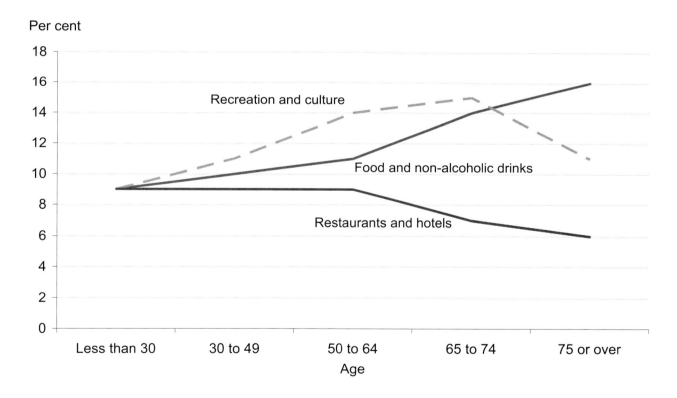

Household expenditure by economic activity and socio-economic classification

This analysis uses the National Statistics Socio-Economic Classification (NS-SEC), see Appendix B, page 222.

Household spending varied with the economic status of the HRP. The average weekly expenditure of households where the HRP was in employment (£587.20 per week) was just under twice that of households where the HRP was unemployed (£309.60 per week), and of households where the HRP was economically inactive (£309.50 per week) (Table A17).

In households where the HRP was in employment, spending was greatest on transport (£84.80 per week) and recreation and culture (£69.90 per week). Among households where the HRP was unemployed, spending on housing, fuel and power was greatest (£54.50 per week), followed by transport (£47.40 per week) (Table A17).

Average weekly expenditure was highest among households where the HRP was in the 'large employers and higher managerial' occupational group, at £856.10 per week. An average weekly expenditure of £418.10 was recorded for households where the HRP was in a 'routine' occupation (Table A22).

Household expenditure by household composition

Generally, household expenditure increased with the size of the household. Thus, average weekly household expenditure was lowest among retired one-person households who were mainly dependent on the state pension (£161.10) and highest among households containing three or more adults with children (£747.30) (Table A23).

Household expenditure by region

Overall, average household expenditure in the UK was £466.50 per week for the years 2008–10 combined. There were five regions in which expenditure over this period was higher than the UK average: expenditure was highest in London (£577.80), followed by the South East (£523.20 per week), the East (£493.40), the South West (£482.60) and Northern Ireland (£482.80). Spending was lowest among households in the North East (£372.70), Wales (£394.00), and Yorkshire and the Humber (£405.50) (Figure 1.3, Table A33).

Figure 1.3   **Household expenditure by region, 2008 to 2010**
United Kingdom

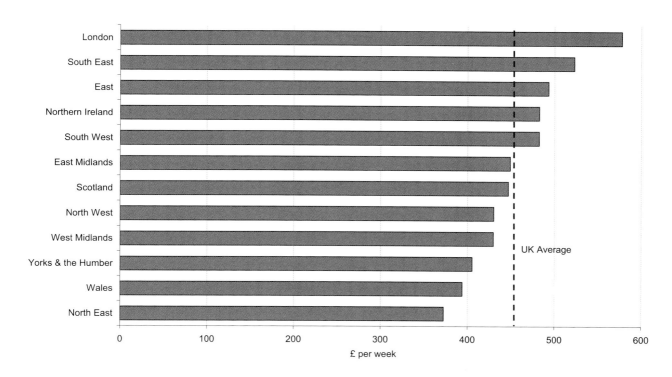

During the three-year period, 2008–2010, spending on transport was highest among households in the South East (£73.50 per week) and lowest among those in the North East (£49.10 per week). Households in London spent the most on housing, fuel and power (£87.00 per week) compared with the UK national average of £56.90 per week (Table A35). Housing expenditure is looked at in more detail in chapter two.

Households in rural areas had higher overall expenditure (£506.30 per week) than those in urban areas (£454.60 per week). This was reflected in expenditure on transport, where spending was highest (£76.00 in rural areas and £58.30 in urban areas), and recreation and culture (£68.00 in rural areas and £55.80 in urban areas). However, expenditure on housing, fuel and power was slightly higher in urban areas (£58.00 per week) than in rural areas (£54.50 per week) (Table A36).

Household income

Income is defined within the survey as the gross weekly cash income current at the time of interview. Income includes salaries and wages, income from self employment, benefits and pensions. See Appendix B for further details on income.

Average gross weekly household income in the UK in 2010 was £700.00, £17.00 more than in 2009 (£683.00 per week). Besides wages and salaries (65 per cent), social security benefits formed the largest proportion of income (14 per cent), followed by self-employment income (10 per cent), and income from annuities and pensions (8 per cent) (Figure 1.4, Table A37).

Figure 1.4   **Percentage of gross weekly household income by source
of income, 2010**
United Kingdom

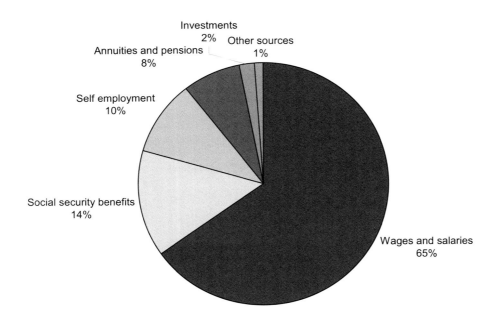

Household income by age

Households with an HRP aged 30 to 49 recorded the highest gross weekly income at £894 per week. Of this, 80 per cent was acquired from wages and salaries. The lowest gross weekly income was recorded by households with an HRP aged 75 years or over (£350, an increase of £19.00 from the previous year) with 52 per cent of their income gained through social security benefits.

Households with a household reference person (HRP) aged less than 65 years of age had a higher average weekly gross income than those with an HRP aged 65 years and over. However, the older age groups had a higher proportion of disposable income: 92 per cent for households with an HRP aged 65 to 74 years and 93 per cent for households with an HRP aged 75 years or over. This compares with households with an HRP aged 30-49, where 80 per cent of income was disposable (Table A38).

Household income by region

There were three English regions that exceeded the 2008–2010 UK national average income of £699. They were London (£982), the South East (£810) and the East (£749). Income was lowest among households in the North East (£543 per week) and Yorkshire and the Humber (£570 per week).

Among UK countries, households in England had the highest average gross weekly income (£711), whereas those in Wales had the lowest average income at £604 per week (Table A41).

Household income by economic activity and socio-economic classification

Households where the HRP was in the 'large employers and higher managerial' occupational group had an average gross weekly income of £1,653, almost three times the income of households where the HRP worked in a 'routine' occupation (£612). Incomes for these occupational groups were acquired mainly from wages and salaries (92 and 84 per cent respectively).

Households with an HRP in the 'large employers and higher managerial' occupational group also received the highest proportion of their average gross weekly income (£1,653) from wages and salaries (92 per cent). By contrast, those households with an HRP in the 'long-term unemployed' occupational group obtained 75 per cent of their average gross weekly income (£258) from social security benefits (Table A43).

Ownership of durable goods

Overall, 77 per cent of households had a home computer and 73 per cent had an internet connection at home, an increase of 2 percentage points for both from 2009 (Table A45). Among households in the highest income decile 98 per cent had a home computer and 97 per cent an internet connection, compared with only 46 and 39 per cent of households in the lowest income decile. This does, however, represent an 8 percentage point increase in households with a home computer and a 6 percentage point increase on households with an internet connection in the lowest income decile from 2009 continuing the upward trend (Table A46).

In general, households with children were more likely to have an internet connection than those without. Overall, 92 per cent of two-adult, non-retired households owned a home computer, with 89 per cent having an internet connection (Table A46).

Connection to the internet was lowest among households in Northern Ireland (61 per cent) and the North East (64 per cent) and highest in the London (77 per cent). Ownership of a mobile phone was lowest among households in Wales (48 per cent) and highest in the East Midlands at 86 per cent (Table A48).

Three-quarters (75 per cent) of all households owned a car or van, with 30 per cent owning two or more. Ownership of at least one car or van varied from 32 per cent in the lowest income decile, to 96 per cent in the ninth decile (Table A47).

Ownership of a car or van was highest among households in the South West (83 per cent), the East (82 per cent) and the South East (81 per cent), and lowest among households in London (65 per cent) and the North East (68 per cent) (Table A48).

1   From 2001-02, the Classification Of Individual COnsumption by Purpose (COICOP) was introduced as a new coding frame for expenditure items. COICOP is the internationally agreed classification system for reporting household consumption expenditure. Total expenditure is made up from the total of the COICOP expenditure groups (1 to 12) plus 'Other expenditure items (13)'. Other expenditure items are those items excluded from the narrower COICOP classifications, such as mortgage interest payments, council tax, domestic rates, holiday spending, cash gifts and charitable donations.

Housing expenditure

Background

This chapter presents housing-related costs such as rent, mortgage payments, repair and maintenance, and home improvements. The first section outlines the definitions of housing expenditure: the Classification Of Individual COnsumption by Purpose (COICOP) definition, followed by the definition used in the analysis of this chapter, which includes expenditure not present in COICOP. This chapter also examines housing expenditure over time and by income, region, and household characteristics. The final section explores housing costs for renters, and for mortgage holders in more depth.

Definitions of housing expenditure

The COICOP system has been used to classify expenditure on the Living Costs and Food Survey (LCF) and previously the Expenditure and Food Survey (EFS) since 2001/02. COICOP is an internationally agreed system of classification for reporting consumption expenditure within National Accounts and is used by other household budget surveys across the European Union. Further information on COICOP can be found on the United Nations Statistics Division website: **http://unstats.un.org/unsd/cr/registry/regct.asp?Lg=1**.

Under COICOP, household consumption expenditure is categorised into the following 12 headings:
1. Food & non-alcoholic drinks
2. Alcoholic drinks, tobacco & narcotics
3. Clothing & footwear
4. Housing (net), fuel & power
5. Household goods & services
6. Health
7. Transport
8. Communication
9. Recreation   & culture
10. Education
11. Restaurants & hotels
12. Miscellaneous goods & services

It is important to note that COICOP classified housing costs do not include what is considered to be non-consumption expenditure, for example, mortgage interest, mortgage capital repayments, mortgage protection premiums, council tax and domestic rates.

In addition to the 12 COICOP expenditure categories, the tables contained in Appendix A include a category called 'other expenditure items' under which certain non-consumption expenditures can be found. This category includes the following housing-related costs: mortgage interest payments; mortgage protection premiums; council tax; and domestic rates. Housing costs that are not included in either the COICOP definition of housing or the 'other expenditure item' category are captured within the 'other items recorded' category that can be viewed in Table A1 in Appendix A.

For the purpose of this chapter all data relating to housing expenditure have been combined to facilitate an understanding of total housing costs. This comprehensive definition of housing expenditure is made up from three types of expenditure detailed in Table 2.1: expenditure included in COICOP, housing costs included in the 'other expenditure items' and 'other items recorded' categories of *Family Spending*.

It should also be noted that throughout *Family Spending*, including this chapter, rent excluding service charges and benefit receipts associated with housing (net rent) has been used when calculating total expenditure. This convention ensures that rebates, benefits and allowances are excluded from the calculation of total household expenditure on rent.

Table 2.1 Definition of total housing expenditure

Housing costs which are included in the COICOP classification:	Housing costs which are included as 'other expenditure items' but excluded from COICOP classification:

• Actual rentals for housing
  – net rent (gross rent *less* housing benefit, rebates and allowances received)
  – second dwelling rent
• Maintenance and repair of dwelling
  – central heating maintenance and repair
  – house maintenance and repair
  – paint, wallpaper, timber
  – equipment hire, small materials
• Water supply and miscellaneous services relating to dwelling
  – water charges
  – other regular housing payments including service charge for rent
  – refuse collection, including skip hire.
• Household Insurances
  – structural insurance
  – contents insurance
  – insurance for household appliances.
	• Housing: mortgage interest payments etc
	  – mortgage interest payments
	  – mortgage protection premiums
	  – council tax, domestic rates
	  – council tax, mortgage, insurance (second dwelling).
	Housing costs which are included as 'other items recorded' and are excluded from COICOP classification:
	• Purchase or alteration of dwellings (contracted out), mortgages
	  – outright purchase of houses, flats etc. including deposits
	  – capital repayment of mortgage
	  – central heating installation
	  – DIY improvements: double glazing, kitchen units, sheds etc
	  – home improvements (contracted out)
	  – bathroom fittings
	  – purchase of materials for capital improvements
	  – purchase of second dwelling.

Housing expenditure

Table 2.2 shows expenditure on the items included in the comprehensive definition of housing expenditure. It also displays total household expenditure, which includes all expenditure items covered by the survey. The total expenditure figure reported here is therefore greater than the expenditure totals shown in the tables in Appendix A, as these exclude certain non-consumption costs.

Under the comprehensive definition of housing expenditure, UK households spent on average £134.70 a week on housing in 2010, which equates to a fifth (20 per cent) of total weekly expenditure. The COICOP definition of housing expenditure (excluding fuel and power) on the other hand, gave an average of £39.00 per week for each household (see Table A1).

In 2010 spending was highest on mortgages (interest payments, protection premiums and capital repayments) at £44.90 a week. The next highest expenditure was on charges (council tax or domestic rates, water charges, refuse collection and other regular services) at £27.20 a week. This was followed by net rent at £24.40 a week and household alterations and improvements at £19.70 per week. Figure 2.1 provides a breakdown of housing expenditure items as a proportion of housing expenditure.

Figure 2.1   Housing expenditure items as a percentage of total housing expenditure, 2010

United Kingdom

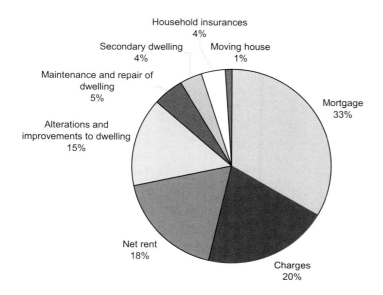

Housing expenditure over time

Overall, expenditure on housing has decreased over the last three years, from £143.40 in 2008 to £134.70 per week in 2010. There was a decrease of £4.10 per week from 2008 to 2009, which was followed by a further decrease of £4.60 between 2009 and 2010. However, housing expenditure as a percentage of total expenditure has remained stable, decreasing by only 1 percentage point from 2008 (21 per cent) to 2010 (20 per cent). See Table 2.2 for a comparison of housing expenditure from 2008 to 2010.

Figure 2.2 presents the average weekly spend on each category of housing expenditure from 2008 to 2010. The largest decrease was seen in the mortgages category, which has decreased from £57.20 in 2008 to £44.90 in 2010. Spending has remained relatively consistent for most other categories, with slight downward trends in alterations and improvements to dwellings and upward trends in charges such as council tax and water.

Figure 2.2 Housing expenditure 2008 to 2010
United Kingdom

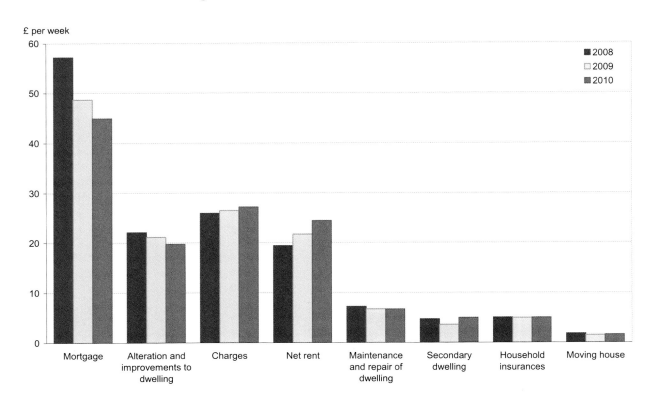

Expenditure by gross income

Table 2.3 presents housing expenditure by gross income decile group (a decile is one-tenth of the distribution). Overall, spending on housing increased with income. The highest income group spent £318.20 per week, more than twice the average for all income groups (£134.70) and more than six times that of the lowest income group (£49.30).

The categories that showed the greatest variation by income are mortgages, and alteration and improvements to dwellings. Figure 2.3 shows expenditure on mortgages to be consistently higher through income deciles, up to a weekly average of £129.60 in the highest income decile. Expenditure on alteration and improvements to dwellings increased with income, and displayed a sharp increase in the tenth decile group to £71.40 per week, almost double that of the ninth decile group (£36.20).

Figure 2.3 **Expenditure on selected items by gross income decile group, 2010**
United Kingdom

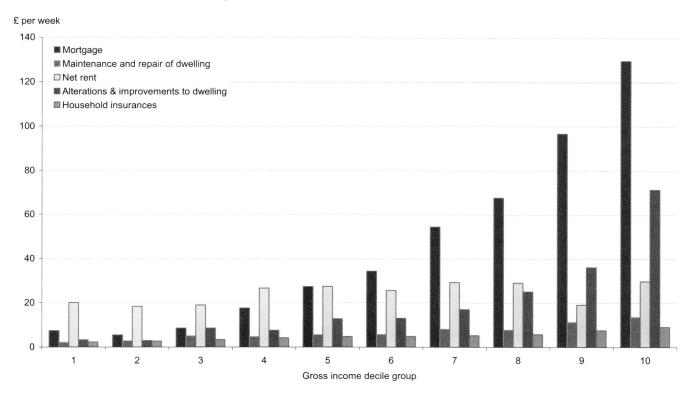

A different pattern can be seen when looking at net rent by income decile group; households in the second decile group spent the least on net rent at £18.30. Expenditure on rent was relatively consistent for higher income deciles, with the highest decile showing just the highest expenditure at £29.80. The exception was the ninth decile where expenditure was substantially lower at £19.10.

Expenditure on household insurances and on maintenance and repairs of dwellings increased slightly with income, but the increase was far less pronounced than for other categories.

Expenditure by age of the household reference person

Table 2.4 presents average weekly expenditure by age of household reference person (HRP, defined in Appendix B). Figure 2.4 presents spending on three key housing expenditure categories by age of HRP. Average weekly expenditure on mortgages peaked at £79.70 for households with an HRP aged 30 to 49. Average weekly expenditure for households with an HRP within the age range 50 to 64 was £42.20, and households with an HRP under 30 was even lower at £35.30. However, the average weekly spend for household alterations and improvements was highest for households with an HRP aged 50 to 64 at £26.20. Net rent expenditure decreased as the age of the HRP increased. Average weekly expenditure for households with an HRP under the age of 30 was £65.90, compared with £31.90 for households with an HRP aged between 30 and 49, and £7.00 for households with an HRP aged over 75.

Figure 2.4   **Expenditure on selected items by age of household reference person, 2010**
United Kingdom

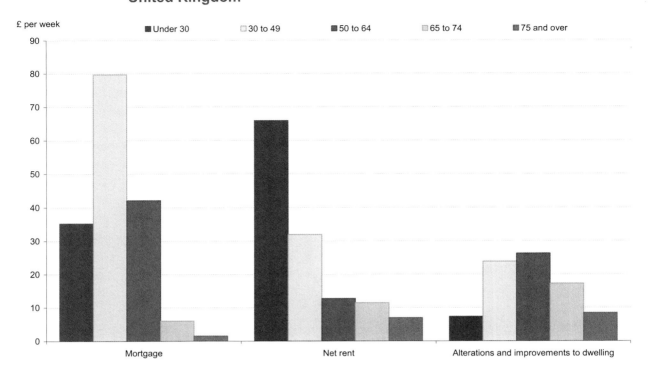

Figure 2.5 shows that expenditure on net rent for households with an HRP aged under 30 has decreased from £72.00 in 2009 to £65.90 in 2010. There were slight increases for HRP age groups over 30 and below 75.

Figure 2.5   **Expenditure on net rent by age of household reference person, 2009 and 2010**
United Kingdom

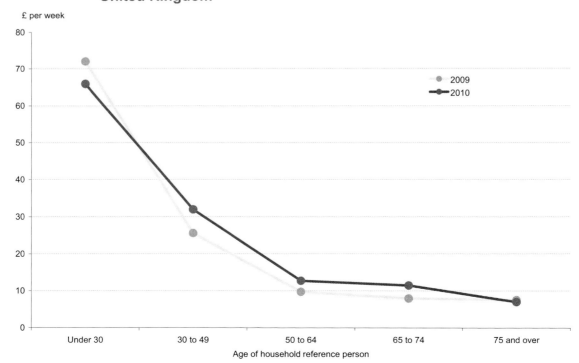

Expenditure by region

Table 2.5, Figure 2.6, and Figure 2.7 show housing expenditure by UK country and region. Looking first at expenditure by country (Figure 2.6), households in England spent the most on housing at £139.20 a week, followed by Scotland (£121.80), Wales (£112.80) and Northern Ireland (£80.10).

Figure 2.6 **Housing expenditure by Country, 2010**
United Kingdom

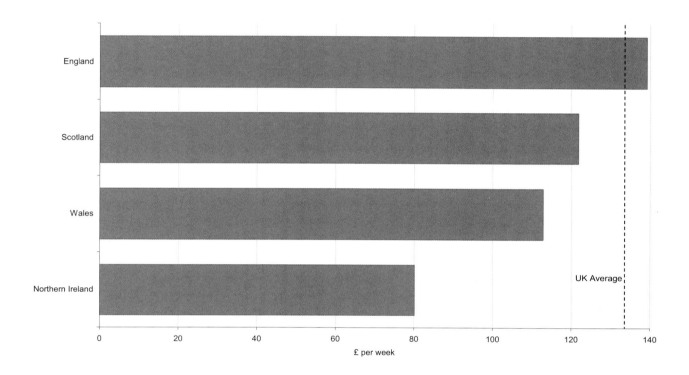

Figure 2.7 shows housing expenditure by region and country (excluding England). Five English regions had average weekly household spends that were greater than the UK average. Housing expenditure was greatest in London at £199.00 a week, followed by the South East (£155.70) and South West (£153.10). Expenditure was lowest in Northern Ireland with an average spend of £80.10. The lowest spending in England was in the North East, where average weekly housing costs were £98.00.

Figure 2.7 Housing expenditure by UK regions and Countries, 2010

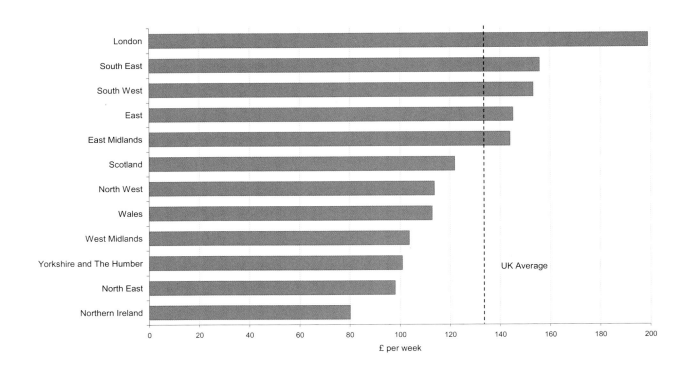

Variations in the total spending on housing are largely due to differences between regions in the average amount spent on rent and mortgages. Table 2.5 presents housing expenditure by country and region, while Figures 2.8 and 2.9 present the percentage difference in each region from the UK average for net rent and mortgages, respectively. It is important to note that these figures include all households. Average expenditure on rent only by renters, and mortgages only by mortgage holders is examined later in the chapter.

Average weekly expenditure on net rent was £24.40 (Table 2.5). Figure 2.8 shows that expenditure was more than twice the average in London (£54.60), and slightly above average in the South East (£25.50); these were the only regions to exceed the UK average. Spending on net rent was lowest in Northern Ireland with an average weekly expenditure of £13.10, closely followed by the North East of England where the average weekly net rent was £15.80

Figure 2.8 **Percentage difference compared with UK average for net rent by UK Countries and regions, 2010**

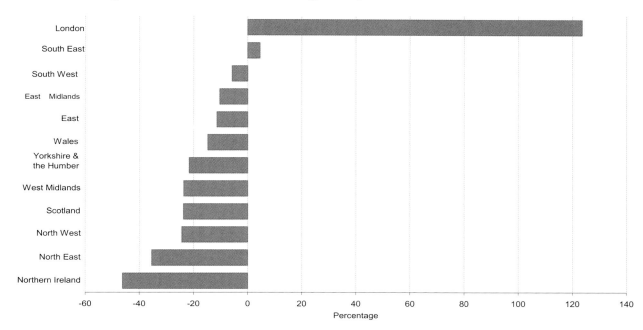

Figure 2.9 shows that London, the South East and the East of England had average mortgage payments above the UK average of £44.90; all other regions spent less on mortgages than the UK average. Expenditure on mortgages was lowest in Wales with average weekly mortgage payments of £34.90.

Figure 2.9 **Percentage difference compared with UK average for mortgage payments by UK Countries and regions, 2010**

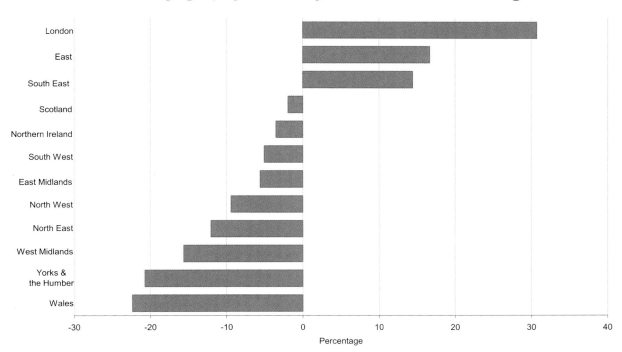

Expenditure by socio-economic classification

Figure 2.10 and Table 2.6 present housing expenditure by socio-economic classification of the household reference person (HRP). Households with an HRP in the 'large employer and higher managerial' occupation group spent the most, at £309.30 per week: more than twice that of households with an HRP in the 'routine' occupation group, where the average weekly spend was £114.30.

Figure 2.10 **Housing expenditure by socio-economic classification of household reference person, 2010**
United Kingdom

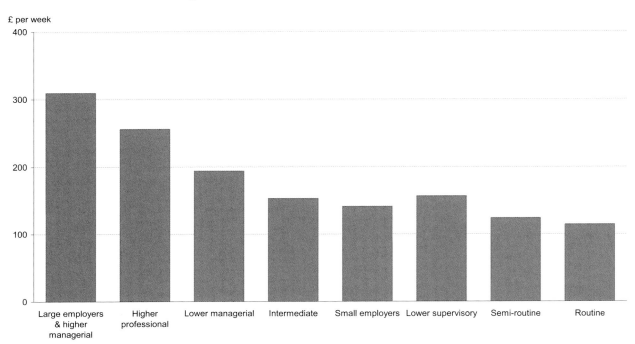

Figure 2.11 presents expenditure on selected items by the socio-economic classification of the HRP. Overall, expenditure followed a similar pattern as described above, with those with an HRP classified as 'large employer and higher managerial' spending more than those in 'routine' occupations. The exception to this was net rent where the opposite pattern is observed and the highest expenditure was for the 'routine' occupation group, reflecting the high number of renters in this group.

Figure 2.11   **Expenditure on selected items by socio-economic classification of household reference person, 2010**
United Kingdom

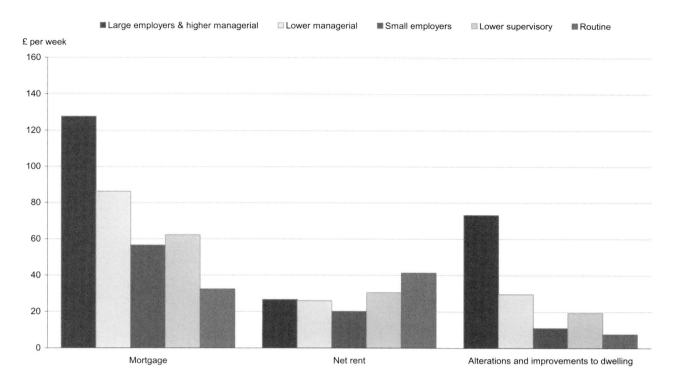

Analysis of housing costs for renters and mortgage holders

The following analysis looks at average expenditure on net rent for households that report spending on net rent and average expenditure on mortgages for mortgage holders. This is the only place in *Family Spending* where averages are not across all households. Excluding households with nil expenditure for net rent and mortgages provides a more informative picture of expenditure on these items.

Table 2.8 provides expenditure on rent over the last three years. In 2010 renters spent on average £74.40 per week on net rent. Table 2.9 provides expenditure on mortgages over the last three years. In 2010 the average weekly expenditure on mortgages by mortgage holders was £130.80.

Table 2.10 and Figure 2.12 present average weekly expenditure for the relevant households by income decile group for mortgage holders and renters.

Figure 2.12 shows a steady increase in net rent as the income decile increases. Households in the first income decile spent £31.70 on net rent, compared with £146.80 in the ninth income decile and £255.20 in the tenth income decile. It should be noted, however, that a relatively small number of households in the highest income group paid rent. The estimate of net costs for this income group should therefore be viewed with caution. Average expenditure for mortgages followed a roughly similar pattern to net rent, increasing towards the higher income deciles. The fluctuations in the lower decile groups may be due to a low number of mortgage holders in this decile group, and should be viewed with caution.

Figure 2.12   **Expenditure on net rent by renters, and mortgages by mortgage holders, by gross income decile group, 2010**
United Kingdom

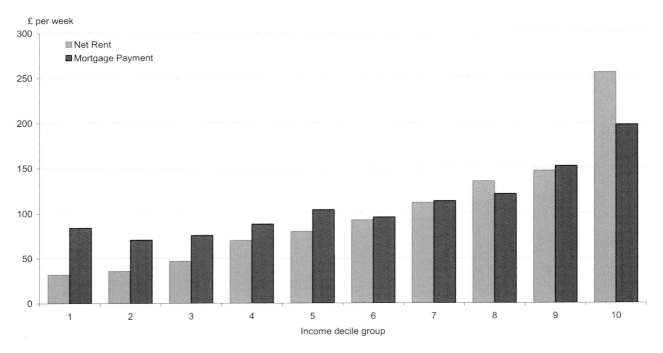

Figure 2.13, and Figure 2.14 show net rent and mortgage expenditure, respectively, averaged across renters and mortgage holders, by country and region. The figures are presented in Table 2.11.

Figure 2.13   **Expenditure on net rent by renters, by UK Countries and regions, 2010**

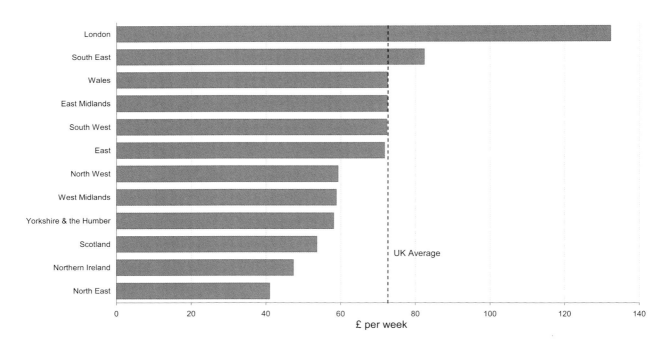

Figure 2.14 **Expenditure on mortgages by mortgage holders by UK Countries and regions, 2010**

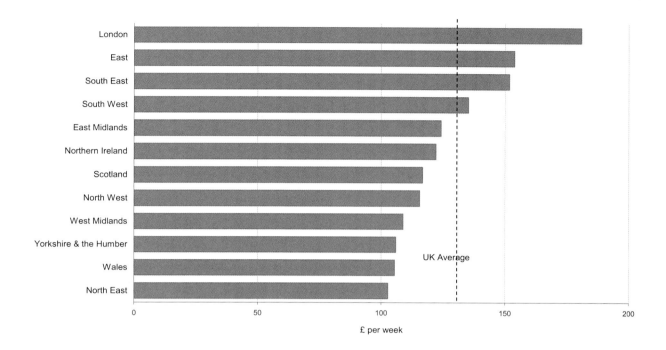

Table 2.11 shows the country with the highest average spend on net rent was England, which at £77.50 a week was above the UK average of £74.40. Wales had the next highest average weekly expenditure among UK countries on net rent at £72.50, followed by Scotland (£53.60) and Northern Ireland with the lowest average weekly spend (£47.40).

There were two regions that had average weekly expenditures on net rent greater than the UK average of £74.40. London spent the most on net rent by a substantial margin (£132.30), followed by the South East (£82.40). All the other regions had average weekly spends lower than the UK average. The region with the lowest average weekly expenditure on net rent was North East (£41.00).

Looking at expenditure on mortgages by country (Table 2.11), England at £134.20 was the only country that had an average weekly expenditure on mortgages greater than the UK average of £130.80 per week. Closely following England was Northern Ireland (£122.10), Scotland (£116.70) and finally Wales (£105.40). The differences between average weekly expenditure by country are much smaller for mortgages compared with net rent.

Four regions had a greater average weekly spend on mortgage payments than the UK average (Figure 2.14). London had the highest average spend at £180.90 per week, although the difference was less pronounced than for net rent. The next highest regions were the East (£153.80), the South East (£151.80) and the South West (£135.20). The region with the lowest weekly spend was North East at £102.70 per week; just under £30 a week lower than the UK average.

Table 2.2   Housing expenditure, 2008 to 2010
United Kingdom

	2008			2009			2010		
	£ per week	% of total expend-iture	% of housing expend-iture	£ per week	% of total expend-iture	% of housing expend-iture	£ per week	% of total expend-iture	% of housing expend-iture
Weighted number of households (thousands)	25,690			25,980			26,320		
Total number of households in sample	5,850			5,830			5,260		
Total number of persons in sample	13,830			13,740			12,180		
Total number of adults in sample	10,640			10,650			9,430		
Weighted average number of persons per household	2.4			2.3			2.3		
Commodity or service	Average weekly household expenditure (£)								
Primary dwelling									
Rent	**31.50**	**5**	**22**	**35.40**	**5**	**25**	**39.90**	**6**	**30**
Gross rent	31.50	5	22	35.40	5	25	39.90	6	30
less housing benefit, rebates and allowances received	12.10	2	8	13.70	2	10	15.50	2	11
Net rent[1]	19.40	3	14	21.70	3	16	24.40	4	18
Mortgage	**57.20**	**8**	**40**	**48.70**	**7**	**35**	**44.90**	**7**	**33**
Mortgage interest payments	37.50	6	26	26.70	4	19	23.80	4	18
Mortgage protection premiums	1.90	0	1	1.60	0	1	1.40	0	1
Capital repayment of mortgage	17.80	3	12	20.30	3	15	19.70	3	15
Outright purchase, including deposits	**[0.10]**	**0**	**0**	**[4.50]**	**1**	**3**	**[0.20]**	**0**	**0**
Secondary dwelling	**4.70**	**1**	**3**	**3.60**	**1**	**3**	**5.00**	**1**	**4**
Rent	[0.10]	0	0	[0.00]	0	0	[0.00]	0	0
Council tax, mortgage, insurance (secondary dwelling)	0.50	0	0	1.80	0	1	1.60	0	1
Purchase of second dwelling	4.10	1	3	1.80	0	1	3.40	1	2
Charges	**25.90**	**4**	**18**	**26.50**	**4**	**19**	**27.20**	**4**	**20**
Council tax, domestic rates	18.50	3	13	18.90	3	14	19.40	3	14
Water charges	6.30	1	4	6.50	1	5	6.60	1	5
Other regular housing payments including service charge for rent	1.10	0	1	1.10	0	1	1.10	0	1
Refuse collection, including skip hire	[0.10]	0	0	[0.00]	0	0	[0.10]	0	0
Moving house	**1.80**	**0**	**1**	**1.50**	**0**	**1**	**1.60**	**0**	**1**
Property transaction - purchase and sale	0.90	0	1	0.60	0	0	0.70	0	1
Property transaction - sale only	0.40	0	0	0.40	0	0	0.50	0	0
Property transaction - purchase only	0.30	0	0	0.30	0	0	0.30	0	0
Property transaction - other payments	0.20	0	0	0.20	0	0	0.20	0	0
Maintenance and repair of dwelling	**7.20**	**1**	**5**	**6.70**	**1**	**5**	**6.70**	**1**	**5**
Central heating repairs	1.50	0	1	1.30	0	1	1.30	0	1
House maintenance etc.	3.90	1	3	3.70	1	3	3.60	1	3
Paint, wallpaper, timber	0.90	0	1	1.00	0	1	1.00	0	1
Equipment hire, small materials	0.90	0	1	0.70	0	1	0.80	0	1
Alterations and improvements to dwelling	**22.10**	**3**	**15**	**21.10**	**3**	**15**	**19.70**	**3**	**15**
Central heating installation	1.20	0	1	0.80	0	1	1.30	0	1
DIY improvements: double glazing, kitchen units, sheds etc.	1.60	0	1	0.90	0	1	0.70	0	1
Home improvements - contracted out	18.10	3	13	18.90	3	14	16.80	3	12
Bathroom fittings	0.50	0	0	0.30	0	0	0.60	0	0
Purchase of materials for Capital Improvements	0.80	0	1	0.30	0	0	0.20	0	0
Household insurances	**5.00**	**1**	**4**	**5.00**	**1**	**4**	**5.00**	**1**	**4**
Structure	2.50	0	2	2.50	0	2	2.50	0	2
Contents	2.50	0	2	2.40	0	2	2.50	0	2
Household appliances	0.10	0	0	0.10	0	0	0.10	0	0
Housing expenditure	**143.40**	**21**	**100**	**139.30**	**21**	**100**	**134.70**	**20**	**100**
Total expenditure[2]	**674.10**			**653.90**			**664.00**		

Note: Please see page xiii for symbols and conventions used in this report.

1 The figure included in total expenditure is net rent as opposed to gross rent.

2 This total includes all categories recorded in the LCF, including those outside the 'COICOP' total expenditure

Table 2.3 **Housing expenditure by gross income decile group, 2010**
United Kingdom

	Gross income decile group										
	1	2	3	4	5	6	7	8	9	10	All
Weighted number of households (thousands)	2,630	2,640	2,630	2,630	2,630	2,640	2,630	2,640	2,630	2,630	26,320
Total number of households in sample	510	530	540	550	550	530	530	520	510	500	5,260
Total number of persons in sample	690	860	960	1,150	1,280	1,290	1,410	1,430	1,530	1,580	12,180
Total number of adults in sample	570	670	780	900	970	1,010	1,090	1,090	1,170	1,170	9,430
Weighted average number of persons per household	1.3	1.6	1.8	2.1	2.4	2.4	2.7	2.8	3.0	3.2	2.3
Commodity or service	Average weekly household expenditure (£)										
Primary dwelling											
Rent	**71.80**	**58.10**	**46.80**	**43.70**	**38.80**	**28.90**	**31.70**	**29.50**	**19.80**	**29.80**	**39.90**
Gross rent	71.80	58.10	46.80	43.70	38.80	28.90	31.70	29.50	19.80	29.80	39.90
less housing benefit, rebates and allowances received	51.80	39.70	27.80	17.00	11.40	3.30	2.50	[0.50]	[0.70]	[0.00]	15.50
Net rent[1]	20.10	18.30	19.00	26.60	27.40	25.60	29.30	29.00	19.10	29.80	24.40
Mortgage	**7.40**	**5.50**	**8.60**	**17.60**	**27.40**	**34.40**	**54.40**	**67.60**	**96.70**	**129.60**	**44.90**
Mortgage interest payments	5.00	2.80	4.50	10.20	12.80	18.20	28.40	35.00	51.60	69.20	23.80
Mortgage protection premiums	[0.20]	[0.10]	0.40	0.40	0.80	1.40	1.60	2.60	3.20	3.40	1.40
Capital repayment of mortgage	2.20	2.50	3.70	7.00	13.80	14.80	24.40	30.10	41.90	56.90	19.70
Outright purchase, including deposits	**-**	**[0.10]**	**[0.20]**	**-**	**[0.00]**	**[0.10]**	**-**	**[0.20]**	**[0.80]**	**[0.40]**	**[0.20]**
Secondary dwelling	**[0.30]**	**[0.10]**	**[0.50]**	**[0.40]**	**[1.50]**	**[1.20]**	**[1.90]**	**[6.00]**	**19.20**	**18.70**	**5.00**
Rent	-	[0.00]	-	[0.00]	[0.10]	-	[0.10]	-	-	[0.20]	[0.00]
Council tax, mortgage, insurance (secondary dwelling)	[0.30]	-	[0.30]	[0.40]	[1.30]	[1.10]	[1.30]	[1.60]	[5.00]	4.30	1.60
Purchase of second dwelling	-	[0.10]	[0.10]	-	[0.10]	[0.10]	[0.50]	[4.00]	[14.00]	[14.20]	3.40
Charges	**13.30**	**16.50**	**21.30**	**25.00**	**27.10**	**29.10**	**30.80**	**32.80**	**34.70**	**41.50**	**27.20**
Council tax, domestic rates	6.60	9.60	13.80	17.20	19.80	21.60	22.80	24.20	26.40	31.80	19.40
Water charges	5.50	5.70	5.70	6.40	6.30	6.90	7.00	7.00	7.40	8.40	6.60
Other regular housing payments including service charge for rent	1.30	1.20	1.80	1.40	1.00	0.60	0.80	1.40	0.70	1.10	1.10
Refuse collection, including skip hire	-	-	-	[0.00]	[0.00]	-	[0.30]	[0.20]	[0.20]	[0.10]	[0.10]
Moving house	**[0.40]**	**[0.60]**	**[1.00]**	**[1.20]**	**[1.40]**	**[1.10]**	**[1.30]**	**1.60**	**3.30**	**4.20**	**1.60**
Property transaction - purchase and sale	[0.00]	[0.00]	[0.50]	[0.50]	[1.00]	[0.60]	[0.50]	[0.60]	[1.30]	[1.60]	0.70
Property transaction - sale only	-	[0.50]	[0.20]	[0.70]	[0.00]	-	[0.30]	[0.50]	[0.70]	[1.70]	0.50
Property transaction - purchase only	-	-	[0.20]	[0.10]	[0.20]	[0.30]	[0.20]	[0.30]	[0.90]	[0.60]	0.30
Property transaction - other payments	-	[0.10]	[0.10]	[0.10]	[0.20]	[0.10]	[0.20]	[0.20]	[0.40]	[0.40]	0.20
Maintenance and repair of dwelling	**2.10**	**2.80**	**5.10**	**4.70**	**5.60**	**5.80**	**8.20**	**7.80**	**11.30**	**13.60**	**6.70**
Central heating repairs	0.50	0.70	0.70	1.10	1.10	1.40	1.70	1.90	1.90	2.20	1.30
House maintenance etc.	0.80	1.50	3.20	2.70	2.50	2.80	3.50	3.80	6.40	8.50	3.60
Paint, wallpaper, timber	0.60	0.40	0.80	0.60	0.80	0.80	2.10	0.90	1.70	1.70	1.00
Equipment hire, small materials	0.20	0.20	0.40	0.30	1.20	0.80	0.80	1.20	1.30	1.20	0.80
Alterations and improvements to dwelling	**3.30**	**2.90**	**8.60**	**7.60**	**12.80**	**13.00**	**17.10**	**25.10**	**36.20**	**71.40**	**19.80**
Central heating installation	[0.70]	[0.90]	[0.60]	[1.70]	[1.40]	[0.80]	[1.10]	[1.70]	[1.50]	3.30	1.40
DIY improvements: double glazing, kitchen units, sheds etc.	[0.10]	[0.00]	[0.10]	[0.70]	[0.30]	[0.10]	[1.10]	[0.60]	[3.10]	[1.00]	0.70
Home improvements - contracted out	2.40	1.70	7.80	4.80	10.00	9.60	14.20	21.90	29.70	66.10	16.80
Bathroom fittings	[0.10]	[0.20]	[0.00]	[0.00]	[0.60]	[2.50]	[0.10]	[0.40]	[1.70]	[0.70]	0.60
Purchase of materials for capital improvements	[0.10]	[0.10]	[0.00]	[0.40]	[0.50]	[0.10]	[0.40]	[0.50]	[0.10]	[0.30]	0.20
Household insurances	**2.30**	**2.70**	**3.40**	**4.20**	**4.80**	**4.90**	**5.30**	**5.80**	**7.60**	**9.10**	**5.00**
Structure	1.00	1.30	1.60	1.90	2.40	2.40	2.60	2.90	3.80	4.70	2.50
Contents	1.20	1.40	1.70	2.10	2.40	2.40	2.60	2.80	3.60	4.30	2.50
Household appliances	[0.00]	[0.00]	[0.10]	[0.10]	[0.10]	[0.10]	[0.10]	[0.10]	[0.20]	[0.10]	0.10
Housing expenditure	**49.30**	**49.60**	**67.60**	**87.30**	**107.90**	**115.10**	**148.10**	**175.80**	**228.80**	**318.20**	**134.70**
Total expenditure[2]	**203.10**	**231.20**	**306.50**	**398.70**	**491.50**	**588.00**	**711.00**	**857.20**	**1098.30**	**1757.00**	**664.00**

Note: Please see page xiii for symbols and conventions used in this report.
1 The figure included in total expenditure is net rent as opposed to gross rent.
2 This total includes all categories recorded in the LCF, including those outside the 'COICOP' total expenditure.

Table 2.4

Housing expenditure by age of household reference person, 2010

United Kingdom

	Under 30	30 to 49	50 to 64	65 to 74	75 or over	All
Weighted number of households (thousands)	2,810	9,540	7,020	3,420	3,530	26,320
Total number of households in sample	450	1,890	1,490	800	630	5,260
Total number of persons in sample	1,060	5,640	3,200	1,380	910	12,180
Total number of adults in sample	760	3,500	2,900	1,360	910	9,430
Weighted average number of persons per household	2.4	3.0	2.2	1.8	1.4	2.3
Commodity or service	Average weekly household expenditure (£)					
Primary dwelling						
Rent	**91.00**	**48.60**	**23.50**	**25.40**	**22.30**	**39.90**
Gross rent	91.00	48.60	23.50	25.40	22.30	39.90
less housing benefit, rebates and allowances received	25.10	16.70	10.80	14.00	15.30	15.50
Net rent[1]	65.90	31.90	12.70	11.40	7.00	24.40
Mortgage	**35.30**	**79.70**	**42.20**	**6.00**	**1.60**	**44.90**
Mortgage interest payments	22.20	43.60	18.70	4.10	0.70	23.80
Mortgage protection premiums	1.10	2.40	1.40	[0.10]	[0.00]	1.40
Capital repayment of mortgage	12.00	33.70	22.00	1.80	[0.80]	19.70
Outright purchase, including deposits	**[0.70]**	**[0.20]**	**[0.10]**	**[0.10]**	**-**	**[0.20]**
Secondary dwelling	**[1.30]**	**7.60**	**6.00**	**[3.70]**	**[0.00]**	**5.00**
Rent	-	[0.00]	[0.10]	-	-	[0.00]
Council tax, mortgage, insurance (secondary dwelling)	[0.80]	2.70	1.50	[0.70]	[0.00]	1.60
Purchase of second dwelling	[0.50]	4.80	[4.40]	[3.00]	-	3.40
Charges	**21.40**	**28.00**	**29.70**	**26.80**	**25.20**	**27.20**
Council tax, domestic rates	13.90	20.10	21.80	19.80	16.60	19.40
Water charges	6.00	7.00	7.10	6.20	5.70	6.60
Other regular housing payments including service charge for ren	1.60	0.80	0.60	0.90	2.80	1.10
Refuse collection, including skip hire	-	[0.10]	[0.10]	[0.00]	-	[0.10]
Moving house	**1.50**	**2.30**	**1.30**	**[1.50]**	**[0.50]**	**1.60**
Property transaction - purchase and sale	[0.40]	0.90	[0.60]	[0.90]	[0.40]	0.70
Property transaction - sale only	[0.20]	0.70	[0.40]	[0.60]	-	0.50
Property transaction - purchase only	[0.60]	0.40	[0.10]	[0.00]	[0.00]	0.30
Property transaction - other payments	[0.30]	0.30	[0.10]	[0.00]	[0.00]	0.20
Maintenance and repair of dwelling	**2.50**	**6.90**	**7.70**	**7.10**	**7.00**	**6.70**
Central heating repairs	0.40	1.20	1.70	1.80	1.20	1.30
House maintenance etc.	0.90	3.90	3.30	3.80	5.10	3.60
Paint, wallpaper, timber	1.00	1.10	1.40	0.60	0.50	1.00
Equipment hire, small materials	0.30	0.70	1.40	0.80	0.20	0.80
Alterations and improvements to dwelling	**7.30**	**23.80**	**26.20**	**17.20**	**8.40**	**19.80**
Central heating installation	[0.90]	1.60	1.80	[0.80]	[1.00]	1.40
DIY improvements: double glazing, kitchen units, sheds etc.	[0.30]	0.40	1.50	[0.90]	[0.10]	0.70
Home improvements - contracted out	6.10	20.80	21.40	15.20	7.20	16.80
Bathroom fittings	[0.00]	0.70	1.20	[0.20]	[0.10]	0.60
Purchase of materials for capital improvements	[0.00]	[0.30]	[0.40]	[0.00]	[0.00]	0.20
Household insurances	**2.30**	**5.30**	**6.00**	**5.20**	**4.40**	**5.00**
Structure	0.90	2.60	3.00	2.60	2.10	2.50
Contents	1.40	2.60	2.80	2.50	2.20	2.50
Household appliances	[0.00]	0.10	0.10	0.20	[0.10]	0.10
Housing expenditure	**138.10**	**185.70**	**131.80**	**79.00**	**54.00**	**134.70**
Total expenditure[2]	**578.00**	**850.90**	**747.90**	**437.60**	**279.40**	**664.00**

Note: Please see page xiii for symbols and conventions used in this report.

1 The figure included in total expenditure is net rent as opposed to gross rent.

2 This total includes all categories recorded in the LCF, including those outside the 'COICOP' total expenditure.

Table 2.5   Household expenditure by UK Countries and regions, 2010

	North East	North West	Yorks & the Humber	East Midlands	West Midlands	East	London
Weighted number of households (thousands)	1,190	3,040	2,280	1,950	2,270	2,460	3,010
Total number of households in sample	260	600	490	410	470	520	480
Total number of persons in sample	570	1,340	1,130	940	1,100	1,190	1,210
Total number of adults in sample	450	1,050	860	740	850	940	890
Weighted average number of persons per household	2.2	2.3	2.3	2.3	2.4	2.3	2.6
Commodity or service	Average weekly household expenditure (£)						
Primary dwelling							
Rent	**35.40**	**32.60**	**32.40**	**30.10**	**34.60**	**35.60**	**82.50**
Gross rent	35.40	32.60	32.40	30.10	34.60	35.60	82.50
less housing benefit, rebates and allowances received	19.60	14.10	13.30	8.20	15.90	14.00	27.90
Net rent[1]	15.80	18.50	19.20	21.90	18.70	21.70	54.60
Mortgage	**39.50**	**40.70**	**35.60**	**42.40**	**37.90**	**52.40**	**58.70**
Mortgage interest payments	18.30	21.30	18.20	22.80	17.90	28.20	33.10
Mortgage protection premiums	1.60	1.30	1.40	1.30	1.10	1.40	1.50
Capital repayment of mortgage	19.50	18.10	16.00	18.30	18.90	22.80	24.10
Outright purchase, including deposits	-	-	[0.00]	[0.10]	[0.50]	[0.00]	[0.40]
Secondary dwelling	**[0.90]**	**[1.30]**	**[1.50]**	**[9.40]**	**[0.60]**	**[2.40]**	**[4.00]**
Rent	-	-	[0.00]	-	-	-	[0.20]
Council tax, mortgage, insurance (secondary dwelling)	[0.70]	[1.20]	[1.10]	[0.80]	[0.60]	[1.40]	[2.80]
Purchase of second dwelling	[0.30]	[0.10]	[0.50]	[8.60]	[0.00]	[1.10]	[1.10]
Charges	**21.40**	**26.20**	**24.00**	**27.10**	**23.70**	**31.20**	**31.50**
Council tax, domestic rates	15.10	18.00	16.50	19.90	16.60	22.30	22.30
Water charges	6.10	7.20	6.60	6.80	6.00	7.30	6.40
Other regular housing payments including service charge for rent	[0.20]	0.90	0.90	0.40	1.10	1.50	2.70
Refuse collection, including skip hire	-	[0.20]	[0.10]	-	-	-	[0.20]
Moving house	**[1.00]**	**1.50**	**[0.60]**	**[1.20]**	**[1.40]**	**[2.60]**	**[1.70]**
Property transaction - purchase and sale	[0.70]	[1.20]	[0.00]	[0.30]	[0.80]	[0.90]	[0.50]
Property transaction - sale only	-	[0.00]	[0.20]	[0.60]	[0.30]	[1.10]	[0.40]
Property transaction - purchase only	[0.20]	[0.30]	[0.20]	[0.20]	[0.10]	[0.40]	[0.50]
Property transaction - other payments	[0.10]	[0.00]	[0.20]	[0.10]	[0.20]	[0.20]	[0.30]
Maintenance and repair of dwelling	**4.50**	**6.50**	**6.00**	**7.90**	**4.90**	**7.00**	**8.50**
Central heating repairs	0.50	1.70	1.20	1.60	0.90	1.60	1.30
House maintenance etc.	1.70	3.00	2.40	3.70	2.80	3.90	6.00
Paint, wallpaper, timber	2.00	0.70	1.20	1.10	0.70	1.00	0.60
Equipment hire, small materials	[0.40]	1.20	1.20	1.50	0.50	0.60	0.60
Alterations and improvements to dwelling	**10.50**	**14.20**	**9.10**	**29.00**	**11.70**	**22.60**	**33.50**
Central heating installation	[1.00]	[1.90]	[0.60]	[1.20]	[0.60]	[1.60]	[2.70]
DIY improvements: double glazing, kitchen units, sheds etc.	[0.40]	[0.40]	[0.40]	[0.30]	[0.30]	[1.20]	[2.20]
Home improvements - contracted out	8.10	11.20	4.70	27.30	9.90	19.30	27.40
Bathroom fittings	[0.00]	[0.60]	[2.90]	[0.20]	[0.70]	[0.10]	[1.20]
Purchase of materials for Capital Improvements	[1.00]	[0.10]	[0.50]	[0.10]	[0.20]	[0.40]	[0.00]
Household insurances	**4.40**	**4.80**	**4.90**	**4.90**	**4.30**	**5.20**	**6.20**
Structure	2.20	2.40	2.30	2.40	2.00	2.60	3.10
Contents	2.20	2.30	2.40	2.40	2.30	2.50	3.00
Household appliances	[0.00]	[0.10]	[0.20]	[0.10]	[0.10]	[0.10]	[0.10]
Housing expenditure	**98.00**	**113.70**	**100.90**	**143.90**	**103.70**	**145.10**	**199.00**
Total expenditure[2]	**510.30**	**585.40**	**568.00**	**670.70**	**537.40**	**712.70**	**893.00**

Please see page xiii for symbols and conventions used in this report.

1 The figure included in total expenditure is net rent as opposed to gross rent.

2 This total includes all categories recorded in the LCF, including those outside the 'COICOP' total expenditure.

Table 2.5   Household expenditure by UK Countries and region, 2010 (cont.)

	South East	South West	England	Wales	Scotland	Northern Ireland	United Kingdom
Weighted number of households (thousands)	3,470	2,340	22,010	1,260	2,320	720	26,320
Total number of households in sample	680	500	4,390	260	470	150	5,260
Total number of persons in sample	1,560	1,120	10,160	620	1,040	360	12,180
Total number of adults in sample	1,200	860	7,860	480	830	270	9,430
Weighted average number of persons per household	2.4	2.3	2.3	2.4	2.2	2.5	2.3
Commodity or service	Average weekly household expenditure (£)						
Primary dwelling							
Rent	**42.80**	**36.10**	**41.80**	**32.40**	**30.60**	**23.80**	**39.90**
Gross rent	42.80	36.10	41.80	32.40	30.60	23.80	39.90
less housing benefit, rebates and allowances received	17.30	13.10	16.20	11.50	12.00	10.60	15.50
Net rent[1]	25.50	23.00	25.60	20.80	18.60	13.10	24.40
Mortgage	**51.40**	**42.60**	**45.60**	**34.90**	**44.00**	**43.30**	**44.90**
Mortgage interest payments	28.20	24.00	24.40	17.70	22.20	21.20	23.80
Mortgage protection premiums	1.50	1.10	1.40	1.00	1.50	2.80	1.40
Capital repayment of mortgage	21.60	17.50	19.90	16.10	20.30	19.30	19.70
Outright purchase, including deposits	**[0.30]**	**[0.00]**	**[0.20]**	-	**[0.30]**	-	**[0.20]**
Secondary dwelling	**2.90**	**25.60**	**5.30**	**[0.60]**	**[5.70]**	**[0.40]**	**5.00**
Rent	[0.10]	[0.00]	[0.00]	-	-	[0.20]	[0.00]
Council tax, mortgage, insurance (secondary dwelling)	[1.90]	[4.70]	1.80	[0.20]	[0.80]	-	1.60
Purchase of second dwelling	[0.90]	[20.80]	3.50	[0.40]	[4.90]	[0.20]	3.40
Charges	**30.80**	**30.40**	**28.00**	**24.80**	**26.70**	**9.20**	**27.20**
Council tax, domestic rates	22.60	21.20	19.80	17.00	19.40	8.90	19.40
Water charges	6.80	7.90	6.80	7.70	6.30	-	6.60
Other regular housing payments including service charge for rent	1.30	1.00	1.20	[0.10]	0.90	0.30	1.10
Refuse collection, including skip hire	[0.00]	[0.40]	[0.10]	-	-	-	[0.10]
Moving house	**3.00**	**[1.30]**	**1.70**	**[0.30]**	**[1.80]**	**[0.10]**	**1.60**
Property transaction - purchase and sale	[1.20]	[0.60]	0.70	-	[1.10]	[0.10]	0.70
Property transaction - sale only	[1.10]	[0.40]	0.50	[0.00]	[0.50]	-	0.50
Property transaction - purchase only	[0.30]	[0.30]	0.30	[0.10]	[0.20]	[0.00]	0.30
Property transaction - other payments	[0.30]	[0.00]	0.20	[0.20]	[0.10]	-	0.20
Maintenance and repair of dwelling	**7.60**	**7.50**	**6.90**	**5.20**	**6.20**	**4.40**	**6.70**
Central heating repairs	1.70	1.00	1.30	0.90	1.30	1.20	1.30
House maintenance etc.	4.70	4.40	3.80	2.80	2.30	1.30	3.60
Paint, wallpaper, timber	0.70	1.60	1.00	[0.60]	2.10	[0.30]	1.00
Equipment hire, small materials	0.50	0.50	0.80	0.90	0.40	[1.60]	0.80
Alterations and improvements to dwelling	**28.80**	**17.30**	**20.70**	**21.60**	**14.00**	**5.70**	**19.80**
Central heating installation	[1.40]	[0.90]	1.40	[2.00]	[0.90]	[0.30]	1.40
DIY improvements: double glazing, kitchen units, sheds etc.	[0.30]	[0.50]	0.70	[0.20]	[1.10]	[0.10]	0.70
Home improvements - contracted out	26.50	15.80	17.70	18.60	11.30	[4.80]	16.80
Bathroom fittings	[0.20]	[0.10]	0.70	[0.70]	[0.20]	[0.00]	0.60
Purchase of materials for Capital Improvements	[0.30]	[0.00]	0.20	[0.10]	[0.40]	[0.40]	0.20
Household insurances	**5.50**	**5.30**	**5.10**	**4.60**	**4.50**	**3.70**	**5.00**
Structure	2.70	2.60	2.50	2.20	2.10	1.80	2.50
Contents	2.60	2.50	2.50	2.20	2.30	1.90	2.50
Household appliances	[2.00]	[0.10]	0.10	[0.20]	[0.10]	[0.00]	0.10
Housing expenditure	**155.70**	**153.10**	**139.20**	**112.80**	**121.80**	**80.10**	**134.70**
Total expenditure[2]	**739.50**	**705.50**	**675.50**	**550.60**	**639.20**	**590.30**	**664.00**

Please see page xiii for symbols and conventions used in this report.

1 The figure included in total expenditure is net rent as opposed to gross rent.

2 This total includes all categories recorded in the LCF, including those outside the 'COICOP' total expenditure.

Table 2.6   **Housing expenditure by socio-economic classification of household reference person, 2010**
United Kingdom

Commodity or service	Large employers & higher managerial	Higher professional	Lower managerial & professional	Intermediate	Small employers	Lower supervisory
Weighted number of households (thousands)	1,260	1,770	4,620	1,350	1,580	1,690
Total number of households in sample	260	350	910	270	330	320
Total number of persons in sample	710	930	2,430	640	870	860
Total number of adults in sample	520	680	1,790	480	650	650
Weighted average number of persons per household	2.8	2.6	2.7	2.4	2.7	2.7
Commodity or service	Average weekly household expenditure (£)					
Primary dwelling						
Rent	**27.10**	**33.40**	**27.70**	**34.50**	**27.90**	**33.60**
Gross rent	27.10	33.40	27.70	34.50	27.90	33.60
less housing benefit, rebates and allowances received	0.40	2.60	1.70	4.80	7.80	3.10
Net rent[3]	26.70	30.80	26.00	29.70	20.20	30.50
Mortgage	**127.70**	**91.10**	**86.20**	**61.40**	**56.60**	**62.30**
Mortgage interest payments	71.10	48.50	45.80	31.70	31.50	28.40
Mortgage protection premiums	3.80	2.40	2.70	1.90	1.90	2.80
Capital repayment of mortgage	52.80	40.20	37.60	27.70	23.20	31.20
Outright purchase, including deposits	**[0.40]**	**[0.20]**	**[0.20]**	**[0.80]**	**[0.10]**	**[0.30]**
Secondary dwelling	**20.20**	**28.40**	**2.30**	**[0.70]**	**[9.20]**	**[3.20]**
Rent	-	[0.30]	[0.10]	-	-	[0.10]
Council tax, mortgage, insurance (secondary dwelling)	[4.90]	[6.80]	2.00	[0.20]	[2.80]	[2.10]
Purchase of second dwelling	[15.30]	[21.40]	[0.20]	[0.60]	[6.50]	[1.00]
Charges	**37.70**	**35.10**	**32.50**	**29.50**	**30.80**	**27.50**
Council tax, domestic rates	28.60	27.00	24.20	21.30	23.10	20.50
Water charges	8.10	7.30	7.00	6.70	7.10	6.60
Other regular housing payments including service charge for rent	0.70	0.80	1.00	1.50	[0.60]	0.40
Refuse collection, including skip hire	[0.30]	-	[0.30]	-	-	-
Moving house	**[4.20]**	**[2.40]**	**2.20**	**[1.90]**	**[1.30]**	**[2.10]**
Property transaction - purchase and sale	[2.10]	[0.80]	[0.90]	[0.30]	[0.10]	[1.10]
Property transaction - sale only	[0.30]	[0.80]	[0.50]	[0.60]	[0.80]	[0.70]
Property transaction - purchase only	[0.90]	[0.40]	0.60	[0.80]	[0.20]	[0.20]
Property transaction - other payments	[0.90]	[0.40]	0.20	[0.20]	[0.10]	[0.10]
Maintenance and repair of dwelling	**11.10**	**12.20**	**8.50**	**6.00**	**6.20**	**6.20**
Central heating repairs	1.80	2.70	1.70	1.00	0.70	1.50
House maintenance etc.	6.50	7.20	4.70	2.40	2.50	2.70
Paint, wallpaper, timber	1.80	1.50	1.50	[0.80]	1.80	1.00
Equipment hire, small materials	1.00	0.80	0.60	1.90	1.20	1.00
Alterations and improvements to dwelling	**73.20**	**47.60**	**29.60**	**18.10**	**10.80**	**19.30**
Central heating installation	[2.20]	[3.80]	1.80	[1.20]	[1.10]	[0.60]
DIY improvements: double glazing, kitchen units, sheds etc.	[1.30]	[0.90]	[1.60]	-	[0.20]	[1.40]
Home improvements - contracted out	68.60	42.40	24.40	11.60	9.30	16.20
Bathroom fittings	[0.20]	[0.50]	1.10	[5.30]	[0.10]	[1.00]
Purchase of materials for capital improvements	[0.70]	[0.00]	[0.70]	[0.00]	[0.10]	[0.20]
Household insurances	**8.10**	**7.60**	**6.40**	**5.10**	**6.00**	**5.20**
Structure	4.20	3.90	3.30	2.40	3.10	2.50
Contents	3.80	3.70	3.10	2.60	2.90	2.50
Household appliances	[0.10]	[0.10]	0.10	[0.10]	[0.10]	[0.20]
Housing expenditure	**309.30**	**255.60**	**193.80**	**153.20**	**141.20**	**156.60**
Total expenditure[4]	**1,516.70**	**1,202.80**	**970.60**	**704.90**	**688.00**	**731.60**

Please see page xiii for symbols and conventions used in this report.
1 Includes those who have never worked.
2 Includes those who are economically inactive.
3 The figure included in total expenditure is net rent as opposed to gross rent.
4 This total includes all categories recorded in the LCF, including those outside the 'COICOP' total expenditure.

Table 2.6 Housing expenditure by socio-economic classification of household reference person, 2010 (cont.)
United Kingdom

	Semi-routine	Routine	Long-term unemployed[1]	Students	Occupation not stated[2] & not classifiable	All groups
Weighted number of households (thousands)	1,770	1,520	490	590	9,680	26,320
Total number of households in sample	350	290	100	100	1,990	5,260
Total number of persons in sample	870	770	270	260	3,560	12,180
Total number of adults in sample	650	590	150	180	3,090	9,430
Weighted average number of persons per household	2.6	2.7	2.9	2.7	1.8	2.3
Commodity or service	Average weekly household expenditure (£)					
Primary dwelling						
Rent	**54.30**	**53.80**	**104.80**	**107.40**	**40.20**	**39.90**
Gross rent	54.30	53.80	104.80	107.40	40.20	39.90
less housing benefit, rebates and allowances received	11.70	12.50	83.40	26.80	28.30	15.50
Net rent[3]	42.70	41.30	21.40	80.50	11.80	24.40
Mortgage	**34.70**	**32.40**	**4.20**	**38.40**	**5.00**	**44.90**
Mortgage interest payments	17.50	15.40	[2.50]	23.50	2.90	23.80
Mortgage protection premiums	1.00	1.10	[0.20]	[0.60]	0.10	1.40
Capital repayment of mortgage	16.20	15.90	[1.60]	14.30	1.90	19.70
Outright purchase, including deposits	**[0.30]**	**[0.10]**	-	-	**[0.00]**	**[0.20]**
Secondary dwelling	**[2.50]**	**[1.20]**	**[0.20]**	-	**[1.70]**	**5.00**
Rent	-	-	-	-	[0.00]	[0.00]
Council tax, mortgage, insurance (secondary dwelling)	[0.10]	[0.90]	[0.20]	-	[0.40]	1.60
Purchase of second dwelling	[2.40]	[0.40]	-	-	[1.30]	3.40
Charges	**26.40**	**25.10**	**11.60**	**17.40**	**22.80**	**27.20**
Council tax, domestic rates	18.50	17.90	5.40	8.80	15.10	19.40
Water charges	6.80	6.40	6.10	6.20	6.10	6.60
Other regular housing payments including service charge for rent	0.80	[0.80]	[0.20]	[2.40]	1.50	1.10
Refuse collection, including skip hire	[0.40]	-	-	-	[0.00]	[0.10]
Moving house	**[1.60]**	**[0.30]**	**[0.40]**	**[1.10]**	**1.10**	**1.60**
Property transaction - purchase and sale	[0.80]	[0.00]	[0.20]	[0.40]	[0.60]	0.70
Property transaction - sale only	[0.60]	-	[0.20]	-	[0.40]	0.50
Property transaction - purchase only	[0.00]	[0.30]	-	[0.10]	[0.10]	0.30
Property transaction - other payments	[0.10]	[0.00]	-	[0.60]	[0.00]	0.20
Maintenance and repair of dwelling	**3.20**	**2.80**	**[1.30]**	**3.20**	**6.20**	**6.70**
Central heating repairs	0.80	0.90	[0.10]	[0.30]	1.20	1.30
House maintenance etc.	1.40	[0.50]	[0.20]	[1.50]	3.60	3.60
Paint, wallpaper, timber	0.90	0.50	[1.10]	[0.60]	0.60	1.00
Equipment hire, small materials	[0.10]	[0.90]	[0.10]	[0.80]	0.70	0.80
Alterations and improvements to dwelling	**9.20**	**7.50**	**0.80**	**8.50**	**10.30**	**19.80**
Central heating installation	[2.20]	[0.70]	[0.50]	[0.50]	0.80	1.40
DIY improvements: double glazing, kitchen units, sheds etc.	[0.60]	[0.20]	[0.00]	[0.00]	0.50	0.70
Home improvements - contracted out	6.30	6.20	[0.30]	[8.00]	8.80	16.80
Bathroom fittings	[0.00]	[0.20]	-	[0.00]	[0.10]	0.60
Purchase of materials for capital improvements	[0.00]	[0.20]	-	-	[0.10]	0.20
Household insurances	**3.40**	**3.50**	**1.10**	**2.30**	**4.10**	**5.00**
Structure	1.50	1.60	[0.30]	1.10	2.00	2.50
Contents	1.80	1.90	0.70	1.20	2.00	2.50
Household appliances	[0.10]	[0.00]	-	[0.00]	0.10	0.10
Housing expenditure	**124.00**	**114.30**	**41.10**	**151.40**	**63.00**	**134.70**
Total expenditure[4]	**515.00**	**555.30**	**226.50**	**682.00**	**351.40**	**664.00**

Please see page xiii for symbols and conventions used in this report.

1 Includes those who have never worked.
2 Includes those who are economically inactive.
3 The figure included in total expenditure is net rent as opposed to gross rent.
4 This total includes all categories recorded in the LCF, including those outside the 'COICOP' total expenditure.

Table 2.7   Housing expenditure by household composition, 2010
United Kingdom

	Retired households		Non-retired		Retired and non-retired households			
	One Person	Two adults	One Person	Two adults	One adult with children	Two adults with children	Three or more adults without children	with children
Weighted number of households (thousands)	3,770	2,760	4,030	5,760	1,410	5,160	2,200	1,230
Total number of households in sample	710	640	800	1,170	330	1,060	370	200
Total number of persons in sample	710	1,270	800	2,340	900	3,990	1,220	950
Total number of adults in sample	710	1,270	800	2,340	330	2,110	1,220	650
Weighted average number of persons per household	1.0	2.0	1.0	2.0	2.8	3.7	3.4	4.8

Commodity or service	Average weekly household expenditure (£)							
Primary dwelling								
Rent	**32.30**	**14.20**	**50.50**	**36.90**	**105.30**	**38.80**	**32.30**	**43.30**
Gross rent	32.30	14.20	50.50	36.90	105.30	38.80	32.30	43.30
less housing benefit, rebates & allowances received	22.70	8.50	19.90	3.90	70.80	14.30	4.00	10.90
Net rent[1]	9.70	5.70	30.50	33.10	34.50	24.50	28.20	32.40
Mortgage	**1.70**	**2.70**	**32.10**	**57.90**	**21.30**	**91.40**	**49.90**	**75.60**
Mortgage interest payments	1.10	1.30	17.40	30.00	12.60	51.10	22.30	36.10
Mortgage protection premiums	[0.00]	[0.00]	0.90	1.70	0.70	3.00	2.00	2.10
Capital repayment of mortgage	[0.60]	[1.40]	13.90	26.20	8.00	37.40	25.60	37.40
Outright purchase, including deposits	-	[0.10]	[0.10]	[0.20]	[0.00]	[0.10]	[0.50]	[0.90]
Secondary dwelling	**0.10**	**0.50**	**1.10**	**14.20**	**0.30**	**5.30**	**6.10**	**1.30**
Rent	-	-	[0.10]	[0.10]	[0.00]	-	[0.10]	-
Council tax, mortgage, insurance (secondary dwelling)	[0.10]	[0.50]	[0.80]	2.90	[0.20]	3.30	[0.90]	[0.30]
Purchase of second dwelling	-	-	[0.20]	11.20	-	[2.00]	[5.10]	[1.00]
Charges	**20.10**	**31.10**	**20.40**	**31.60**	**15.30**	**30.70**	**32.10**	**32.40**
Council tax, domestic rates	12.40	23.30	13.30	23.50	8.60	22.50	23.70	23.70
Water charges	5.20	6.60	5.40	6.90	6.70	7.50	7.60	8.60
Other regular housing payments including service charge for rent	2.50	1.20	1.70	1.00	[0.10]	0.50	0.50	[0.10]
Refuse collection, including skip hire	-	[0.00]	[0.00]	[0.10]	-	[0.10]	[0.30]	-
Moving house	**0.60**	**1.80**	**1.70**	**2.10**	**1.70**	**2.00**	**0.90**	**1.10**
Property transaction - purchase and sale	[0.40]	[1.10]	[1.00]	[0.50]	[1.00]	[0.80]	[0.40]	[1.00]
Property transaction - sale only	[0.20]	[0.60]	[0.30]	[0.80]	[0.50]	[0.50]	[0.20]	-
Property transaction - purchase only	-	[0.10]	[0.30]	0.60	-	[0.40]	[0.10]	-
Property transaction - other payments	[0.00]	[0.00]	[0.10]	0.20	[0.10]	0.40	[0.20]	[0.10]
Maintenance and repair of dwelling	**5.40**	**8.60**	**3.50**	**8.30**	**2.30**	**8.50**	**6.60**	**6.60**
Central heating repairs	1.10	1.90	0.70	1.50	1.10	1.30	1.80	1.20
House maintenance etc.	3.80	5.10	2.10	4.00	0.60	4.80	2.80	2.20
Paint, wallpaper, timber	0.40	0.60	0.50	1.30	[0.50]	1.40	1.20	2.80
Equipment hire, small materials	[0.10]	1.00	0.20	1.50	[0.10]	1.00	0.70	[0.50]
Alterations and improvements to dwelling	**5.40**	**17.50**	**11.20**	**28.10**	**4.00**	**35.70**	**15.70**	**16.60**
Central heating installation	[0.90]	[0.90]	1.40	1.70	[0.50]	2.10	[1.20]	[0.50]
DIY improvements: double glazing, kitchen units, sheds etc.	[0.00]	[1.00]	[1.70]	1.00	[0.20]	[0.40]	[0.60]	-
Home improvements - contracted out	4.30	15.50	7.90	22.80	[3.20]	32.30	13.60	15.90
Bathroom fittings	[0.10]	[0.10]	[0.10]	2.00	[0.10]	[0.60]	[0.30]	[0.20]
Purchase of materials for capital improvements	[0.00]	[0.00]	[0.00]	[0.70]	[0.10]	[0.40]	[0.00]	[0.10]
Household insurances	**3.70**	**5.60**	**3.50**	**5.70**	**2.60**	**5.90**	**6.10**	**6.50**
Structure	1.70	2.80	1.80	2.80	1.00	2.90	3.20	3.30
Contents	1.90	2.70	1.80	2.90	1.50	2.90	2.70	3.20
Household appliances	[0.10]	[0.10]	[0.00]	0.10	[0.10]	[0.10]	[0.20]	[0.10]
Housing expenditure	**46.60**	**73.60**	**104.20**	**181.20**	**82.00**	**204.10**	**146.10**	**173.50**
Total expenditure[2]	**230.20**	**441.60**	**411.80**	**867.60**	**358.70**	**949.40**	**946.70**	**1011.30**

Note: Please see page xiii for symbols and conventions used in this report.

1 The figure included in total expenditure is net rent as opposed to gross rent.

2 This total includes all categories recorded in the LCF, including those outside the 'COICOP' total expenditure.

Table 2.8 Expenditure on rent[1] by renters, 2008 to 2010
United Kingdom

	2008		2009		2010	
	£[2]	% of total expenditure	£[2]	% of total expenditure	£[2]	% of total expenditure
Weighted number of households (thousands)	7,520		7,980		8,640	
Total number of households in sample	1,610		1,680		1,620	
Total number of persons in sample	3,610		3,780		3,620	
Total number of adults in sample	2,570		2,710		2,580	
Weighted average number of persons per household	2.3		2.2		2.3	
Total expenditure for renters	**420.90**		**412.20**		**435.10**	
Rent	**107.70**	**25.6**	**115.40**	**28.0**	**121.50**	**27.9**
Gross rent	107.70	25.6	115.40	28.0	121.50	27.9
less housing benefit, rebates and allowances received	41.50	9.9	44.70	10.8	47.10	10.8
Net rent[3]	66.30	15.7	70.70	17.2	74.40	17.1

Note: Please see page xiii for symbols and conventions used in this report.

1 Primary dwelling.

2 Average weekly household expenditure (£).

3 The figure included in total expenditure is net rent as opposed to gross rent.

Table 2.9 Expenditure on mortgages[1] by mortgage holders, 2008 to 2010
United Kingdom

	2008		2009		2010	
	£[2]	% of total expenditure	£[2]	% of total expenditure	£[2]	% of total expenditure
Weighted number of households (thousands)	9,830		9,460		8,970	
Total number of households in sample	2,210		2,100		1,810	
Total number of persons in sample	6,330		5,960		5,030	
Total number of adults in sample	4,450		4,210		3,620	
Weighted average number of persons per household	2.8		2.8		2.8	
Total expenditure for mortgage payers	**985.30**		**941.30**		**974.50**	
Mortgage	**148.50**	**15.1**	**133.00**	**14.1**	**130.80**	**13.4**
Mortgage interest payments	97.40	9.9	73.00	7.8	69.20	7.1
Mortgage protection premiums	4.80	0.5	4.40	0.5	4.10	0.4
Capital repayment of mortgage	46.30	4.7	55.60	5.9	57.50	5.9

Note: Please see page xiii for symbols and conventions used in this report.

1 Primary dwelling.

2 Average weekly household expenditure (£).

Table 2.10 Expenditure on rent and mortgages[1] by renters and mortgage holders by gross income decile group, 2010
United Kingdom

| | Gross income decile group | | | | | | | | | | |
	1	2	3	4	5	6	7	8	9	10	All
Weighted number of households (thousands)	1,670	1,360	1,070	1,010	910	730	690	570	340	310	8,640
Total number of households in sample	320	270	210	200	160	130	120	100	60	50	1,620
Total number of persons in sample	440	520	420	470	480	380	340	270	170	150	3,620
Total number of adults in sample	340	350	300	310	300	260	260	210	140	120	2,580
Weighted average number of persons per household	1.3	1.9	2.0	2.4	2.8	2.8	2.8	2.8	3.0	3.4	2.3
Commodity or service	Average weekly household expenditure (£)										
Rent for renters	**113.60**	**112.90**	**114.50**	**114.20**	**112.90**	**104.10**	**121.00**	**137.60**	**152.20**	**255.40**	**121.50**
Gross rent	113.60	112.90	114.50	114.20	112.90	104.10	121.00	137.60	152.20	255.40	121.50
less housing benefit, rebates and allowances received	81.90	77.20	68.00	44.60	33.30	11.90	[9.40]	[2.30]	[5.40]	[0.30]	47.10
Net rent[2]	31.70	35.60	46.60	69.60	79.60	92.20	111.60	135.30	146.80	255.20	74.40
Weighted number of households (thousands)	230	200	300	530	680	930	1,240	1,470	1,680	1,720	8,970
Total number of households in sample	50	40	70	110	150	190	250	290	330	330	1,810
Total number of persons in sample	70	70	150	220	350	480	710	840	1,040	1,090	5,030
Total number of adults in sample	60	60	100	170	250	350	520	600	740	780	3,620
Weighted average number of persons per household	1.4	1.7	2.1	2.0	2.4	2.6	2.9	2.9	3.1	3.2	2.8
Commodity or service	Average weekly household expenditure (£)										
Mortgage for mortgage holders	**84.00**	**70.40**	**75.30**	**87.80**	**103.70**	**95.40**	**113.40**	**121.20**	**151.80**	**197.20**	**130.80**
Mortgage interest payments	57.10	36.20	39.90	50.50	48.30	50.50	58.90	62.70	81.00	105.50	69.20
Mortgage protection premiums	[2.20]	[1.70]	3.20	2.10	2.90	3.80	3.30	4.60	5.00	5.20	4.10
Capital repayment of mortgage	24.70	32.50	32.20	35.20	52.60	41.10	51.10	53.90	65.80	86.50	57.50

Note: Please see page xiii for symbols and conventions used in this report.
1 Primary dwelling.
2 The figure included in total expenditure is net rent as opposed to gross rent.

Table 2.11   Expenditure on rent and mortgages[1] by renters and mortgage holders by UK Countries and region, 2010

	North East	North West	Yorks & the Humber	East Midlands	West Midlands	East	London
Weighted number of households (thousands)	460	950	750	590	720	740	1,240
Total number of households in sample	90	180	150	110	140	150	190
Total number of persons in sample	190	370	330	250	310	330	490
Total number of adults in sample	150	270	230	190	220	250	330
Weighted average number of persons per household	2.1	2.1	2.1	2.3	2.3	2.2	2.7
Commodity or service	Average weekly household expenditure (£)						
Rent by renters	**92.00**	**104.50**	**98.30**	**99.70**	**109.00**	**118.00**	**199.90**
Gross rent	92.00	104.50	98.30	99.70	109.00	118.00	199.90
less housing benefit, rebates and allowances received	51.00	45.20	40.20	27.20	50.20	46.20	67.60
Net rent[2]	41.00	59.30	58.10	72.50	58.80	71.70	132.30
Weighted number of households (thousands)	450	1,060	770	660	780	840	960
Total number of households in sample	100	200	170	150	160	170	150
Total number of persons in sample	250	570	480	400	460	490	420
Total number of adults in sample	190	410	340	290	330	350	310
Weighted average number of persons per household	2.6	2.8	2.8	2.7	2.8	2.8	2.7
Commodity or service	Average weekly household expenditure (£)						
Mortgage by mortgage holders	**102.70**	**115.50**	**105.80**	**124.20**	**108.80**	**153.80**	**180.90**
Mortgage interest payments	47.30	60.30	54.10	66.80	51.30	82.80	101.90
Mortgage protection premiums	4.30	3.80	4.20	3.80	3.20	4.10	4.70
Capital repayment of mortgage	51.10	51.40	47.50	53.60	54.30	67.00	74.30

	South East	South West	England	Wales	Scotland	Northern Ireland	United Kingdom
Weighted number of households (thousands)	1,080	740	7,270	360	810	200	8,640
Total number of households in sample	200	150	1,350	70	150	40	1,620
Total number of persons in sample	460	330	3,060	180	300	80	3,620
Total number of adults in sample	310	220	2,160	120	240	60	2,580
Weighted average number of persons per household	2.4	2.1	2.3	2.4	1.9	2.0	2.3
Commodity or service	Average weekly household expenditure (£)						
Rent by renters	**138.20**	**113.40**	**126.70**	**112.60**	**88.00**	**85.80**	**121.50**
Gross rent	138.20	113.40	126.70	112.60	88.00	85.80	121.50
less housing benefit, rebates and allowances received	55.80	41.00	49.10	40.10	34.40	38.40	47.10
Net rent[2]	82.40	72.40	77.50	72.50	53.60	47.40	74.40
Weighted number of households (thousands)	1,170	730	7,420	420	880	260	8,970
Total number of households in sample	230	160	1,490	90	180	50	1,810
Total number of persons in sample	650	450	4,150	240	480	160	5,030
Total number of adults in sample	470	320	2,990	180	350	100	3,620
Weighted average number of persons per household	2.9	2.9	2.8	2.7	2.7	3.1	2.8
Commodity or service	Average weekly household expenditure (£)						
Mortgage by mortgage holders	**151.80**	**135.20**	**134.20**	**105.40**	**116.70**	**122.10**	**130.80**
Mortgage interest payments	83.50	76.20	71.70	53.60	58.80	59.80	69.20
Mortgage protection premiums	4.50	3.40	4.00	3.10	4.00	7.90	4.10
Capital repayment of mortgage	63.70	55.60	58.60	48.70	53.80	54.40	57.50

Note: Please see page xiii for symbols and conventions used in this report.

1 Primary dwelling.

2 The figure included in total expenditure is net rent as opposed to gross rent.

Equivalised income

Background

Equivalisation is a standard methodology that adjusts household income to account for different demands on resources, by considering the household size and composition. The purpose of this chapter is to show the impact of using this methodology on Living Costs and Food Survey (LCF) data. This is the only chapter that presents equivalised income data; other tables included in Family Spending are available on an equivalised income basis on request from the Office for National Statistics (ONS) (see page xvi Introduction).

Equivalisation Methodology

When the incomes of households are compared, income is often adjusted in order to take different demands on resources into account. Household size is an important factor to consider because larger households usually need a higher income than smaller households in order to achieve a comparable standard of living. The composition of a household also affects resource needs, for example living costs for adults are normally higher than those for children.

Equivalisation scales are used to adjust household income in such a way that both household size and composition are taken into account. There are various scales available, which differ in their complexity and methodology. For example, the Organisation for Economic Cooperation and Development (OECD) modified equivalence scale is used widely across Europe: it adjusts household income to reflect the different resource needs of single adults, any additional adults in the household, and children in various age groups.

The OECD-modified equivalence scale is the standard scale for the Statistical Office of the European Union (EUROSTAT) and several government departments in the UK use it for key household income statistics. For example, the Department for Work and Pensions (DWP) use the OECD-modified scale for their Households Below Average Income (HBAI) publication and ONS use it for the Effects of Taxes and Benefits on Household Income (ETB) analysis.

To calculate equivalised income using the OECD-modified equivalence scale, each member of the household is first given an equivalence value. The OECD-modified equivalence values are shown in the table below. Single adult households are taken as the reference group and are given a value of one. For larger households, each additional adult is given a smaller value of 0.5 to reflect the economies of scale achieved when people live together. Economies of scale arise when households share resources such as water and electricity, which reduces the living costs per person. Children under the age of 14 are given a value of 0.3 to take account of their lower living costs while children aged 14 and over are given a value of 0.5 because their living costs are assumed to be the same as those of an adult.

OECD-modified equivalence scale	
Type of Household Member	**Equivalence value**
First adult	1.0
Additional adult	0.5
Child aged: 14 and over	0.5
Child aged: 0–13	0.3

In the next stage of the calculation, the equivalence values for each household member are summed to give a total equivalence number for the household. For example, the total equivalence value for a household containing a married couple with two children aged 10 and 14 is calculated as follows:

1 (first adult) + 0.5 (second adult) + 0.5 (14-year-old child) + 0.3 (10-year-old child) = 2.3

The total equivalence value of 2.3 shows that the household needs more than twice the income of a single adult household in order to achieve a comparable standard of living.

In the final step of the calculation the total income for the household is divided by the equivalence value. For example, if the household described in the example above has an annual income of £30,000, their equivalised income is calculated as follows:

£30,000/2.3 = £13,043

For a single adult household with an actual income of £30,000 the equivalised income remains at £30,000, because the equivalence value for this household is equal to one. This demonstrates that a single adult household will have a higher standard of living than a larger household with the same level of income.

Results

Equivalised household incomes were calculated for each household using the OECD-modified equivalence scale. Household equivalised incomes were then ranked in ascending order and divided into ten equally sized (decile) groups, with households having the lowest equivalised income in the first decile group. Gross (non-equivalised) income data are presented in Tables 3.1 to 3.12; equivalised gross income data based on the OECD-modified scale are shown in Tables 3.2E to 3.12E.

The income decile groups were as follows:

Income decile	Gross weekly income	Gross weekly equivalised income (OECD-modified scale)
1	Up to £159	Up to £131
2	£160 to £237	£132 to £179
3	£238 to £314	£180 to £226
4	£315 to £412	£227 to £276
5	£413 to £521	£277 to £332
6	£522 to £650	£333 to £399
7	£651 to £800	£400 to £483
8	£801 to £1,014	£484 to £594
9	£1,015 to £1,367	£595 to £793
10	£1,368 and over	£794 and over

Household composition by income groups

Table 3.1 shows the household composition of the gross (non-equivalised) income decile groups and the OECD-equivalised income decile groups. Equivalisation has a large impact on those groups of households containing one adult without children. The effects of equivalisation were particularly noticeable for households containing one retired adult. These households accounted for just under two-fifths (38 per cent) of households in the lowest gross income decile group but when income was equivalised they accounted for only 12 per cent of the lowest income group. These households tended to move to a higher income decile group after income was equivalised. For example, households containing one retired adult made up 7 per cent of the fifth gross income decile group, but after income was equivalised this group accounted for 15 per cent of the fifth decile group, which was due to households moving up the income distribution. Households containing one non-retired adult also moved up the income distribution after income was equivalised. These results demonstrate how equivalisation increases relatively the incomes of one-adult households.

Table 3.1 also shows how equivalisation affects the average household size for each income decile group. As gross income increases the average number of people in each household also increases: the average household size for the highest income group (3.2 people) was almost two and a half times that of the lowest income group (1.3 people). After income was equivalised, the average number of people in each household was more similar for each income decile group, with the average varying between 2.0 and 2.6. This pattern of results occurs because the equivalisation process scales up the income of households containing one adult (relative to other households) and scales down the income of households with more people.

Figures 3.1 and 3.1E show the percentage of households with children in each income group before and after income equivalisation. As gross income increases, the proportion of households with children generally increases: from 15 per cent of households in the bottom gross income decile group to 44 per cent in the top gross income decile group. In contrast, after equivalisation households with children are most likely to be found in the bottom income decile group; just over two-fifths (42 per cent) of households in this group contained at least one child. The proportion of households with children was only 23 per cent in the second equivalised income decile group compared with 35 per cent in the fifth income group. After the fifth decile group, the proportion of households with children fell slightly. These results demonstrate how factoring in living costs for children as part of the equivalisation process can bring about large changes in the income distribution.

Figure 3.1 **Percentage of households with children in each gross income decile group, 2010,**
United Kingdom

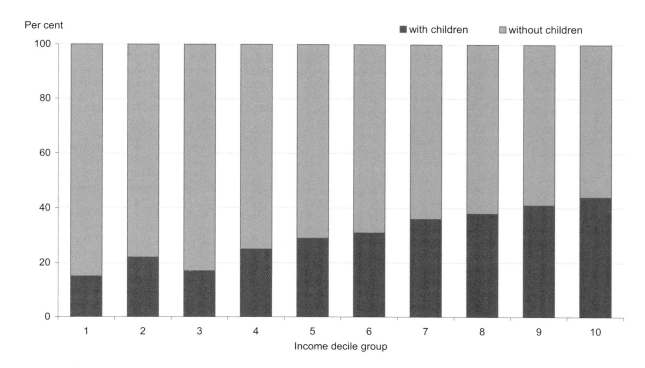

Figure 3.1E **Percentage of households with children in each gross OECD-modified equivalised income decile group, 2010**
United Kingdom

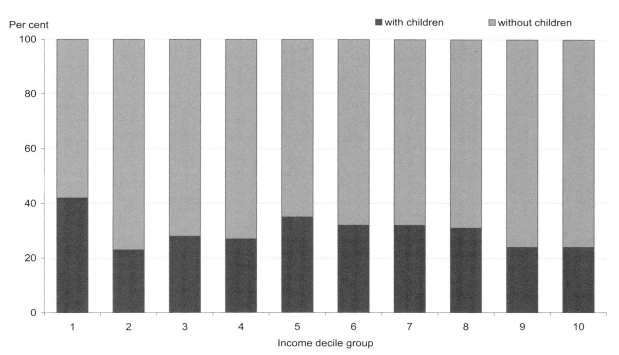

The proportion of households containing at least one retired person by income decile group before and after income equivalisation is shown in Figures 3.2 and 3.2E. Equivalisation has a large effect on the proportion of retired households in the lowest income decile group. Retired households accounted for just under two-fifths (39 per cent) of households in the bottom gross income group but after equivalisation they accounted for only 17 per cent of households in the bottom income group. This result can largely be explained by the fact that a relatively high proportion of retired households contain only one adult and, as explained above, the incomes of single adult households are scaled up (relative to other households) when income is equivalised. The proportion of retired households in the second lowest income decile also decreased after equivalisation, although the effect was much smaller. The opposite was true of the higher income decile groups; the proportion of retired households increased slightly after income was equivalised.

Figure 3.2   **Percentage of retired and non-retired households by gross income decile group, 2010**
United Kingdom

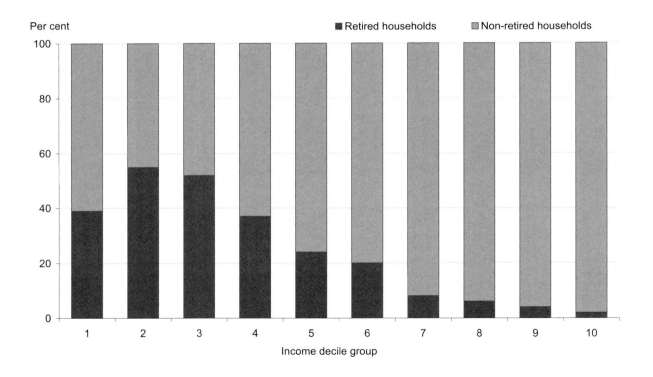

Figure 3.2E   **Percentage of retired and non-retired households by
OECD-modified equivalised income decile group, 2010**
United Kingdom

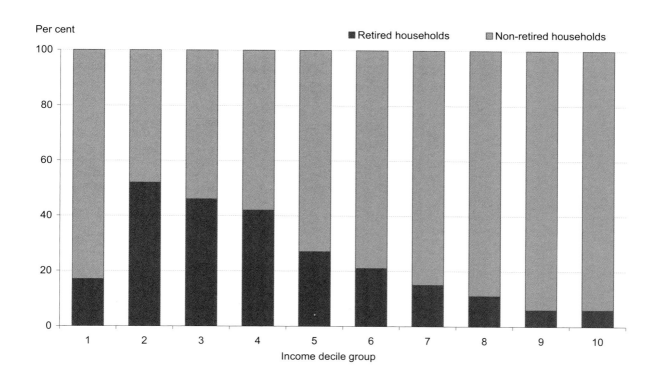

Household expenditure by income

Tables 3.2 and 3.2E show expenditure, in total and for each of the Classification of Individual COnsumption by Purpose (COICOP) categories, by gross and equivalised income decile groups respectively. As incomes increase with both measures of income, total expenditure also increased, although the gap in spending between the top and the bottom income group was slightly smaller when the equivalised income measure was used. Households in the top gross income decile group spent an average of £1,018.50 per week; five and a half times that of households in the bottom gross income group (£185.60). In comparison, the top equivalised income group spent £906.20 per week, which was just under four times higher than that of the lowest equivalised income group (£236.00). For each COICOP category spending rose consistently with income, although as with total expenditure, the difference in spending between the top and the bottom income groups tended to be smaller when looking at equivalised income.

For most COICOP categories, expenditure in the lower part of the income distribution was lower for the gross income decile groups than for the corresponding equivalised income groups. The opposite was true for the higher income decile groups. Therefore equivalisation flattens the distribution of household expenditure. This pattern of results can be illustrated using the examples of expenditure on food and non-alcoholic drinks, and on clothing and footwear. As shown in Figure 3.3, average weekly expenditure on food and non-alcoholic drinks for the bottom gross income group was £27.20 compared with £35.30 for the bottom equivalised income group. Spending was also higher for the equivalised income groups than for the gross income groups in the next deciles, from the second decile group (the second lowest) up to the sixth. For the top two decile groups spending was higher in the gross income groups than the equivalised income group. For the top income group, for example, the average weekly spend on food and non-alcoholic drinks was £83.40 per week compared with £66.80 for the top equivalised income group. Figure 3.4 shows that the pattern of results was similar for expenditure on clothing and footwear.

Figure 3.3   **Expenditure on food and non-alcoholic drinks by gross and OECD-modified equivalised income decile group, 2010**
United Kingdom

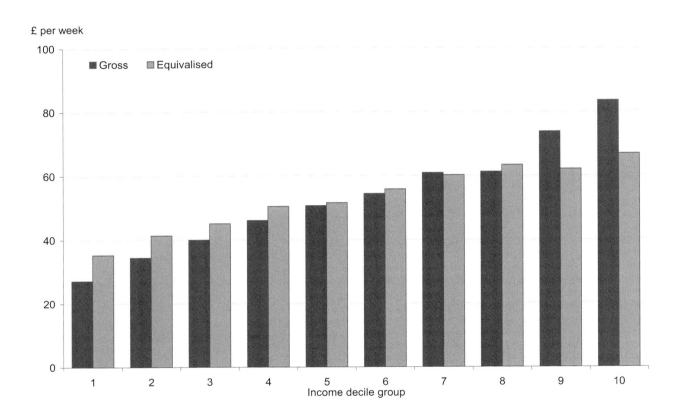

£ per week

■ Gross   ▨ Equivalised

Income decile group

Figure 3.4   **Expenditure on clothing and footwear by gross and OECD-modified equivalised income decile group, 2010**
United Kingdom

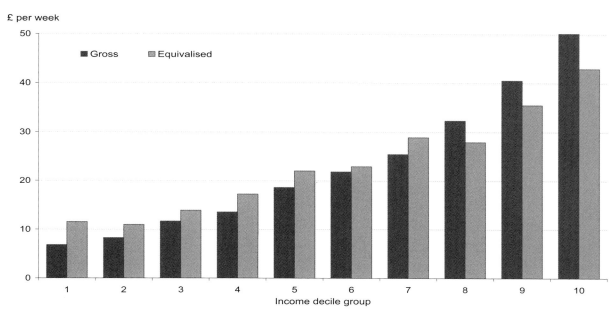

£ per week

Tables 3.3 and Table 3.3E show the share of total expenditure on each COICOP category, by gross income group and equivalised income group respectively. There was notable variation in spending with income for some COICOP categories. As shown in Figure 3.5, the proportion of total expenditure spent on food and non-alcoholic drinks, and on housing, fuel and power, when these categories are combined, decreased steadily as equivalised income increased. Spending on these categories combined accounted for just under two-fifths (36 per cent) of total expenditure for the bottom equivalised income decile group, compared with just under a fifth (16 per cent) for households in the top equivalised income decile group. In contrast, the proportion of total expenditure spent on transport increased slightly with equivalised income, from 10 per cent for the lowest decile group to 16 per cent in the highest income group. The pattern of results was similar when looking at spending by gross (unequivalised) income group.

Figure 3.5 **Percentage of total expenditure on selected items by OECD-modified equivalised income decile group, 2010**
United Kingdom

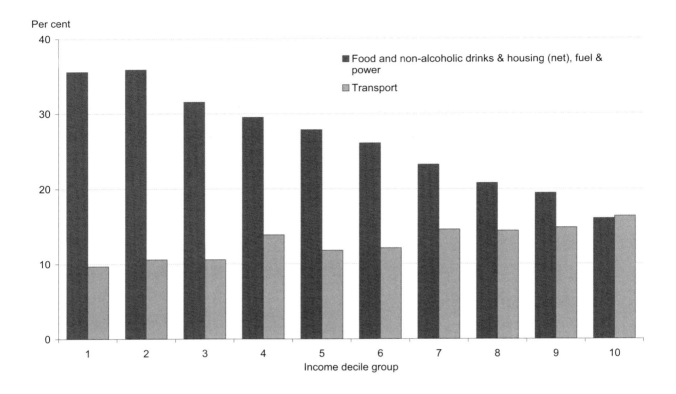

Household expenditure by household composition and income

This section describes the effect that equivalisation has when looking at the expenditure in the income quintile groups of different household types (see Tables 3.4 to 3.11 and Tables 3.4E to 3.11E). The analysis focuses on one and two adult households, with and without children. It should be noted that the sample for some groups, particularly retired households who are mainly dependent on the state pension, contain a small number of households and the results should therefore be treated with caution.

Equivalisation had a large impact on households containing one adult without children (see Tables 3.4, 3.5, 3.4E, and 3.5E). Expenditure for each income quintile group decreased after equivalisation and the effect was greatest for the highest income groups. This effect is due to the movements between income groups before and after equivalisation, as discussed earlier in the chapter. This was particularly noticeable for households containing one non-retired adult. Among these households, total expenditure was £169.80 for the bottom gross income group but after equivalisation total expenditure fell to £154.60 for this group. Equivalisation had a much larger impact on households in the higher income quintile groups; total expenditure for households containing one non-retired household in the top gross income quintile was £628.00 compared with £476.50 for the top equivalised income group.

Equivalisation had the opposite effect on expenditure for two adult households with children (Tables 3.6 and 3.6E). Expenditure for households containing two adults with children increased from £305.50 for the lowest gross income group to £372.30 after income was equivalised. There was a similar increase in expenditure after income was equivalised for each of the remaining quintiles.

Sources of income

Tables 3.12 and 3.12E, and Figures 3.6 and 3.6E show the breakdown of income sources for each income quintile, by gross household income and equivalised household income respectively. For both measures of income, the proportion of income obtained from self employment decreased for higher income groups, while, conversely, the proportion of income from wages and salaries increased.

Equivalisation mainly affected the distribution of income sources for the lowest income quintile groups. Annuities and pensions made up 10 per cent of the income received by households in the lowest gross income quintile groups, but after income was equivalised this income source accounted for only 5 per cent of total income received by the lowest income group. In contrast, the proportion of total income provided by wages and salaries increased after equivalisation. Among households in the bottom gross income quintile, 8 per cent of income came from salaries and wages compared with 12 per cent for households in the lowest equivalised income quintile. These results largely reflect the fact that after income was equivalised, the lowest quintile groups contained fewer pensioner households and more working households.

Figure 3.6   Sources of income by gross income decile group, 2010
United Kingdom

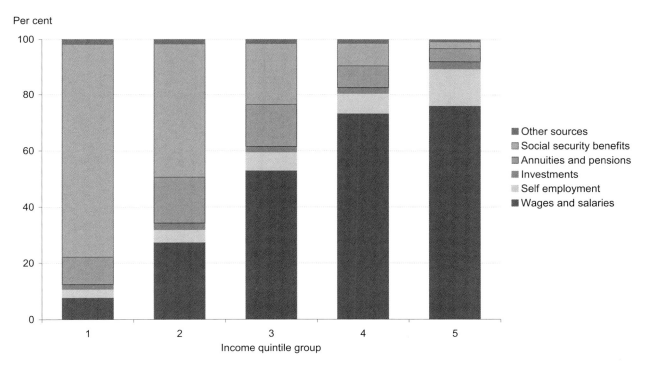

Figure 3.6E Sources of income by gross OECD-modified equivalised quintile group, 2010

United Kingdom

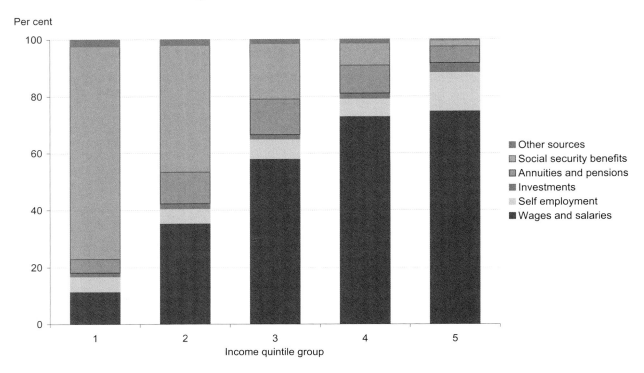

Table 3.1 **Percentage of households by composition in each gross and equivalised income decile group (OECD –modified scale), 2010**
United Kingdom

Percentages

	Income decile group									
	Lowest ten per cent		Second		Third		Fourth		Fifth	
	Gross	Equivalised	Gross	Equivalised	Gross	Equivalised	Gross	Equivalised	Gross	Equivalised
Lower boundary of group (£ per week)			160	132	238	180	315	227	413	277
Average size of household	**1.3**	**2.1**	**1.6**	**2.0**	**1.8**	**2.2**	**2.1**	**2.2**	**2.4**	**2.4**
One adult retired mainly dependent on state pensions[1]	15	6	9	13	5	4	[1]	4	-	[2]
One adult, other retired	23	6	36	25	26	23	13	21	7	13
One adult, non-retired	40	32	16	10	17	10	20	11	17	13
One adult, one child	10	11	4	4	4	3	4	[2]	[3]	[3]
One adult, two or more children	[2]	12	7	3	5	4	6	3	4	[1]
Two adults, retired mainly dependent on state pensions[1]	[0]	[2]	5	6	7	6	3	[2]	[1]	[1]
Two adults, other retired	[1]	[3]	5	7	13	13	20	15	17	12
Two adults, non-retired	6	6	6	7	12	10	14	10	23	15
Two adults, one child	[2]	8	7	4	[3]	8	8	6	7	10
Two adults, two children	[0]	5	[3]	[3]	[3]	5	[3]	9	7	12
Two adults, three children	[1]	[2]	[0]	[3]	[1]	[2]	[1]	[1]	[3]	[2]
Two adults, four or more children	[0]	[1]	[0]	[2]	[0]	[1]	[1]	[1]	[2]	[1]
Three adults	[0]	[1]	[1]	[3]	[1]	[3]	[3]	6	6	6
Three adults, one or more children	-	[2]	[1]	[2]	[1]	[3]	[1]	[2]	[3]	6
All other households without children	[0]	[1]	-	5	-	[3]	[1]	5	[1]	[4]
All other households with children	[0]	[1]	-	[1]	[0]	[2]	[0]	[1]	[1]	[1]

	Income decile group									
	Sixth		Seventh		Eighth		Ninth		Highest ten per cent	
	Gross	Equivalised	Gross	Equivalised	Gross	Equivalised	Gross	Equivalised	Gross	Equivalised
Lower boundary of group (£ per week)	522	333	651	400	801	484	1,015	595	1,368	794
Average size of household	**2.4**	**2.5**	**2.7**	**2.6**	**2.8**	**2.5**	**3.0**	**2.4**	**3.2**	**2.3**
One adult retired mainly dependent on state pensions[1]	-	[0]	-	-	-	-	-	-	-	-
One adult, other retired	5	9	[1]	6	[1]	6	[0]	[3]	[0]	[2]
One adult, non-retired	17	13	11	16	7	13	5	19	[4]	16
One adult, one child	[1]	[1]	[0]	[1]	[0]	[0]	-	[0]	[0]	[0]
One adult, two or more children	[2]	[1]	[1]	[1]	[0]	[1]	[1]	[1]	[0]	[0]
Two adults, retired mainly dependent on state pensions[1]	[0]	[0]	-	-	-	-	-	-	-	-
Two adults, other retired	15	11	7	9	5	4	[3]	4	[2]	4
Two adults, non-retired	24	20	33	21	35	31	34	36	31	41
Two adults, one child	12	10	11	11	14	13	12	9	11	10
Two adults, two children	10	9	13	10	12	10	14	9	17	10
Two adults, three children	[1]	[3]	[2]	[3]	[3]	[2]	4	[1]	4	[2]
Two adults, four or more children	[1]	[0]	[1]	-	[1]	[0]	[0]	-	[1]	[0]
Three adults	6	7	9	9	10	7	10	9	13	8
Three adults, one or more children	[3]	5	8	6	6	[4]	7	[3]	7	[2]
All other households without children	[2]	7	[3]	7	[4]	8	7	6	7	6
All other households with children	[2]	[1]	[0]	[2]	[1]	[1]	[3]	[1]	[4]	[0]

Note: Please see page xiii for symbols and conventions used in this report.

1 Mainly dependent on state pension and not economically active - see Appendix B.

Table 3.2   Household expenditure by gross income decile group, 2010
United Kingdom

	Lowest ten per cent	Second decile group	Third decile group	Fourth decile group	Fifth decile group	Sixth decile group
Lower boundary of group (£ per week)		160	238	315	413	522
Weighted number of households (thousands)	2,630	2,640	2,620	2,630	2,630	2,640
Total number of households in sample	510	530	540	550	550	530
Total number of persons in sample	690	860	960	1,150	1,280	1,290
Total number of adults in sample	570	670	780	900	970	1,010
Weighted average number of persons per household	1.3	1.6	1.8	2.1	2.4	2.4
Commodity or service	Average weekly household expenditure (£)					
1 Food & non-alcoholic drinks	27.20	34.50	40.00	46.10	50.70	54.40
2 Alcoholic drinks, tobacco & narcotics	7.10	7.90	8.60	10.30	12.10	13.20
3 Clothing & footwear	6.80	8.30	11.70	13.50	18.60	21.90
4 Housing (net)[1], fuel & power	42.50	45.10	50.30	58.10	61.20	60.70
5 Household goods & services	14.40	16.20	17.90	25.40	22.50	31.40
6 Health	1.40	3.00	3.40	3.60	4.00	5.00
7 Transport	16.40	19.60	29.20	43.30	48.40	60.70
8 Communication	6.80	7.60	9.00	10.60	11.30	13.40
9 Recreation & culture	18.70	25.40	31.30	39.90	52.40	54.80
10 Education	[2.70]	[1.60]	[4.60]	5.70	8.50	3.90
11 Restaurants & hotels	10.50	12.40	16.60	23.30	30.30	35.80
12 Miscellaneous goods & services	12.20	14.10	19.60	23.50	29.00	38.40
1-12 All expenditure groups	166.90	195.60	242.20	303.10	349.00	393.40
13 Other expenditure items	18.70	19.10	30.50	41.30	51.00	65.40
Total expenditure	185.60	214.80	272.70	344.40	400.00	458.80
Average weekly expenditure per person (£)						
Total expenditure	141.00	130.90	154.50	166.50	169.90	188.30

Note: The commodity and service categories are not comparable to those in publications before 2001-02.

Please see page xiii for symbols and conventions used in this report.

1 Excluding mortgage interest payments, council tax and Northern Ireland rates.

Table 3.2   **Household expenditure by gross income decile group, 2010 (cont.)**
United Kingdom

	Seventh decile group	Eighth decile group	Ninth decile group	Highest ten per cent	All house-holds
Low er boundary of group (£ per w eek)	651	801	1,015	1,368	
Weighted number of households (thousands)	2,630	2,640	2,630	2,630	26,320
Total number of households in sample	530	520	510	500	5,260
Total number of persons in sample	1,410	1,430	1,530	1,580	12,180
Total number of adults in sample	1,090	1,090	1,170	1,170	9,430
Weighted average number of persons per household	2.7	2.8	3.0	3.2	2.3
Commodity or service	Average w eekly household expenditure (£)				
1 Food & non-alcoholic drinks	60.90	61.20	73.70	83.40	53.20
2 Alcoholic drinks, tobacco & narcotics	12.10	13.60	15.70	17.60	11.80
3 Clothing & footwear	25.40	32.30	40.50	54.90	23.40
4 Housing (net)[1], fuel & power	67.90	68.70	64.70	84.70	60.40
5 Household goods & services	38.60	31.10	37.30	79.60	31.40
6 Health	4.60	5.50	7.50	12.60	5.00
7 Transport	77.90	87.40	111.00	155.00	64.90
8 Communication	14.60	16.70	18.00	22.20	13.00
9 Recreation & culture	59.80	73.10	92.80	132.60	58.10
10 Education	9.30	7.80	19.30	37.00	10.00
11 Restaurants & hotels	44.80	54.10	67.60	97.10	39.20
12 Miscellaneous goods & services	37.20	48.30	54.20	82.60	35.90
1-12 All expenditure groups	453.00	499.90	602.20	859.20	406.30
13 Other expenditure items	78.20	90.90	118.50	159.30	67.30
Total expenditure	**531.30**	**590.80**	**720.70**	**1018.50**	**473.60**
Average weekly expenditure per person (£)					
Total expenditure	**194.80**	**211.10**	**237.80**	**320.30**	**203.10**

Note: The commodity and service categories are not comparable to those in publications before 2001-02.
Please see page xiii for symbols and conventions used in this report.
1 Excluding mortgage interest payments, council tax and Northern Ireland rates.

Table 3.2E   Household expenditure by gross equivalised income decile group (OECD-modified scale), 2010
United Kingdom

	Lowest ten per cent	Second decile group	Third decile group	Fourth decile group	Fifth decile group	Sixth decile group
Lower boundary of group (£ per week)		132	180	227	277	333
Weighted number of households (thousands)	2,630	2,630	2,640	2,630	2,630	2,630
Total number of households in sample	520	530	530	540	520	540
Total number of persons in sample	1,120	1,070	1,150	1,200	1,230	1,320
Total number of adults in sample	700	800	880	940	950	1,040
Weighted average number of persons per household	2.1	2.0	2.2	2.2	2.4	2.5
Commodity or service	Average weekly household expenditure (£)					
1 Food & non-alcoholic drinks	35.30	41.40	45.10	50.50	51.60	55.80
2 Alcoholic drinks, tobacco & narcotics	10.30	8.90	10.40	9.50	11.50	11.40
3 Clothing & footwear	11.50	11.00	13.90	17.20	22.00	22.90
4 Housing (net)[1], fuel & power	48.60	47.60	51.50	58.20	60.10	65.30
5 Household goods & services	17.10	14.10	20.00	26.90	22.70	30.80
6 Health	1.60	2.50	3.30	4.10	5.40	4.20
7 Transport	22.80	26.20	32.40	51.10	47.40	56.30
8 Communication	8.40	8.30	10.30	11.50	12.50	13.80
9 Recreation & culture	24.20	27.80	41.10	37.50	44.60	58.90
10 Education	4.00	4.70	3.00	8.80	9.90	5.80
11 Restaurants & hotels	15.00	17.10	20.80	25.00	30.90	37.30
12 Miscellaneous goods & services	15.10	15.90	21.70	26.00	31.40	36.20
1-12 All expenditure groups	213.90	225.60	273.50	326.20	349.80	398.70
13 Other expenditure items	22.10	22.50	33.00	42.10	51.50	66.10
Total expenditure	**236.00**	**248.10**	**306.50**	**368.30**	**401.30**	**464.80**
Average weekly expenditure per person (£)						
Total expenditure	**110.80**	**124.30**	**139.40**	**164.90**	**164.50**	**184.00**

Note: The commodity and service categories are not comparable to those in publications before 2001-02.

Please see page xiii for symbols and conventions used in this report.

1 Excluding mortgage interest payments, council tax and Northern Ireland Rates.

Table 3.2E **Household expenditure by gross equivalised income decile group (OECD-modified scale), 2010 (cont.)**
United Kingdom

	Seventh decile group	Eighth decile group	Ninth decile group	Highest ten per cent	All house- holds
Low er boundary of group (£ per w eek)	400	484	595	794	
Weighted number of households (thousands)	2,630	2,630	2,630	2,630	26,320
Total number of households in sample	540	520	510	520	5,260
Total number of persons in sample	1,360	1,300	1,200	1,230	12,180
Total number of adults in sample	1,080	1,050	990	1,010	9,430
Weighted average number of persons per household	2.6	2.5	2.4	2.3	2.3
Commodity or service	Average w eekly household expenditure (£)				
1 Food & non-alcoholic drinks	60.20	63.30	62.00	66.80	53.20
2 Alcoholic drinks, tobacco & narcotics	11.20	13.90	14.30	16.80	11.80
3 Clothing & footwear	28.90	27.90	35.50	43.00	23.40
4 Housing (net)[1], fuel & power	62.50	61.90	69.20	78.80	60.40
5 Household goods & services	27.80	40.70	42.20	72.10	31.40
6 Health	5.50	5.50	7.10	11.20	5.00
7 Transport	77.10	86.90	100.20	148.30	64.90
8 Communication	14.30	14.80	17.10	19.20	13.00
9 Recreation & culture	64.50	75.90	93.80	112.40	58.10
10 Education	15.90	7.20	11.70	29.20	10.00
11 Restaurants & hotels	42.20	54.80	64.80	84.80	39.20
12 Miscellaneous goods & services	40.50	45.70	52.10	74.50	35.90
1-12 All expenditure groups	450.60	498.70	570.10	757.00	406.30
13 Other expenditure items	77.70	104.00	104.70	149.20	67.30
Total expenditure	**528.30**	**602.70**	**674.80**	**906.20**	**473.60**
Average weekly expenditure per person (£)					
Total expenditure	**205.10**	**240.40**	**285.60**	**386.40**	**203.10**

Note: The commodity and service categories are not comparable to those in publications before 2001-02.
 Please see page xiii for symbols and conventions used in this report.
1 Excluding mortgage interest payments, council tax and Northern Ireland Rates.

Table 3.3 Household expenditure as a percentage of total expenditure by gross income decile group, 2010
United Kingdom

	Lowest ten per cent	Second decile group	Third decile group	Fourth decile group	Fifth decile group	Sixth decile group
Lower boundary of group (£ per week)		160	238	315	413	522
Weighted number of households (thousands)	2,630	2,640	2,620	2,630	2,630	2,640
Total number of households in sample	510	530	540	550	550	530
Total number of persons in sample	690	860	960	1,150	1,280	1,290
Total number of adults in sample	570	670	780	900	970	1,010
Weighted average number of persons per household	1.3	1.6	1.8	2.1	2.4	2.4
Commodity or service	Percentage of total expenditure					
1 Food & non-alcoholic drinks	15	16	15	13	13	12
2 Alcoholic drinks, tobacco & narcotics	4	4	3	3	3	3
3 Clothing & footwear	4	4	4	4	5	5
4 Housing (net)[1], fuel & power	23	21	18	17	15	13
5 Household goods & services	8	8	7	7	6	7
6 Health	1	1	1	1	1	1
7 Transport	9	9	11	13	12	13
8 Communication	4	4	3	3	3	3
9 Recreation & culture	10	12	11	12	13	12
10 Education	[1]	[1]	[2]	2	2	1
11 Restaurants & hotels	6	6	6	7	8	8
12 Miscellaneous goods & services	7	7	7	7	7	8
1-12 All expenditure groups	90	91	89	88	87	86
13 Other expenditure items	10	9	11	12	13	14
Total expenditure	100	100	100	100	100	100

Note: The commodity and service categories are not comparable to those in publications before 2001-02.
Please see page xiii for symbols and conventions used in this report.
1 Excluding mortgage interest payments, council tax and Northern Ireland rates.

Table 3.3 **Household expenditure as a percentage of total expenditure by gross income decile group, 2010 (cont.)**
United Kingdom

	Seventh decile group	Eighth decile group	Ninth decile group	Highest ten per cent	All house-holds
Low er boundary of group (£ per w eek)	651	801	1,015	1,368	
Weighted number of households (thousands)	2,630	2,640	2,630	2,630	26,320
Total number of households in sample	530	520	510	500	5,260
Total number of persons in sample	1,410	1,430	1,530	1,580	12,180
Total number of adults in sample	1,090	1,090	1,170	1,170	9,430
Weighted average number of persons per household	2.7	2.8	3.0	3.2	2.3
Commodity or service	Percentage of total expenditure				
1 Food & non-alcoholic drinks	11	10	10	8	11
2 Alcoholic drinks, tobacco & narcotics	2	2	2	2	2
3 Clothing & footwear	5	5	6	5	5
4 Housing (net)[1], fuel & power	13	12	9	8	13
5 Household goods & services	7	5	5	8	7
6 Health	1	1	1	1	1
7 Transport	15	15	15	15	14
8 Communication	3	3	2	2	3
9 Recreation & culture	11	12	13	13	12
10 Education	2	1	3	4	2
11 Restaurants & hotels	8	9	9	10	8
12 Miscellaneous goods & services	7	8	8	8	8
1-12 All expenditure groups	85	85	84	84	86
13 Other expenditure items	15	15	16	16	14
Total expenditure	**100**	**100**	**100**	**100**	**100**

Note: The commodity and service categories are not comparable to those in publications before 2001-02.
Please see page xiii for symbols and conventions used in this report.
1 Excluding mortgage interest payments, council tax and Northern Ireland rates.

Table 3.3E   **Household expenditure as a percentage of total expenditure by gross equivalised income decile group (OECD-modified scale), 2010**
United Kingdom

	Lowest ten per cent	Second decile group	Third decile group	Fourth decile group	Fifth decile group	Sixth decile group
Lower boundary of group (£ per week)		132	180	227	277	333
Weighted number of households (thousands)	2,630	2,630	2,640	2,630	2,630	2,630
Total number of households in sample	520	530	530	540	520	540
Total number of persons in sample	1,120	1,070	1,150	1,200	1,230	1,320
Total number of adults in sample	700	800	880	940	950	1,040
Weighted average number of persons per household	2.1	2.0	2.2	2.2	2.4	2.5
Commodity or service	Percentage of total expenditure					
1 Food & non-alcoholic drinks	15	17	15	14	13	12
2 Alcoholic drinks, tobacco & narcotics	4	4	3	3	3	2
3 Clothing & footwear	5	4	5	5	5	5
4 Housing (net)[1], fuel & power	21	19	17	16	15	14
5 Household goods & services	7	6	7	7	6	7
6 Health	1	1	1	1	1	1
7 Transport	10	11	11	14	12	12
8 Communication	4	3	3	3	3	3
9 Recreation & culture	10	11	13	10	11	13
10 Education	2	2	1	2	2	1
11 Restaurants & hotels	6	7	7	7	8	8
12 Miscellaneous goods & services	6	6	7	7	8	8
1-12 All expenditure groups	91	91	89	89	87	86
13 Other expenditure items	9	9	11	11	13	14
Total expenditure	**100**	**100**	**100**	**100**	**100**	**100**

Note: The commodity and service categories are not comparable to those in publications before 2001-02.
Please see page xiii for symbols and conventions used in this report.
1 Excluding mortgage interest payments, council tax and Northern Ireland Rates.

Table 3.3E   **Household expenditure as a percentage of total expenditure by gross equivalised income decile group (OECD-modified scale), 2010 (cont.)**
United Kingdom

	Seventh decile group	Eighth decile group	Ninth decile group	Highest ten per cent	All house-holds
Lower boundary of group (£ per week)	400	484	595	794	
Weighted number of households (thousands)	2,630	2,630	2,630	2,630	26,320
Total number of households in sample	540	520	510	520	5,260
Total number of persons in sample	1,360	1,300	1,200	1,230	12,180
Total number of adults in sample	1,080	1,050	990	1,010	9,430
Weighted average number of persons per household	2.6	2.5	2.4	2.3	2.3
Commodity or service	Percentage of total expenditure				
1 Food & non-alcoholic drinks	11	11	9	7	11
2 Alcoholic drinks, tobacco & narcotics	2	2	2	2	2
3 Clothing & footwear	5	5	5	5	5
4 Housing (net)[1], fuel & power	12	10	10	9	13
5 Household goods & services	5	7	6	8	7
6 Health	1	1	1	1	1
7 Transport	15	14	15	16	14
8 Communication	3	2	3	2	3
9 Recreation & culture	12	13	14	12	12
10 Education	3	1	2	3	2
11 Restaurants & hotels	8	9	10	9	8
12 Miscellaneous goods & services	8	8	8	8	8
1-12 All expenditure groups	85	83	84	84	86
13 Other expenditure items	15	17	16	16	14
Total expenditure	100	100	100	100	100

Note: The commodity and service categories are not comparable to those in publications before 2001-02.
Please see page xiii for symbols and conventions used in this report.
1 Excluding mortgage interest payments, council tax and Northern Ireland Rates.

Table 3.4   Expenditure of one person non-retired households by gross income quintile group, 2010
United Kingdom

	Lowest twenty per cent	Second quintile group	Third quintile group	Fourth quintile group	Highest twenty per cent	All house-holds
Lower boundary of group (£ per week)		238	413	651	1,015	
Weighted number of households (thousands)	1,470	980	890	470	230	4,030
Total number of households in sample	290	200	180	90	40	800
Total number of persons in sample	290	200	180	90	40	800
Total number of adults in sample	290	200	180	90	40	800
Weighted average number of persons per household	1.0	1.0	1.0	1.0	1.0	1.0

Commodity or service	Average weekly household expenditure (£)					
1 Food & non-alcoholic drinks	21.30	27.40	28.00	29.40	38.30	26.20
2 Alcoholic drinks, tobacco & narcotics	7.20	7.20	7.60	7.20	11.60	7.60
3 Clothing & footwear	3.60	8.60	15.40	11.70	23.10	9.40
4 Housing (net)[1], fuel & power	40.40	59.90	57.10	73.80	85.70	55.30
5 Household goods & services	12.20	11.30	28.40	35.10	67.40	21.30
6 Health	1.10	1.40	2.00	3.30	15.80	2.40
7 Transport	15.90	31.00	58.80	60.30	87.00	38.20
8 Communication	6.70	8.80	9.20	11.10	15.10	8.70
9 Recreation & culture	16.30	23.10	49.10	41.20	72.50	31.20
10 Education	[2.90]	[0.40]	[1.30]	[7.80]	[9.00]	2.80
11 Restaurants & hotels	11.40	16.90	23.60	40.50	49.00	20.90
12 Miscellaneous goods & services	8.20	18.80	24.60	19.80	53.20	18.30
1-12 All expenditure groups	147.40	214.80	305.00	341.20	527.60	242.50
13 Other expenditure items	22.30	39.50	70.00	77.60	100.50	47.80
Total expenditure	**169.80**	**254.30**	**374.90**	**418.80**	**628.00**	**290.30**
Average weekly expenditure per person (£)						
Total expenditure	**169.80**	**254.30**	**374.90**	**418.80**	**628.00**	**290.30**

Note: The commodity and service categories are not comparable to those in publications before 2001-02.
Please see page xiii for symbols and conventions used in this report.
1 Excluding mortgage interest payments, council tax and Northern Ireland rates.

Table 3.4E   **Expenditure of one adult non-retired households by gross equivalised income quintile group (OECD-modified scale), 2010**

United Kingdom

	Lowest twenty per cent	Second quintile group	Third quintile group	Fourth quintile group	Highest twenty per cent	All house- holds
Lower boundary of group (£ per week)		180	277	400	595	
Weighted number of households (thousands)	1,130	550	700	760	900	4,030
Total number of households in sample	220	120	140	150	170	800
Total number of persons in sample	220	120	140	150	170	800
Total number of adults in sample	220	120	140	150	170	800
Weighted average number of persons per household	1.0	1.0	1.0	1.0	1.0	1.0
Commodity or service	Average weekly household expenditure (£)					
1 Food & non-alcoholic drinks	20.00	27.00	26.60	28.00	31.50	26.20
2 Alcoholic drinks, tobacco & narcotics	7.50	7.20	7.00	6.80	9.00	7.60
3 Clothing & footwear	3.40	5.00	9.20	14.50	15.70	9.40
4 Housing (net)[1], fuel & power	38.40	51.70	58.70	60.70	71.40	55.30
5 Household goods & services	7.30	20.20	12.80	20.30	47.10	21.30
6 Health	0.90	1.80	1.40	2.00	6.00	2.40
7 Transport	15.10	20.90	31.90	54.20	69.10	38.20
8 Communication	5.90	8.70	9.10	8.90	11.90	8.70
9 Recreation & culture	15.30	18.20	24.00	46.70	51.60	31.20
10 Education	[2.70]	[2.30]	[0.40]	[1.20]	[6.70]	2.80
11 Restaurants & hotels	10.70	14.40	17.30	21.70	39.90	20.90
12 Miscellaneous goods & services	7.30	15.00	18.60	22.20	30.40	18.30
1-12 All expenditure groups	134.50	192.20	217.00	287.00	390.30	242.50
13 Other expenditure items	20.20	27.80	41.80	63.30	86.20	47.80
Total expenditure	**154.60**	**220.00**	**258.80**	**350.30**	**476.50**	**290.30**
Average weekly expenditure per person (£)						
Total expenditure	**154.60**	**220.00**	**258.80**	**350.30**	**476.50**	**290.30**

Note: The commodity and service categories are not comparable to those in publications before 2001-02.

Please see page xiii for symbols and conventions used in this report.

1 Excluding mortgage interest payments, council tax and Northern Ireland Rates.

Table 3.5 **Expenditure of one person retired households not mainly dependent on state pensions[1] by gross income quintile group, 2010**
United Kingdom

	Lowest twenty per cent	Second quintile group	Third quintile group	Fourth quintile group	Highest twenty per cent	All house-holds
Lower boundary of group (£ per week)		238	413	651	1,015	
Weighted number of households (thousands)	1,540	1,040	310	80	10	2,980
Total number of households in sample	290	190	60	..	..	570
Total number of persons in sample	290	190	60	..	..	570
Total number of adults in sample	290	190	60	..	..	570
Weighted average number of persons per household	1.0	1.0	1.0	1.0	1.0	1.0
Commodity or service	Average weekly household expenditure (£)					
1 Food & non-alcoholic drinks	27.60	30.00	36.30	[37.40]	[42.10]	29.70
2 Alcoholic drinks, tobacco & narcotics	4.10	5.50	7.00	[13.40]	[9.50]	5.10
3 Clothing & footwear	5.40	8.20	8.90	[9.10]	-	6.80
4 Housing (net)[2], fuel & power	36.00	46.20	41.70	[91.10]	[53.30]	41.60
5 Household goods & services	12.00	24.50	24.50	[22.50]	[101.50]	18.30
6 Health	4.00	4.20	2.70	[27.10]	-	4.50
7 Transport	11.00	21.80	44.00	[39.50]	[143.80]	19.50
8 Communication	6.00	7.70	7.00	[7.50]	[6.70]	6.70
9 Recreation & culture	19.40	24.70	38.00	[90.40]	[223.50]	25.90
10 Education	[0.10]	[0.10]	[0.10]	-	-	[0.10]
11 Restaurants & hotels	7.00	13.60	21.70	[31.00]	[12.00]	11.50
12 Miscellaneous goods & services	11.30	17.00	36.50	[47.60]	[62.90]	17.10
1-12 All expenditure groups	143.80	203.40	268.50	[416.60]	[655.20]	186.80
13 Other expenditure items	14.60	32.30	65.10	[63.90]	[54.70]	27.50
Total expenditure	**158.40**	**235.60**	**333.60**	**[480.50]**	**[709.90]**	**214.20**
Average weekly expenditure per person (£)						
Total expenditure	**158.40**	**235.60**	**333.60**	**[480.50]**	**[709.90]**	**214.20**

Note: The commodity and service categories are not comparable to those in publications before 2001-02.

Please see page xiii for symbols and conventions used in this report.

1 Mainly dependent on state pension and not economically active - see appendix B.

2 Excluding mortgage interest payments, council tax and Northern Ireland rates.

Table 3.5E   **Expenditure of one person retired households not mainly dependent on state pensions[1] by gross equivalised income quintile group (OECD-modified scale), 2010**
United Kingdom

	Lowest twenty per cent	Second quintile group	Third quintile group	Fourth quintile group	Highest twenty per cent	All house-holds
Lower boundary of group (£ per week)		180	277	400	595	
Weighted number of households (thousands)	810	1,160	560	330	120	2,980
Total number of households in sample	150	220	110	60	30	570
Total number of persons in sample	150	220	110	60	30	570
Total number of adults in sample	150	220	110	60	30	570
Weighted average number of persons per household	1.0	1.0	1.0	1.0	1.0	1.0
Commodity or service	Average weekly household expenditure (£)					
1 Food & non-alcoholic drinks	25.40	30.00	29.80	35.40	38.80	29.70
2 Alcoholic drinks, tobacco & narcotics	3.20	5.30	5.20	6.90	[10.30]	5.10
3 Clothing & footwear	4.10	7.30	7.80	10.30	[7.90]	6.80
4 Housing (net)[2], fuel & power	37.80	39.40	42.10	45.70	76.00	41.60
5 Household goods & services	9.40	18.80	22.60	26.10	32.90	18.30
6 Health	2.70	4.60	4.40	3.90	[17.20]	4.50
7 Transport	6.60	18.80	19.90	39.80	57.60	19.50
8 Communication	6.00	6.40	7.90	7.30	7.40	6.70
9 Recreation & culture	16.00	23.00	22.70	41.50	93.60	25.90
10 Education	-	[0.10]	[0.10]	[0.10]	-	[0.10]
11 Restaurants & hotels	6.00	9.30	14.40	19.90	33.30	11.50
12 Miscellaneous goods & services	9.70	13.90	19.00	32.50	48.00	17.10
1-12 All expenditure groups	127.00	176.80	195.80	269.30	423.00	186.80
13 Other expenditure items	11.40	24.20	30.20	60.20	66.50	27.50
Total expenditure	**138.30**	**201.10**	**226.00**	**329.50**	**489.50**	**214.20**
Average weekly expenditure per person (£)						
Total expenditure	**138.30**	**201.10**	**226.00**	**329.50**	**489.50**	**214.20**

Note: The commodity and service categories are not comparable to those in publications before 2001-02.

Please see page xiii for symbols and conventions used in this report.

1 Mainly dependent on state pension and not economically active - see Appendix B.

2 Excluding mortgage interest payments, council tax and Northern Ireland Rates.

Table 3.6   **Expenditure of two adult households with children by gross income quintile group, 2010**
United Kingdom

	Lowest twenty per cent	Second quintile group	Third quintile group	Fourth quintile group	Highest twenty per cent	All house- holds
Lower boundary of group (£ per week)		238	413	651	1,015	
Weighted number of households (thousands)	340	550	1,140	1,500	1,640	5,160
Total number of households in sample	60	100	230	310	350	1,060
Total number of persons in sample	220	390	880	1,160	1,340	3,990
Total number of adults in sample	120	210	460	620	710	2,110
Weighted average number of persons per household	3.5	3.7	3.8	3.7	3.8	3.7
Commodity or service	Average weekly household expenditure (£)					
1 Food & non-alcoholic drinks	49.40	58.30	62.60	67.80	85.70	70.10
2 Alcoholic drinks, tobacco & narcotics	13.20	15.90	14.00	10.60	14.30	13.20
3 Clothing & footwear	18.60	24.90	23.80	32.80	50.20	34.50
4 Housing (net)[1], fuel & power	62.00	66.40	67.40	61.90	69.10	65.90
5 Household goods & services	14.80	26.10	24.20	39.00	64.60	40.90
6 Health	[0.60]	1.90	3.30	4.50	9.40	5.30
7 Transport	27.30	52.80	56.60	76.60	124.70	81.70
8 Communication	10.50	13.80	15.30	16.70	20.10	16.80
9 Recreation & culture	29.00	46.30	51.30	64.90	120.40	75.20
10 Education	[14.10]	[6.60]	6.30	5.90	46.60	19.50
11 Restaurants & hotels	16.90	29.70	35.00	45.10	81.10	50.80
12 Miscellaneous goods & services	24.80	30.80	36.60	50.80	87.40	55.50
1-12 All expenditure groups	281.20	373.60	396.30	476.50	773.60	529.50
13 Other expenditure items	24.30	49.30	61.20	100.50	172.40	104.30
Total expenditure	**305.50**	**422.90**	**457.60**	**577.00**	**946.00**	**633.80**
Average weekly expenditure per person (£) **Total expenditure**	**86.80**	**115.10**	**120.60**	**155.20**	**249.80**	**169.50**

Note: The commodity and service categories are not comparable to those in publications before 2001-02.

Please see page xiii for symbols and conventions used in this report.

1 Excluding mortgage interest payments, council tax and Northern Ireland rates.

Table 3.6E **Expenditure of two adult households with children by gross equivalised income quintile group (OECD-modified scale), 2010**

United Kingdom

	Lowest twenty per cent	Second quintile group	Third quintile group	Fourth quintile group	Highest twenty per cent	All house-holds
Lower boundary of group (£ per week)		180	277	400	595	
Weighted number of households (thousands)	740	860	1,230	1,280	1,060	5,160
Total number of households in sample	140	170	250	270	230	1,060
Total number of persons in sample	590	660	930	970	840	3,990
Total number of adults in sample	280	340	500	530	460	2,110
Weighted average number of persons per household	4.0	3.8	3.7	3.6	3.7	3.7
Commodity or service	Average weekly household expenditure (£)					
1　Food & non-alcoholic drinks	58.50	63.30	65.20	70.70	88.90	70.10
2　Alcoholic drinks, tobacco & narcotics	15.20	13.00	11.50	12.10	15.50	13.20
3　Clothing & footwear	25.60	24.00	27.70	35.60	56.10	34.50
4　Housing (net)[1], fuel & power	58.40	71.50	65.10	60.00	74.50	65.90
5　Household goods & services	21.60	23.00	37.10	35.20	80.10	40.90
6　Health	1.30	1.90	5.10	4.20	12.20	5.30
7　Transport	46.20	53.90	63.40	92.20	137.80	81.70
8　Communication	12.40	14.60	16.30	16.90	22.00	16.80
9　Recreation & culture	36.00	54.90	59.90	73.90	138.30	75.20
10　Education	[10.20]	5.70	5.90	12.10	62.10	19.50
11　Restaurants & hotels	28.60	28.80	42.00	53.20	91.70	50.80
12　Miscellaneous goods & services	27.20	34.70	42.20	59.60	102.50	55.50
1-12　All expenditure groups	341.10	389.30	441.20	525.80	881.70	529.50
13　Other expenditure items	31.20	63.00	83.60	130.10	181.50	104.30
Total expenditure	**372.30**	**452.30**	**524.80**	**655.80**	**1063.20**	**633.80**
Average weekly expenditure per person (£)						
Total expenditure	**92.70**	**119.40**	**140.50**	**181.00**	**291.20**	**169.50**

Note: The commodity and service categories are not comparable to those in publications before 2001-02.
Please see page xiii for symbols and conventions used in this report.
1 Excluding mortgage interest payments, council tax and Northern Ireland Rates.

Table 3.7   **Expenditure of one adult households with children by gross income quintile group, 2010**
United Kingdom

	Lowest twenty per cent	Second quintile group	Third quintile group	Fourth quintile group	Highest twenty per cent	All house-holds
Low er boundary of group (£ per w eek)		238	413	651	1,015	
Weighted number of households (thousands)	630	480	240	40	30	1,410
Total number of households in sample	140	110	60	..	..	330
Total number of persons in sample	350	340	160	30	20	900
Total number of adults in sample	140	110	60	..	..	330
Weighted average number of persons per household	2.5	3.1	2.8	2.9	2.9	2.8
Commodity or service	Average w eekly household expenditure (£)					
1 Food & non-alcoholic drinks	37.80	53.00	51.70	[54.80]	[85.40]	46.70
2 Alcoholic drinks, tobacco & narcotics	10.60	9.20	12.10	[6.60]	[11.10]	10.30
3 Clothing & footwear	14.10	16.90	28.50	[20.20]	[25.90]	17.80
4 Housing (net)[1], fuel & power	59.90	59.40	65.70	[53.50]	[59.20]	60.50
5 Household goods & services	27.20	18.10	20.00	[78.20]	[45.40]	24.80
6 Health	0.80	3.50	3.00	[2.20]	[15.10]	2.40
7 Transport	18.50	31.30	46.30	[33.70]	[98.00]	29.50
8 Communication	8.10	11.10	14.50	[12.50]	[20.60]	10.60
9 Recreation & culture	23.40	36.20	54.10	[53.20]	[73.10]	34.70
10 Education	[2.40]	[5.40]	[2.50]	[0.70]	[26.10]	3.80
11 Restaurants & hotels	13.00	21.00	37.90	[36.10]	[100.30]	22.30
12 Miscellaneous goods & services	14.50	21.70	29.00	[58.50]	[64.50]	21.70
1-12 All expenditure groups	230.20	286.80	365.20	[410.10]	[624.50]	285.10
13 Other expenditure items	12.00	31.50	48.40	[72.00]	[147.20]	29.10
Total expenditure	**242.20**	**318.30**	**413.60**	**[482.20]**	**[771.70]**	**314.20**
Average weekly expenditure per person (£)						
Total expenditure	**96.60**	**104.30**	**148.70**	**[165.90]**	**[262.00]**	**114.00**

Note: The commodity and service categories are not comparable to those in publications before 2001-02.

Please see page xiii for symbols and conventions used in this report.

1 Excluding mortgage interest payments, council tax and Northern Ireland rates.

Table 3.7E   **Expenditure of one adult households with children by gross equivalised income quintile group (OECD-modified scale), 2010**
United Kingdom

	Lowest twenty per cent	Second quintile group	Third quintile group	Fourth quintile group	Highest twenty per cent	All house-holds
Low er boundary of group (£ per w eek)		180	277	400	595	
Weighted number of households (thousands)	810	360	170	50	30	1,410
Total number of households in sample	180	90	40	..	..	330
Total number of persons in sample	520	240	90	30	30	900
Total number of adults in sample	180	90	40	..	..	330
Weighted average number of persons per household	2.8	2.8	2.4	2.6	2.7	2.8

Commodity or service	Average w eekly household expenditure (£)					
1 Food & non-alcoholic drinks	42.90	52.10	47.30	[54.10]	[69.70]	46.70
2 Alcoholic drinks, tobacco & narcotics	10.60	9.60	11.90	[5.20]	[10.10]	10.30
3 Clothing & footwear	16.00	18.70	23.30	[17.10]	[27.70]	17.80
4 Housing (net)[1], fuel & power	55.90	71.70	57.40	[64.90]	[59.90]	60.50
5 Household goods & services	23.90	24.30	13.20	[21.90]	[121.40]	24.80
6 Health	1.00	4.10	3.70	[3.20]	[12.00]	2.40
7 Transport	18.60	38.30	47.20	[45.70]	[93.40]	29.50
8 Communication	8.40	12.80	13.70	[12.50]	[20.20]	10.60
9 Recreation & culture	24.60	41.30	58.20	[45.70]	[80.50]	34.70
10 Education	[3.00]	[4.40]	[2.80]	[4.20]	[23.20]	3.80
11 Restaurants & hotels	14.90	22.70	38.90	[64.50]	[55.30]	22.30
12 Miscellaneous goods & services	14.70	25.00	33.00	[30.20]	[92.80]	21.70
1-12 All expenditure groups	234.40	324.90	350.70	[369.20]	[666.30]	285.10
13 Other expenditure items	17.60	28.70	54.70	[80.90]	[118.70]	29.10
Total expenditure	**252.00**	**353.60**	**405.40**	**[450.10]**	**[785.00]**	**314.20**
Average weekly expenditure per person (£)						
Total expenditure	**89.10**	**127.80**	**166.60**	**[171.50]**	**[287.40]**	**114.00**

Note: The commodity and service categories are not comparable to those in publications before 2001-02.
Please see page xiii for symbols and conventions used in this report.
1 Excluding mortgage interest payments, council tax and Northern Ireland Rates.

Table 3.8   **Expenditure of two adult non-retired households by gross income quintile group, 2010**
United Kingdom

	Lowest twenty per cent	Second quintile group	Third quintile group	Fourth quintile group	Highest twenty per cent	All house- holds
Low er boundary of group (£ per w eek)		238	413	651	1,015	
Weighted number of households (thousands)	300	700	1,260	1,800	1,700	5,760
Total number of households in sample	60	150	260	360	340	1,170
Total number of persons in sample	120	290	530	730	670	2,340
Total number of adults in sample	120	290	530	730	670	2,340
Weighted average number of persons per household	2.0	2.0	2.0	2.0	2.0	2.0

Commodity or service	Average w eekly household expenditure (£)					
1 Food & non-alcoholic drinks	48.10	49.00	52.20	54.80	58.40	54.30
2 Alcoholic drinks, tobacco & narcotics	16.80	13.10	13.60	15.00	17.70	15.40
3 Clothing & footwear	13.20	12.40	18.60	25.90	36.00	25.00
4 Housing (net)[1], fuel & power	62.00	61.90	73.90	70.70	75.70	71.40
5 Household goods & services	33.30	26.80	23.20	35.60	58.80	38.60
6 Health	2.50	1.80	4.20	5.20	7.90	5.20
7 Transport	51.10	60.00	52.10	89.80	137.90	90.10
8 Com m unication	11.90	12.00	11.50	15.40	17.30	14.50
9 Recreation & culture	57.90	61.50	54.40	68.40	102.60	74.10
10 Education	[2.00]	[21.80]	[15.90]	3.00	6.90	9.20
11 Restaurants & hotels	23.00	24.00	34.80	51.00	81.90	51.90
12 Miscellaneous goods & services	24.90	23.70	38.90	39.90	55.00	41.40
1-12 All expenditure groups	346.70	368.00	393.40	474.80	656.20	490.90
13 Other expenditure item s	28.60	39.70	59.60	85.40	136.20	86.30
Total expenditure	**375.30**	**407.70**	**452.90**	**560.20**	**792.40**	**577.20**
Average weekly expenditure per person (£)						
Total expenditure	**187.70**	**203.80**	**226.50**	**280.10**	**396.20**	**288.60**

Note: The commodity and service categories are not comparable to those in publications before 2001-02.

Please see page xiii for symbols and conventions used in this report.

1 Excluding mortgage interest payments, council tax and Northern Ireland rates.

Table 3.8E

Expenditure of two adult non-retired households by gross equivalised income quintile group (OECD-modified scale), 2010

United Kingdom

	Lowest twenty per cent	Second quintile group	Third quintile group	Fourth quintile group	Highest twenty per cent	All house-holds
Lower boundary of group (£ per week)		180	277	400	595	
Weighted number of households (thousands)	450	570	1,040	1,480	2,230	5,760
Total number of households in sample	90	120	220	310	440	1,170
Total number of persons in sample	180	240	430	620	870	2,340
Total number of adults in sample	180	240	430	620	870	2,340
Weighted average number of persons per household	2.0	2.0	2.0	2.0	2.0	2.0

Commodity or service	Average weekly household expenditure (£)					
1 Food & non-alcoholic drinks	47.20	51.20	50.70	55.50	57.30	54.30
2 Alcoholic drinks, tobacco & narcotics	14.70	14.40	12.40	14.90	17.40	15.40
3 Clothing & footwear	11.90	13.00	18.80	23.00	34.90	25.00
4 Housing (net)[1], fuel & power	67.60	58.00	75.60	70.60	74.00	71.40
5 Household goods & services	24.90	31.20	23.00	36.60	51.70	38.60
6 Health	2.10	1.90	4.10	4.90	7.40	5.20
7 Transport	45.90	66.00	52.00	79.30	130.00	90.10
8 Communication	11.40	12.40	11.20	14.40	17.30	14.50
9 Recreation & culture	60.60	59.50	53.90	63.30	97.00	74.10
10 Education	[16.90]	[14.50]	[18.70]	[2.70]	6.10	9.20
11 Restaurants & hotels	19.60	26.80	34.60	46.30	76.40	51.90
12 Miscellaneous goods & services	21.00	26.30	41.40	35.30	53.30	41.40
1-12 All expenditure groups	343.90	375.20	396.50	446.80	622.90	490.90
13 Other expenditure items	**28.10**	**43.10**	**59.60**	**77.90**	**126.80**	**86.30**
Total expenditure	**371.90**	**418.30**	**456.10**	**524.70**	**749.70**	**577.20**
Average weekly expenditure per person (£)						
Total expenditure	**186.00**	**209.20**	**228.10**	**262.40**	**374.90**	**288.60**

Note: The commodity and service categories are not comparable to those in publications before 2001-02.

Please see page xiii for symbols and conventions used in this report.

1 Excluding mortgage interest payments, council tax and Northern Ireland Rates.

Table 3.9   **Expenditure of one person retired households mainly dependent on state pensions[1] by gross income quintile group, 2010**
United Kingdom

	Lowest twenty per cent	Second quintile group	Third quintile group	Fourth quintile group	Highest twenty per cent	All house-holds
Lower boundary of group (£ per week)		238	413	651	1,015	
Weighted number of households (thousands)	640	160	0	0	0	780
Total number of households in sample	120	20	0	0	0	140
Total number of persons in sample	120	20	0	0	0	140
Total number of adults in sample	120	20	0	0	0	140
Weighted average number of persons per household	1.0	1.0	0	0	0	1.0
Commodity or service	Average weekly household expenditure (£)					
1 Food & non-alcoholic drinks	26.70	33.40	-	-	-	28.10
2 Alcoholic drinks, tobacco & narcotics	4.10	[7.20]	-	-	-	4.70
3 Clothing & footwear	4.80	[7.90]	-	-	-	5.50
4 Housing (net)[2], fuel & power	36.50	39.10	-	-	-	37.00
5 Household goods & services	8.70	27.50	-	-	-	12.40
6 Health	2.40	[2.10]	-	-	-	2.30
7 Transport	9.60	[9.20]	-	-	-	9.50
8 Communication	5.40	6.40	-	-	-	5.60
9 Recreation & culture	14.70	23.90	-	-	-	16.50
10 Education	-	[13.30]	-	-	-	2.60
11 Restaurants & hotels	6.80	[7.80]	-	-	-	7.00
12 Miscellaneous goods & services	13.10	12.90	-	-	-	13.10
1-12 All expenditure groups	132.80	190.80	-	-	-	144.30
13 Other expenditure items	15.90	20.50	-	-	-	16.90
Total expenditure	**148.70**	**211.30**	**-**	**-**	**-**	**161.10**
Average weekly expenditure per person (£)						
Total expenditure	**148.70**	**211.30**	**-**	**-**	**-**	**161.10**

Note: The commodity and service categories are not comparable to those in publications before 2001-02.

Please see page xiii for symbols and conventions used in this report.

1 Mainly dependent on state pension and not economically active - see appendix B.

2 Excluding mortgage interest payments, council tax and Northern Ireland rates.

Table 3.9E **Expenditure of one person retired households mainly dependent on state pensions[1] by gross equivalised income quintile group (OECD-modified scale), 2010**
United Kingdom

	Lowest twenty per cent	Second quintile group	Third quintile group	Fourth quintile group	Highest twenty per cent	All house-holds
Lower boundary of group (£ per week)		180	277	400	595	
Weighted number of households (thousands)	510	230	50	0	0	790
Total number of households in sample	100	40	..	0	0	150
Total number of persons in sample	100	40	..	0	0	150
Total number of adults in sample	100	40	..	0	0	150
Weighted average number of persons per household	1.0	1.0	1.0	0	0	1.0
Commodity or service	Average weekly household expenditure (£)					
1 Food & non-alcoholic drinks	26.80	29.30	[35.00]	-	-	28.10
2 Alcoholic drinks, tobacco & narcotics	3.30	6.10	[12.30]	-	-	4.70
3 Clothing & footwear	5.60	[6.00]	[1.60]	-	-	5.50
4 Housing (net)[2], fuel & power	35.80	39.40	[37.80]	-	-	37.00
5 Household goods & services	7.20	24.00	[13.00]	-	-	12.40
6 Health	2.20	[2.80]	[0.70]	-	-	2.30
7 Transport	9.30	10.60	[7.00]	-	-	9.50
8 Communication	5.40	5.70	[6.50]	-	-	5.60
9 Recreation & culture	13.10	24.60	[15.60]	-	-	16.50
10 Education	-	-	[41.90]	-	-	2.60
11 Restaurants & hotels	6.10	10.00	[3.00]	-	-	7.00
12 Miscellaneous goods & services	14.20	11.20	[10.40]	-	-	13.10
1-12 All expenditure groups	129.10	169.70	[184.80]	-	-	144.30
13 Other expenditure items	15.30	19.80	[19.00]	-	-	16.90
Total expenditure	**144.40**	**189.50**	**[203.80]**	**-**	**-**	**161.10**
Average weekly expenditure per person (£)						
Total expenditure	**144.40**	**189.50**	**[203.80]**	**-**	**-**	**161.10**

Note: The commodity and service categories are not comparable to those in publications before 2001-02.
Please see page xiii for symbols and conventions used in this report.
1 Mainly dependent on state pension and not economically active - see Appendix B.
2 Excluding mortgage interest payments, council tax and Northern Ireland Rates.

Table 3.10 **Expenditure of two adult retired households mainly dependent on state pensions[1] by gross income quintile group, 2010**
United Kingdom

	Lowest twenty per cent	Second quintile group	Third quintile group	Fourth quintile group	Highest twenty per cent	All house- holds
Lower boundary of group (£ per week)		238	413	651	1,015	
Weighted number of households (thousands)	150	290	20	0	0	450
Total number of households in sample	30	70	..	0	0	110
Total number of persons in sample	60	140	10	0	0	220
Total number of adults in sample	60	140	10	0	0	220
Weighted average number of persons per household	2.0	2.0	2.0	0	0	2.0

Commodity or service	Average weekly household expenditure (£)					
1 Food & non-alcoholic drinks	43.50	51.00	[35.00]	-	-	47.80
2 Alcoholic drinks, tobacco & narcotics	[5.70]	9.80	[13.30]	-	-	8.70
3 Clothing & footwear	[5.70]	11.10	[5.60]	-	-	9.10
4 Housing (net)[2], fuel & power	33.80	44.60	[46.30]	-	-	41.20
5 Household goods & services	12.70	21.00	[61.00]	-	-	20.30
6 Health	[3.30]	5.30	[5.60]	-	-	4.60
7 Transport	22.10	30.30	[33.30]	-	-	27.80
8 Communication	7.00	8.30	[7.40]	-	-	7.80
9 Recreation & culture	32.50	38.50	[46.70]	-	-	37.00
10 Education	-	-	-	-	-	-
11 Restaurants & hotels	[14.20]	21.20	[5.60]	-	-	18.20
12 Miscellaneous goods & services	17.60	26.40	[13.20]	-	-	22.90
1-12 All expenditure groups	198.10	267.40	[272.80]	-	-	245.40
13 Other expenditure items	27.60	34.40	[12.10]	-	-	31.10
Total expenditure	**225.70**	**301.80**	**[284.90]**	**-**	**-**	**276.50**
Average weekly expenditure per person (£)						
Total expenditure	**112.80**	**150.90**	**[142.40]**	**-**	**-**	**138.30**

Note: The commodity and service categories are not comparable to those in publications before 2001-02.
Please see page xiii for symbols and conventions used in this report.
1 Mainly dependent on state pension and not economically active - see appendix B.
2 Excluding mortgage interest payments, council tax and Northern Ireland rates.

Table 3.10E **Expenditure of two adult retired households mainly dependent on state pensions[1] by gross equivalised income quintile group (OECD-modified scale), 2010**
United Kingdom

	Lowest twenty per cent	Second quintile group	Third quintile group	Fourth quintile group	Highest twenty per cent	All house-holds
Lower boundary of group (£ per week)		180	277	400	595	
Weighted number of households (thousands)	210	220	20	0	0	450
Total number of households in sample	50	60	..	0	0	110
Total number of persons in sample	100	110	10	0	0	220
Total number of adults in sample	100	110	10	0	0	220
Weighted average number of persons per household	2.0	2.0	2.0	0	0	2.0
Commodity or service	Average weekly household expenditure (£)					
1 Food & non-alcoholic drinks	45.30	51.50	[35.00]	-	-	47.80
2 Alcoholic drinks, tobacco & narcotics	7.90	9.10	[13.30]	-	-	8.70
3 Clothing & footwear	10.90	7.60	[5.60]	-	-	9.10
4 Housing (net)[2], fuel & power	37.40	44.50	[46.30]	-	-	41.20
5 Household goods & services	12.50	23.70	[61.00]	-	-	20.30
6 Health	5.20	3.90	[5.60]	-	-	4.60
7 Transport	24.80	30.20	[33.30]	-	-	27.80
8 Communication	7.20	8.50	[7.40]	-	-	7.80
9 Recreation & culture	33.50	39.50	[46.70]	-	-	37.00
10 Education	-	-	-	-	-	-
11 Restaurants & hotels	20.50	17.20	[5.60]	-	-	18.20
12 Miscellaneous goods & services	21.20	25.60	[13.20]	-	-	22.90
1-12 All expenditure groups	226.50	261.30	[272.80]	-	-	245.40
13 Other expenditure items	33.40	30.80	[12.10]	-	-	31.10
Total expenditure	**259.90**	**292.10**	**[284.90]**	**-**	**-**	**276.50**
Average weekly expenditure per person (£)						
Total expenditure	**129.90**	**146.10**	**[142.40]**	**-**	**-**	**138.30**

Note: The commodity and service categories are not comparable to those in publications before 2001-02.
Please see page xiii for symbols and conventions used in this report.
1 Mainly dependent on state pension and not economically active - see Appendix B.
2 Excluding mortgage interest payments, council tax and Northern Ireland Rates.

Table 3.11   **Expenditure of two adult retired households not mainly dependent on state pensions[1] by gross income quintile group, 2010**
United Kingdom

	Lowest twenty per cent	Second quintile group	Third quintile group	Fourth quintile group	Highest twenty per cent	All house- holds
Lower boundary of group (£ per week)		238	413	651	1,015	
Weighted number of households (thousands)	160	870	820	300	140	2,300
Total number of households in sample	30	200	190	70	30	530
Total number of persons in sample	70	400	380	140	60	1,050
Total number of adults in sample	70	400	380	140	60	1,050
Weighted average number of persons per household	2.0	2.0	2.0	2.0	2.0	2.0
Commodity or service	Average weekly household expenditure (£)					
1 Food & non-alcoholic drinks	49.20	52.30	56.70	63.50	76.90	56.70
2 Alcoholic drinks, tobacco & narcotics	15.30	8.70	10.70	9.10	20.20	10.60
3 Clothing & footwear	[3.40]	11.90	18.00	18.20	26.00	15.20
4 Housing (net)[2], fuel & power	40.10	40.70	44.20	53.30	81.80	46.10
5 Household goods & services	15.10	23.30	33.10	41.00	38.90	29.50
6 Health	[1.30]	7.10	8.50	6.80	19.80	8.00
7 Transport	24.50	36.60	48.00	74.60	129.70	50.60
8 Communication	9.50	8.40	10.30	9.10	17.50	9.80
9 Recreation & culture	25.70	37.00	71.90	81.70	149.10	61.50
10 Education	[0.60]	[0.20]	[0.20]	[1.50]	[1.00]	[0.40]
11 Restaurants & hotels	14.80	18.40	29.60	54.70	91.40	31.50
12 Miscellaneous goods & services	17.60	20.50	31.70	51.40	81.00	32.10
1-12 All expenditure groups	217.00	265.20	362.90	465.10	733.10	352.10
13 Other expenditure items	22.80	30.10	44.20	81.80	89.10	45.10
Total expenditure	**239.80**	**295.30**	**407.10**	**546.90**	**822.20**	**397.20**
Average weekly expenditure per person (£)						
Total expenditure	**119.90**	**147.60**	**203.60**	**273.40**	**411.10**	**198.60**

Note: The commodity and service categories are not comparable to those in publications before 2001-02.

Please see page xiii for symbols and conventions used in this report.

1 Mainly dependent on state pension and not economically active - see appendix B.

2 Excluding mortgage interest payments, council tax and Northern Ireland rates.

Table 3.11E **Expenditure of two adult retired households not mainly dependent on state pensions[1] by gross equivalised income quintile group (OECD-modified scale), 2010**
United Kingdom

	Lowest twenty per cent	Second quintile group	Third quintile group	Fourth quintile group	Highest twenty per cent	All house-holds
Lower boundary of group (£ per week)		180	277	400	595	
Weighted number of households (thousands)	300	770	650	390	200	2,300
Total number of households in sample	70	180	150	90	50	530
Total number of persons in sample	130	350	300	180	90	1,050
Total number of adults in sample	130	350	300	180	90	1,050
Weighted average number of persons per household	2.0	2.0	2.0	2.0	2.0	2.0
Commodity or service	Average weekly household expenditure (£)					
1 Food & non-alcoholic drinks	47.00	53.70	56.40	61.80	73.60	56.70
2 Alcoholic drinks, tobacco & narcotics	12.70	9.00	10.90	8.40	17.70	10.60
3 Clothing & footwear	6.00	13.60	17.20	19.20	20.60	15.20
4 Housing (net)[2], fuel & power	40.10	41.00	45.20	44.80	80.70	46.10
5 Household goods & services	17.50	23.30	30.90	41.90	42.60	29.50
6 Health	6.10	6.10	8.40	6.80	19.00	8.00
7 Transport	25.50	39.30	47.20	68.20	108.10	50.60
8 Communication	8.70	8.50	10.40	9.60	15.00	9.80
9 Recreation & culture	27.30	39.90	72.50	79.30	125.40	61.50
10 Education	[0.30]	[0.20]	[0.30]	[0.40]	[2.20]	[0.40]
11 Restaurants & hotels	12.70	20.70	30.10	43.20	83.00	31.50
12 Miscellaneous goods & services	16.20	21.80	29.30	40.30	89.10	32.10
1-12 All expenditure groups	220.20	277.00	358.90	423.90	677.00	352.10
13 Other expenditure items	21.40	32.20	44.80	71.70	79.50	45.10
Total expenditure	**241.60**	**309.10**	**403.60**	**495.60**	**756.50**	**397.20**
Average weekly expenditure per person (£)						
Total expenditure	**120.80**	**154.60**	**201.80**	**247.80**	**378.20**	**198.60**

Note: The commodity and service categories are not comparable to those in publications before 2001-02.

Please see page xiii for symbols and conventions used in this report.

1 Mainly dependent on state pension and not economically active - see Appendix B.

2 Excluding mortgage interest payments, council tax and Northern Ireland Rates.

Table 3.12   **Income and source of income by gross income quintile group, 2010**
United Kingdom

	Weighted number of house- holds	Number of house- holds in the sample	Weekly household income		Source of income					
			Dispo- sable	Gross	Wages and salaries	Self employ- ment	Invest- ments	Annuities and pensions[1]	Social security benefits[2]	Other sources
Gross equivalised income quintile group	(000s)	Number	£	£	Percentage of gross weekly household income					
Lowest twenty per cent	5,270	1,040	151	157	8	3	2	10	76	2
Second quintile group	5,260	1,090	297	319	27	4	2	16	48	2
Third quintile group	5,270	1,080	456	525	53	7	2	15	22	1
Fourth quintile group	5,270	1,040	673	815	73	7	2	8	8	1
Highest twenty per cent	5,260	1,020	1,317	1,688	76	13	3	5	2	1

Please see page xiii for symbols and conventions used in this report.
1 Other than social security benefits.
2 Excluding housing benefit and council tax benefit (rates rebate in Northern Ireland) - see Appendix B.

Table 3.12E   **Income and source of income by gross equivalised income quintile group (OECD-modified scale) 2010**
United Kingdom

	Weighted number of house- holds	Number of house- holds in the sample	Weekly household income		Source of income					
			Dispo- sable	Gross	Wages and salaries	Self employ- ment	Invest- ments	Annuities and pensions[1]	Social security benefits[2]	Other sources
Gross equivalised income quintile group	(000s)	Number	£	£	Percentage of gross weekly household income					
Lowest twenty per cent	5,270	1,050	174	180	12	5	1	5	75	2
Second quintile group	5,270	1,070	326	351	35	5	2	11	45	2
Third quintile group	5,260	1,050	480	556	58	7	1	13	19	2
Fourth quintile group	5,270	1,060	681	829	73	6	2	10	8	1
Highest twenty per cent	5,260	1,030	1,232	1,588	75	14	3	6	2	0

Please see page xiii for symbols and conventions used in this report.
1 Other than social security benefits.
2 Excluding housing benefit and council tax benefit (rates rebate in Northern Ireland) - see Appendix B.

Trends in household expenditure over time

Background

This chapter presents household expenditure data over time using two different classifications:

1. Classification Of Individual COnsumption by Purpose (COICOP). COICOP is the internationally agreed standard classification for reporting household consumption expenditure, and has been used since 2001/02, first in the Expenditure and Food Survey (EFS), and subsequently in the Living Costs and Food survey (LCF)

2. The Family Expenditure Survey (FES) classification. This was the main classification prior to 2001/02. Although it is has now been superseded, its use here enables a longer time series to be presented.

The figures and tables in this chapter (except Table 4.5) present figures that have been deflated using the All Items Retail Prices Index (RPI) data. This allows a comparison of expenditure in real terms to be made between the survey years. The commentary refers to the time series produced using these deflated figures. In addition, expenditure over time using COICOP in real terms is shown in Table 4.5, but no commentary is given on this table.

Interpreting EFS/LCF time series data

Before the introduction of the Expenditure and Food Survey (EFS) in 2001/02, expenditure data were collected via the Family Expenditure Survey (FES) and classified using the FES method of classification. These data have been retained and published alongside the COICOP time series and are presented in Tables 4.1 and 4.2.

Time series data based on the FES classification from 2001/02 (Tables 4.1 and 4.2) have been constructed by mapping COICOP data onto the FES classification. As such, the 'all expenditure groups' totals in Table 4.1 may not equal the sum of the component commodities or services as the mapping process is not exact. Due to the differences in the definitions of the classification headings, it is not possible to directly compare the FES data with the COICOP data. For example, 'motoring' in the FES classification includes vehicle insurance, whereas the 'transport' heading under COICOP excludes this expenditure.

As mentioned above, Tables 4.1 to 4.4 contain data that have been deflated to 2010 prices. To produce these data, each year's expenditure figures have been adjusted using the 'All items RPI' to account for price inflation that has occurred since that year. This results in a table of figures displayed in 'real terms' (that is at prices relative to 2010 prices), which allows comparisons to be made between different survey years. (The 'All Items RPI' can be downloaded from the Office for National Statistics website). Data in Table 4.5 have not been deflated to 2010 prices and therefore show the actual expenditure figures for each survey year.

Each year the Living Costs and Food Survey (LCF), previously the EFS, is reviewed and changes are made to keep it up to date. As such, year-on-year changes should be interpreted with caution. A detailed breakdown of the items that feed into each COICOP heading can be found in Table A1, while details of definition changes can be found in Appendix B.

Trends for the categories with lower levels of spending need to be treated with a degree of caution as the standard errors for these categories tend to be higher (standard errors are discussed in more detail in Appendix B). It should also be noted that there may be underreporting on certain items (notably tobacco and alcohol).

COICOP time series data in this publication are not directly comparable with UK National Accounts household expenditure data, which are published in *Consumer Trends*. (This publication can be downloaded from the Office for National Statistics website (http://www.ons.gov.uk/ons/search/index.html?newquery=consumer+trends++). National Accounts figures draw on a number of sources in addition to the LCF (please refer to Appendix B of *Consumer Trends* for details) and may be more appropriate for deriving long term trends on expenditure.

Household expenditure over time

Figure 4.1 and Table 4.3 show total household expenditure at 2010 prices, broken down by COICOP classification, over the period 2004/05 to 2010. Average weekly expenditure was at its highest for this period in 2004/05, a value of £516.30, it has since decreased, and has reached its lowest value of £473.60 in 2010.

Figure 4.1   **Total household expenditure based on COICOP classification, 2004-05 to 2010, at 2010 prices[1]**
United Kingdom

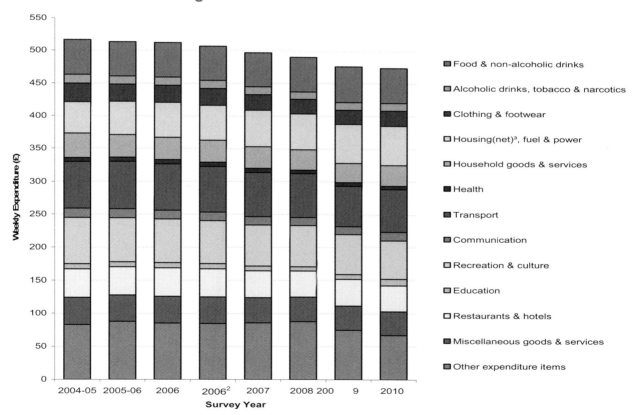

1 Figures have been deflated to 2010 prices using the RPI all items index.

2 Figures from 2006 onwards are based on weighted data using updated weights, with non-response weights and population figs based on 2001 census.

3 Excluding mortgage interest payments, council tax and Northern Ireland rates

Excluding the other items category, transport was the COICOP category which consistently had the highest average weekly spend throughout the time series. Spending levels were greatest towards the start of the time series, with households spending £70.80 per week in 2004/05; this subsequently fell every year to reach the lowest level of £61.10 in 2009. However in 2010 spending on transport has increased to £64.90 per week, Table 4.4 shows this is 14 per cent of overall household spending (slightly higher than the previous three years). The next highest expenditure was on recreation and culture, which followed a similar trend to expenditure on transport. The highest weekly spending on recreation and culture was £70.10 in 2004/05 before steadily declining to its lowest amount of £58.10 in 2010. This category accounted for between 12 and 14 per cent of overall spending throughout the time series.

Among the twelve COICOP headings, food and non–alcoholic drinks, and communication had the most consistent expenditure over the time series at 2010 prices. Food and non-alcoholic drinks expenditure varied between £52.10 and £54.60 per week (between 10 and 11 per cent of total expenditure for each year); communication expenditure varied between £12.20 and £13.90 per week (representing 3 per cent of total expenditure for each year). Housing, fuel and power was the only COICOP heading that has seen a substantial increase over the time series, with average weekly expenditure values increasing from £48.10 in 2004/05 to £60.40 in 2010.

As a proportion of total weekly expenditure, spending on each of the following categories remained relatively stable across the time series: restaurants and hotels (8 per cent), household goods and services (between 6 and 7 per cent) and clothing and footwear (5 per cent). Expenditure on restaurants and hotels was at its highest value of £42.90 per week in 2004/05, and its lowest was £39.20 in 2010. Of these three categories, household goods and services experienced the greatest variation in average weekly spending throughout the time series: the highest expenditure level was £37.50 per week in 2004/05, and the lowest in 2009 at £29.20, but this has since increased to £31.40 per week in 2010. This category is one of four, along with clothing and footwear, transport and communication for which expenditure values in this time series were at their lowest in 2009, but have all increased in 2010.

Table 4.1   **Household expenditure based on the FES classification, 1995-96 to 2010 at 2010 prices[1]**
United Kingdom

Year	1995[2] -96	1995[3] -96	1996 -97	1997 -98	1998 -99	1999 -2000	2000 -01	2001[4] -02	2002 -03
Weighted number of households (thousands)		24,130	24,310	24,560	24,660	25,330	25,030	24,450	24,350
Total number of households in sample	6,800	6,800	6,420	6,410	6,630	7,100	6,640	7,470	6,930
Total number of persons	16,590	16,590	15,730	15,430	16,220	16,790	15,930	18,120	16,590
Average number of persons per household	2.4	2.4	2.5	2.4	2.4	2.3	2.4	2.4	2.4
Commodity or service		Average weekly household expenditure (£)							
1 Housing (net)[7]	71.90	72.50	71.20	72.50	78.00	76.70	83.40	84.70	84.00
2 Fuel and power	19.20	19.10	19.30	17.70	16.00	15.20	15.50	15.10	14.70
3 Food and non-alcoholic drinks	78.80	80.50	81.60	80.40	80.50	80.10	80.80	79.60	81.00
4 Alcoholic drink	17.00	18.20	19.10	19.90	19.10	20.60	19.60	18.40	18.70
5 Tobacco	8.70	8.80	9.00	8.90	8.00	8.10	7.90	7.00	6.80
6 Clothing and footwear	25.50	26.50	27.20	28.60	29.70	28.20	28.70	28.70	27.70
7 Household goods	34.90	35.50	38.70	38.00	40.50	41.30	42.60	42.50	42.60
8 Household services	22.50	22.60	23.50	24.90	25.90	25.40	28.70	30.30	29.40
9 Personal goods and services	17.20	17.50	17.20	17.80	18.20	18.60	19.20	19.20	19.20
10 Motoring	55.10	56.90	61.30	66.50	70.70	70.60	71.90	74.50	77.70
11 Fares and other travel costs	9.20	9.90	11.20	12.10	11.30	12.30	12.30	12.00	12.20
12 Leisure goods	20.50	21.50	23.00	24.40	24.30	24.90	25.80	25.20	25.80
13 Leisure services	47.70	48.80	50.90	55.60	57.30	59.10	66.10	66.70	67.50
14 Miscellaneous	3.50	1.80	1.50	1.50	1.70	1.90	0.90	2.40	2.50
1-14 All expenditure groups	431.80	440.20	454.80	468.90	481.00	483.10	503.40	506.50	509.90
Average weekly expenditure per person (£) **All expenditure groups**	179.90	184.00	181.90	186.70	200.40	210.00	213.90	214.10	214.10
		Average weekly household income (£)[8]							
Gross income (£)	567	581	588	604	624	645	656	695	696
Disposable income (£)	457	467	480	492	506	526	534	569	571

Note: The commodity and service categories are not comparable to the COICOP categories used in Tables 4.3 and 4.4.

Figures are based on FES data between 1984 and 2000-01 and EFS data between 2000-01 and 2007, and LCF thereafter.

1 Figures have been deflated to 2010 prices using the RPI all items index.

2 From 1992 to this version of 1995-96, figures shown are based on unweighted, adult only data.

3 From this version of 1995-96, figures are shown based on weighted data, including children's expenditure. Weighting is based on the population figures from the 1991 and 2001 Censuses.

4 From 2001-02 onwards, commodities and services are based on COICOP codes broadly mapped to FES.

5 From 1995-96 to this version of 2006, figures shown are based on weighted data using non-response weights based on the 1991 Census and population figures from the 1991 and 2001 Censuses.

6 From this version of 2006, figures shown are based on weighted data using updated weights, with non-response weights figures based on the 2001 Census.

7 An improvement to the imputation of mortgage interest payments has been implemented for 2006 data onwards. Also, an error was discovered in the derivation of mortgage capital repayments which was leading to double counting: this has been amended for 2006 data onwards. Both these changes means there is a slight discontinuity between 2006 and earlier years.

8 Does not include imputed income from owner-occupied and rent-free households.

Table 4.1   **Household expenditure based on the FES classification, 1995-96 to 2010 at 2010 prices[1] (cont.)**
United Kingdom

Year	2003 -04	2004 -05	2005 -06	2006[5]	2006[6]	2007	2008	2009	2010
Weighted number of households (thousands)	24,670	24,430	24,800	24,790	25,440	25,350	25,690	25,980	26,320
Total number of households in sample	7,050	6,800	6,790	6,650	6,650	6,140	5,850	5,830	5,260
Total number of persons	16,970	16,260	16,090	15,850	15,850	14,650	13,830	13,740	12,180
Weighted average number of persons per household	2.4	2.4	2.4	2.4	2.3	2.4	2.4	2.3	2.3
Commodity or service	Average weekly household expenditure (£)								
1 Housing (net)[7]	85.60	91.20	93.60	94.20	93.90	99.60	97.80	89.20	85.40
2 Fuel and power	14.70	14.90	16.10	17.90	17.80	18.60	19.70	22.20	21.40
3 Food and non-alcoholic drinks	79.50	80.00	78.70	79.40	78.60	77.20	77.50	79.10	77.50
4 Alcoholic drink	18.10	17.60	17.10	16.80	16.60	15.90	14.00	14.60	14.50
5 Tobacco	6.70	5.90	5.30	5.20	5.30	4.90	4.80	4.60	4.60
6 Clothing and footwear	27.50	27.90	26.00	25.80	25.50	23.40	22.10	21.60	23.10
7 Household goods	43.00	42.30	38.80	38.90	38.40	37.40	35.40	33.30	35.70
8 Household services	30.50	31.30	31.40	30.00	29.80	28.60	28.40	28.00	28.60
9 Personal goods and services	19.80	19.00	19.60	19.90	19.80	19.30	17.90	18.50	17.40
10 Motoring	76.40	74.40	73.90	70.40	69.00	67.20	66.20	62.80	63.50
11 Fares and other travel costs	11.80	11.20	12.80	12.50	12.40	11.70	14.80	11.50	13.60
12 Leisure goods	26.20	25.50	22.50	22.20	21.90	21.70	19.80	19.30	19.00
13 Leisure services	67.40	70.80	73.00	74.90	73.70	66.70	68.60	66.30	62.80
14 Miscellaneous	2.30	2.40	2.50	2.40	2.30	2.10	2.00	2.00	2.30
1-14 All expenditure groups	509.60	514.50	511.10	510.50	505.00	494.40	488.90	473.10	469.30
Average weekly expenditure per person (£)									
All expenditure groups	**216.20**	**215.60**	**215.60**	**215.90**	**216.00**	**209.80**	**207.40**	**202.20**	**201.30**
	Average weekly household income (£)[8]								
Gross income (£)	**699**	**714**	**713**	**724**	**716**	**714**	**742**	**714**	**700**
Disposable income (£)	**569**	**581**	**579**	**588**	**582**	**578**	**606**	**584**	**578**

Note: The commodity and service categories are not comparable to the COICOP categories used in Tables 4.3 and 4.4.

Figures are based on FES data between 1984 and 2000-01 and EFS data thereafter

1 Figures have been deflated to 2010 prices using the RPI all items index.

2 From 1992 to this version of 1995-96, figures shown are based on unweighted, adult only data.

3 From this version of 1995-96, figures are shown based on weighted data, including children's expenditure. Weighting is based on the population figures from the 1991 and 2001 Censuses

4 From 2001-02 onwards, commodities and services are based on COICOP codes broadly mapped to FES.

5 From 1995-96 to this version of 2006, figures shown are based on weighted data using non-response weights based on the 1991 Census and population figures from the 1991 and 2001 Censuses.

6 From this version of 2006, figures shown are based on weighted data using updated weights, with non-response weights figures based on the 2001 Census.

7 An improvement to the imputation of mortgage interest payments has been implemented for 2006 data onwards. Also, an error was discovered in the derivation of mortgage capital repayments which was leading to double counting: this has been amended for 2006 data onwards. Both these changes means there is a slight discontinuity between 2006 and earlier years.

8 Does not include imputed income from owner-occupied and rent-free households.

Table 4.2 **FES household expenditure as a percentage of total expenditure, 1995-96 to 2010**

based on the FES classification at 2010 prices[1]

United Kingdom

Year	1995[2] -96	1995[3] -96	1996 -97	1997 -98	1998 -99	1999 -2000	2000 -01	2001[4] -02	2002 -03
Weighted number of households (thousands)		24,130	24,310	24,560	24,660	25,330	25,030	24,450	24,350
Total number of households in sample	6,800	6,800	6,420	6,410	6,630	7,100	6,640	7,470	6,930
Total number of persons	16,590	16,590	15,730	15,430	16,220	16,790	15,930	18,120	16,590
Weighted average number of persons per household	2.4	2.4	2.5	2.4	2.4	2.3	2.4	2.4	2.3
Commodity or service					Percentage of total expenditure				
1 Housing (net)[7]	17	16	16	15	16	16	17	17	16
2 Fuel and power	4	4	4	4	3	3	3	3	3
3 Food and non-alcoholic drinks	18	18	18	17	17	17	16	16	16
4 Alcoholic drink	4	4	4	4	4	4	4	4	4
5 Tobacco	2	2	2	2	2	2	2	1	1
6 Clothing and footwear	6	6	6	6	6	6	6	6	5
7 Household goods	8	8	9	8	8	9	8	8	8
8 Household services	5	5	5	5	5	5	6	6	6
9 Personal goods and services	4	4	4	4	4	4	4	4	4
10 Motoring	13	13	13	14	15	15	14	15	15
11 Fares and other travel costs	2	2	2	3	2	3	2	2	2
12 Leisure goods	5	5	5	5	5	5	5	5	5
13 Leisure services	11	11	11	12	12	12	13	13	13
14 Miscellaneous	1	0	0	0	0	0	0	0	0
1-14 All expenditure groups	100	100	100	100	100	100	100	100	100

1 Figures have been deflated to 2010 prices using the RPI all items index.
2 From 1992 to this version of 1995-96, figures shown are based on unweighted, adult only data.
3 From this version of 1995-96, figures are shown based on weighted data, including children's expenditure. Weighting is based on the population figures from the 1991 and 2001 Censuses
4 From 2001-02 onwards, commodities and services are based on COICOP codes broadly mapped to FES.
5 From 1995-96 to this version of 2006, figures shown are based on weighted data using non-response weights based on the 1991 Census and population figures from the 1991 and 2001 Censuses.
6 From this version of 2006, figures shown are based on weighted data using updated weights, with non-response weights and population figures based on the 2001 Census.
7 An improvement to the imputation of mortgage interest payments has been implemented for 2006 data onwards. Also, an error was discovered in the derivation of mortgage capital repayments which was leading to double counting: this has been amended for 2006 data onwards. Both these changes means there is a slight discontinuity between 2006 and earlier years.

Table 4.2   **FES household expenditure as a percentage of total expenditure, 1995-96 to 2010 (cont.)**

based on the FES classification at 2010 prices[1]

United Kingdom

Year	2003 -04	2004 -05	2005 -06	2006[5]	2006[6]	2007	2008	2009	2010
Weighted number of households (thousands)	24,670	24,430	24,800	24,790	25,440	25,350	25,690	25,980	26,320
Total number of households in sample	7,050	6,800	6,790	6,650	6,650	6,140	5,850	5,830	5,260
Total number of persons	16,970	16,260	16,090	15,850	15,850	14,650	13,830	13,740	12,180
Weighted average number of persons per household	2.4	2.4	2.4	2.4	2.3	2.4	2.4	2.3	2.3
Commodity or service				Percentage of total expenditure					
1 Housing (net)[7]	17	18	18	18	19	20	20	19	18
2 Fuel and power	3	3	3	4	4	4	4	5	5
3 Food and non-alcoholic drinks	16	16	15	16	16	16	16	17	17
4 Alcoholic drink	4	3	3	3	3	3	3	3	3
5 Tobacco	1	1	1	1	1	1	1	1	1
6 Clothing and footwear	5	5	5	5	5	5	5	5	5
7 Household goods	8	8	8	8	8	8	7	7	8
8 Household services	6	6	6	6	6	6	6	6	6
9 Personal goods and services	4	4	4	4	4	4	4	4	4
10 Motoring	15	14	14	14	14	14	14	13	14
11 Fares and other travel costs	2	2	3	2	2	2	3	2	3
12 Leisure goods	5	5	4	4	4	4	4	4	4
13 Leisure services	13	14	14	15	15	13	14	14	13
14 Miscellaneous	0	0	0	0	0	0	0	0	0
1-14 All expenditure groups	100	100	100	100	100	100	100	100	100

1 Figures have been deflated to 2010 prices using the RPI all items index.

2 From 1992 to this version of 1995-96, figures shown are based on unweighted, adult only data.

3 From this version of 1995-96, figures are shown based on weighted data, including children's expenditure. Weighting is based on the population figures from the 1991 and 2001 Censuses

4 From 2001-02 onwards, commodities and services are based on COICOP codes broadly mapped to FES.

5 From 1995-96 to this version of 2006, figures shown are based on weighted data using non-response weights based on the 1991 Census and population figures from the 1991 and 2001 Censuses.

6 From this version of 2006, figures shown are based on weighted data using updated weights, with non-response weights and population figures based on the 2001 Census.

7 An improvement to the imputation of mortgage interest payments has been implemented for 2006 data onwards. Also, an error was discovered in the derivation of mortgage capital repayments which was leading to double counting: this has been amended for 2006 data onwards. Both these changes means there is a slight discontinuity between 2006 and earlier years.

Table 4.3   **Household expenditure based on COICOP classification, 2004-05 to 2010 at 2010 prices[1]**

United Kingdom

Year	2004-05	2005-06	2006[2]	2006[3]	2007	2008	2009	2010
Weighted number of households (thousands)	24,430	24,800	24,790	25,440	25,350	25,690	25,980	26,320
Total number of households in sample	6,800	6,790	6,650	6,650	6,140	5,850	5,830	5,260
Total number of persons in sample	16,260	16,090	15,850	15,850	14,650	13,830	13,740	12,180
Total number of adults in sample	12,260	12,170	12,000	12,000	11,220	10,640	10,650	9,430
Weighted average number of persons per household	2.4	2.4	2.4	2.3	2.4	2.4	2.3	2.3
Commodity or service	Average weekly household expenditure (£)							
1 Food & non-alcoholic drinks	53.20	52.50	52.90	52.30	52.10	52.80	54.60	53.20
2 Alcoholic drinks, tobacco & narcotics	13.50	12.50	12.50	12.50	12.10	11.20	11.70	11.80
3 Clothing & footwear	28.40	26.30	26.20	25.90	23.80	22.50	21.90	23.40
4 Housing(net)[4], fuel & power	48.10	51.10	53.70	53.60	56.10	55.20	60.00	60.40
5 Household goods & services	37.50	34.70	34.20	33.80	33.30	31.40	29.20	31.40
6 Health	5.90	6.40	6.60	6.60	6.20	5.30	5.50	5.00
7 Transport	70.80	71.50	70.00	68.70	66.70	66.00	61.10	64.90
8 Communication	13.90	13.80	13.30	13.10	12.90	12.40	12.20	13.00
9 Recreation & culture	70.10	66.60	66.00	65.00	62.10	62.50	60.50	58.10
10 Education	7.70	7.60	8.10	7.90	7.30	6.40	7.30	10.00
11 Restaurants & hotels	42.90	42.50	42.80	42.40	40.30	39.30	40.20	39.20
12 Miscellaneous goods & services	41.40	40.10	40.60	40.20	38.20	37.10	36.70	35.90
1-12 All expenditure groups	433.40	425.60	426.90	421.90	411.10	402.20	400.90	406.30
13 Other expenditure items[5]	82.90	87.70	87.60	84.80	85.90	88.10	75.10	67.30
Total expenditure	**516.30**	**513.40**	**514.50**	**506.70**	**496.90**	**490.30**	**476.00**	**473.60**
Average weekly expenditure per person (£)								
Total expenditure	**216.30**	**217.70**	**217.60**	**216.70**	**210.90**	**208.00**	**203.40**	**203.10**
	Average weekly household income (£)							
Gross income (£)	**714**	**713**	**724**	**716**	**714**	**742**	**714**	**700**
Disposable income (£)	**581**	**579**	**588**	**582**	**578**	**606**	**584**	**578**

Note: The commodity and service categories are not comparable to the FES categories used in Tables 4.1 and 4.2

1 Figures have been deflated to 2010 prices using the RPI all items index.

2 From 2002-03 to this version of 2006, figures shown are based on weighted data using non-response weights based on the 1991 Census and population figures from the 1991 and 2001 Censuses.

3 From this version of 2006, figures shown are based on weighted data using updated weights, with non-response weights and population figures based on the 2001 Census.

4 Excluding mortgage interest payments, council tax and Northern Ireland rates.

5 An error was discovered in the derivation of mortgage capital repayments which was leading to double counting. This has been amended for the 2006 data onwards.

Table 4.4   **Household expenditure as a percentage of total expenditure based on COICOP classification, 2004-05 to 2010 at 2010 prices[1]**

United Kingdom

Year	2004-05	2005-06	2006[2]	2006[3]	2007	2008	2009	2010
Weighted number of households (thousands)	24,430	24,800	24,790	25,440	25,350	25,690	25,980	26,320
Total number of households in sample	6,800	6,790	6,650	6,650	6,140	5,850	5,830	5,260
Total number of persons in sample	16,260	16,090	15,850	15,850	14,650	13,830	13,740	12,180
Total number of adults in sample	12,260	12,170	12,000	12,000	11,220	10,640	10,650	9,430
Weighted average number of persons per household	2.4	2.4	2.4	2.3	2.4	2.4	2.3	2.3
Commodity or service				Percentage of total expenditure				
1 Food & non-alcoholic drinks	10	10	10	10	10	11	11	11
2 Alcoholic drinks, tobacco & narcotics	3	2	2	2	2	2	2	2
3 Clothing & footwear	5	5	5	5	5	5	5	5
4 Housing (net)[4], fuel & power	9	10	10	11	11	11	13	13
5 Household goods & services	7	7	7	7	7	6	6	7
6 Health	1	1	1	1	1	1	1	1
7 Transport	14	14	14	14	13	13	13	14
8 Communication	3	3	3	3	3	3	3	3
9 Recreation & culture	14	13	13	13	12	13	13	12
10 Education	1	1	2	2	1	1	2	2
11 Restaurants & hotels	8	8	8	8	8	8	8	8
12 Miscellaneous goods & services	8	8	8	8	8	8	8	8
1-12 **All expenditure groups**	84	83	83	83	83	82	84	86
13 **Other expenditure items[5]**	16	17	17	17	17	18	16	14
Total expenditure	100	100	100	100	100	100	100	100

Note: The commodity and service categories are not comparable to the FES categories used in Tables 4.1 and 4.2

1 Figures have been deflated to 2010 prices using the RPI all items index.

2 From 1995-96 to this version of 2006, figures shown are based on weighted data using non-response weights based on the 1991 Census and population figures from the 1991 and 2001 Censuses.

3 From this version of 2006, figures shown are based on weighted data using updated weights, with non-response weights and population figures based on the 2001 Census.

4 Excluding mortgage interest payments, council tax and Northern Ireland rates.

5 An error was discovered in the derivation of mortgage capital repayments which was leading to double counting. This has been amended for the 2006 data onwards.

Table 4.5   **Household expenditure 2004-05 to 2010 COICOP based on current[1] prices**

United Kingdom

	2004-05	2005-06	2006[2]	2006[3]	2007	2008	2009	2010
Weighted number of households (thousands)	24,430	24,800	24,790	25,440	25,350	25,690	25,980	26,320
Total number of households in sample	6,800	6,790	6,650	6,650	6,140	5,850	5,830	5,260
Total number of persons in sample	16,260	16,090	15,850	15,850	14,650	13,830	13,740	12,180
Total number of adults in sample	12,260	12,170	12,000	12,000	11,220	10,640	10,650	9,430
Weighted average number of persons per household	2.4	2.4	2.4	2.3	2.4	2.4	2.3	2.3

Commodity or service	Average weekly household expenditure (£)							
1 Food & non-alcoholic drinks	44.70	45.30	46.90	46.30	48.10	50.70	52.20	53.20
2 Alcoholic drinks, tobacco & narcotics	11.30	10.80	11.10	11.10	11.20	10.80	11.20	11.80
3 Clothing & footwear	23.90	22.70	23.20	23.00	22.00	21.60	20.90	23.40
4 Housing(net)[4], fuel & power	40.40	44.20	47.60	47.50	51.80	53.00	57.30	60.40
5 Household goods & services	31.60	30.00	30.30	29.90	30.70	30.10	27.90	31.40
6 Health	4.90	5.50	5.90	5.80	5.70	5.10	5.30	5.00
7 Transport	59.60	61.70	62.00	60.80	61.70	63.40	58.40	64.90
8 Communication	11.70	11.90	11.70	11.60	11.90	12.00	11.70	13.00
9 Recreation & culture	59.00	57.50	58.50	57.60	57.40	60.10	57.90	58.10
10 Education	6.50	6.60	7.20	7.00	6.80	6.20	7.00	10.00
11 Restaurants & hotels	36.10	36.70	37.90	37.60	37.20	37.70	38.40	39.20
12 Miscellaneous goods & services	34.90	34.60	36.00	35.70	35.30	35.60	35.00	35.90
1-12 All expenditure groups	364.70	367.60	378.30	373.80	379.80	386.30	383.10	406.30
13 Other expenditure items[5]	69.70	75.80	77.60	75.10	79.30	84.60	71.80	67.30
Total expenditure	**434.40**	**443.40**	**455.90**	**449.00**	**459.20**	**471.00**	**455.00**	**473.6**
Average weekly expenditure per person (£)								
Total expenditure	**182.00**	**188.00**	**192.80**	**192.00**	**194.80**	**199.80**	**194.40**	**203.10**
	Average weekly household income (£)							
Gross income (£)	**601**	**616**	**642**	**635**	**659**	**713**	**683**	**700**
Disposable income (£)	**489**	**500**	**521**	**515**	**534**	**582**	**558**	**578**

Note: The commodity and service categories are not comparable to those in publications before 2001-02

1 Data in Table 4.5 have not been deflated to 2010 prices and therefore show the actual expenditure for the year they were collected. Because inflation is not taken into account, comparisons between the years should be made with caution.

2 From 2002-03 to this version of 2006, figures shown are based on weighted data using non-response weights based on the 1991 Census and population figures from the 1991 and 2001 Censuses.

3 From this version of 2006, figures shown are based on weighted data using updated weights, with non-response weights and population figures based on the 2001 Census.

4 Excluding mortgage interest payments, council tax and Northern Ireland rates.

5 An error was discovered in the derivation of mortgage capital repayments which was leading to double counting. This has been amended for the 2006 data onwards.

Impact of the recession on household expenditure

Executive Summary

Introduction

Household expenditure plays an important part in the UK economy and might also be expected to reflect economic trends. This chapter compares expenditure during the peak of the economic cycle in 2007, with expenditure in 2010, when the UK was recovering from recession. The findings are considered in the context of the macroeconomic changes seen over this period of economic instability.

The chapter follows on from the general trends reported in Chapter 4, and focuses on statistically significant differences observed comparing the two snapshot years. It provides a detailed analysis of differences in expenditure by geographic areas that share common characteristics using the Output Area Code (OAC) categorisation system based on the 2001 Census. The categorisation is based on variables associated with: demographic structure, household composition, housing, socio-economic status and employment (see *Family Spending* 2010[6] for more details).

The results are reported at 2010 prices to remove the effects of inflation and to enable figures for the two years to be compared on a like-for-like basis.

2007 and 2010: UK expenditure and income

When average household expenditure for the two years is considered for the UK, statistically significant decreases were observed for some categories: clothing and footwear; health; recreation and culture; restaurants and hotels; and miscellaneous goods and services. Significant increases were seen for others: education; housing; and water and electricity. While these changes in expenditure were statistically significant, some were small. Interestingly, average disposable income was very similar in 2007 and 2010.

Considered overall, these changes amount to surprisingly little change in average household expenditure and income over the period of recession in the UK. We therefore categorised households according to OAC super group, to consider whether there were any significant changes in expenditure when households are classified according to common characteristics of their local areas. We also disaggregated the broad expenditure categories to enable us to consider spending patterns more fully.

Expenditure by OAC super group

The highest level of OAC categorisation is the super group, which define seven categories of geographic area with common characteristics. (See Table 5.2 for a list of super groups and groups.) Examination of expenditure by OAC super group revealed interesting differences comparing 2007 and 2010:

Transport

Countryside and Prospering Suburbs super groups spent significantly more in 2010 compared with 2007 on operation of personal transport, including fuel, but less on vehicle purchase. In contrast, Typical Traits and Multicultural super groups spent less on private transport but more on public transport. These findings suggest that households reallocate their spending, but how they do this reflects their circumstances. For instance, Countryside or Prospering Suburbs households may find it more difficult to substitute private for public transport than other super groups, but they may be able to make savings by spending less on new vehicles.

Housing, fuel and power

Spending on housing, fuel and power increased across all OAC super groups between 2007 and 2010, but increases in spending on rent were notable for Blue Collar Communities, Countryside, Constrained by Circumstances and Typical Traits.

Spending on electricity, gas and other fuels increased fairly uniformly across super groups, with the exception of Constrained by Circumstances (the lowest spenders in this category), perhaps reflecting this group's economising on fuel.

Recreation and culture; Restaurants and hotels

Changes in expenditure on recreation and culture were variable. For example, Prospering Suburbs and Typical Traits spent less on this category in 2010 than in 2007, but City Living spent significantly more. The patterns by sub-category were complex, with relatively consistent expenditure in some categories suggesting that many households no longer consider some of this spending to be discretionary. Spending on restaurants and hotels gives a similar story, with spending reduced for only some categories.

Expenditure by OAC group

The data were examined at a more detailed level of OAC categorisation: the 21 groups within super groups. This analysis revealed further differences that were masked at the group level:

Alcohol and tobacco showed a significant fall for Prospering Younger Families and a significant increase for Older Workers.

The observed fall in transport expenditure in the Prospering Suburbs super group was driven largely by Prospering Younger Families.

The increase in housing, fuel and power spending among Blue Collar Communities was due to Younger Blue Collar Workers.

Expenditure on food and non-alcoholic drink showed few changes even at this level of breakdown, reflecting the essential nature of purchasing food.

Conclusion

The analysis showed surprisingly little change in average household expenditure between 2007 and 2010, given the scope of the recession. However, when household characteristics are taken into account, using OAC categorisation, a more complex picture emerges: different groups show different changes between 2007 and 2010. It appears that it is difficult to define the impact of the recession on the typical household, with the effect depending greatly on household circumstances and preferences.

Introduction

Household expenditure plays a large part in determining the economic path of the UK. In this chapter, the Living Costs and Food Survey (LCF) data are used to compare 2010 household expenditure when the UK economy was recovering from recession, with expenditure during the peak of the economic cycle in 2007. The macroeconomic environment changed significantly between these two periods; understanding the changes in household expenditure during this period of economic instability is of high general interest.

Context

Between 2008 quarter 1 and 2009 quarter 3, Gross Domestic Product (GDP) fell by 7.1 per cent[1], making this recession more severe than that of the 1990s (where GDP fell by 2.5 per cent) and 1980s (where GDP fell by 5.9 per cent)[2]. Figure 5.1 illustrates that although economic growth returned in the final quarter of 2009, recovery has been slow. Latest estimates continue to show a gradual recovery, with GDP growth not yet returning to pre-recession levels[3]. The impact of the recession on the labour market was gradual and not as severe as in the 1990s recession[4]. Unemployment increased steadily from 5.2 per cent in 2008 quarter 1 to 8 per cent in 2010 quarter 1. However, during the recovery unemployment did not fall immediately. As a consequence, the percentage of workless households, which increased from 17.7 per cent in 2007 to 18.7 per cent in 2009, actually increased again in 2010 to 18.9 per cent. Jenkins (2010) reports that there has also been a fall in the number of full-time workers, illustrating that many households with employed residents have experienced a reduction in hours worked[5]. In 2010 therefore, it is likely that households continued to have been affected by the recession.

Figure 5.1   **UK Quarterly GDP growth and unemployment rate, 2006q4 – 2011q2**

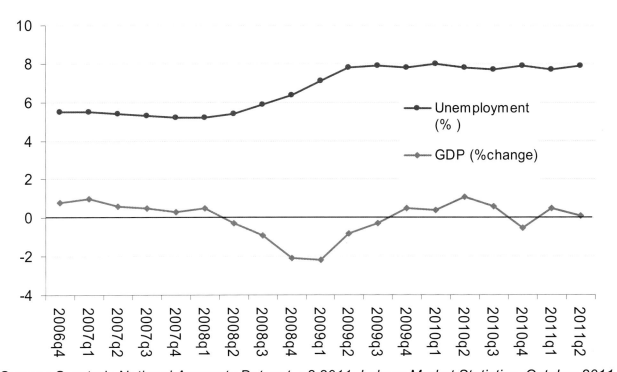

Source: Quarterly National Accounts Dataset, q2 2011; Labour Market Statistics, October 2011

Although the macroeconomic statistics provide us with an overall indication of the impact of the recession for the average household, they are unable to provide the detail on how specific household groups have been affected. To do this, we therefore examine household expenditure by Output Area Classification (OAC) super group and group (described later), following the analysis presented in *Family Spending* 2010[6], to consider the impact of changing economic conditions by household characteristics. This is particularly relevant, as labour market statistics indicate that the young, males, and those with fewer educational qualifications, were most affected by deteriorating employment conditions[7].

We first consider the impact upon households that we may expect. For so-called 'normal goods' and luxury items, we would expect less favourable economic conditions to result in a fall in their consumption. Conversely, we would expect to see the consumption of 'inferior goods', such as supermarket own-brand products for example, to rise as a direct response to constrained income or the expectation of reduced household income in the future. This view however, may be too simplistic for a number of reasons. First, it is important to note that the impact on the household is likely to differ considerably from the impact on the individual. Households can be grouped according to many different circumstances, including for instance, number of working age residents, number of children, and tenure. The impact of rising unemployment may therefore be felt more significantly by households with a greater number of working age adults. In addition, households with a greater number of young adults with few qualifications are likely to have been more affected by rising unemployment. The low interest rates will have had a very different impact upon those households with a large variable-rate mortgage compared with households that have significant savings. Rising inflation will also have had a detrimental impact upon households with large savings. Supporting this view, Howell et al. (2010)[8] showed that while households overall experienced little change in their disposable income during the recession, the impact was uneven. By examining expenditure by OAC super group, and also disaggregating further by OAC group, here we are able to examine changes in household expenditure according to circumstances. This approach enables us to investigate beyond the average impact.

The classification of expenditure into broad Classification Of Individual COnsumption by Purpose (COICOP) items may also mask a degree of substitutability within these categories. Disaggregating expenditure categories may therefore improve our understanding of how households respond to changes in economic conditions.

We should also consider that households may not alter their spending patterns in response to changes in economic conditions. Consumers may determine their spending based upon their long-term expectations of income rather than their current income. This permanent income hypothesis implies that households may not react to any reduction in their household income during the recession, as they expect conditions to improve long-term.

Our view of what constitutes essential item expenditure may also be too simplistic. Certain items that may have traditionally been considered discretionary, such as recreation and culture expenditure for instance, may be considered essential by many households through a habituation effect. Some households may have enjoyed regular holidays and gym subscriptions for many years during economic prosperity and continue with this consumption despite the less favourable economic conditions. Related to this, the spending behaviour of other households may have an impact upon individual household expenditure. Consequently, if friends and neighbours choose not to alter their expenditure and continue to visit restaurants and purchase luxury items, individual households may also be more reluctant to alter their own spending habits. This relative income hypothesis suggests that households are more concerned with relative levels of consumption than absolute levels.

Overall, household expenditure is potentially influenced by many factors, and consequently it is very difficult to predict the household reaction to changing economic conditions. In this chapter, we examine spending behaviour by OAC group and super group to improve our understanding of the impact of the most recent recession upon household expenditure.

Average household COICOP expenditure

This chapter categorises the LCF mean household expenditure estimates using the Classification Of Individual COnsumption by Purpose (COICOP). This is the internationally agreed standard classification for reporting household consumption expenditure.

Table 5.1 presents weekly household COICOP expenditure and weekly household income by year between 2006 and 2010 in 2010 prices[9]. All statistics reported in this article have been deflated to reflect 2010 prices using All Items Retail Price Index data. We should be aware that the price of some goods will have increased at a greater rate relative to others, but this is not accounted for through the use of an All Items deflator. It should be noted that the LCF survey is cross-sectional (the results for each survey year representing a snapshot) and is not primarily intended for analysis over a longer period. Therefore, some caution should be taken when directly looking at trends over time. For this reason, the remainder of the chapter focuses upon comparing snapshots of years 2007 and 2010.

Table 5.1   **Mean COICOP weekly household expenditure and income time series**
weighted and in 2010 prices
United Kingdom

	2006	2007	2008	2009	2010
Weighted number of households	25,440	25,350	25,690	25,980	26,320
Commodity or service	Average weekly household expenditure (£)				
Food & non-alcoholic drinks	52.30	52.10	52.80	54.60	53.20
Alcoholic drinks, tobacco & narcotics	12.50	12.10	11.20	11.70	11.80
Clothing & footwear	25.90	23.80	22.50	21.90	23.40
Housing(net)[1], fuel & power	53.60	56.00	55.20	59.90	60.40
Household goods & services	33.80	33.30	31.40	29.20	31.40
Health	6.60	6.20	5.30	5.50	5.00
Transport	68.60	66.70	66.00	61.10	64.90
Communication	13.10	12.90	12.40	12.20	13.00
Recreation & culture	65.00	62.10	62.50	60.50	58.10
Education	7.90	7.30	6.40	7.30	10.00
Restaurants & hotels	42.40	40.30	39.30	40.10	39.20
Miscellaneous goods & services	40.20	38.20	37.10	36.70	35.90
Total expenditure	**506.70**	**496.90**	**490.30**	**476.00**	**473.60**
Income (£)					
Disposable weekly household income	582	578	605	584	578
Gross weekly household income	716	714	742	714	700

1 Excluding mortgage interest payments, council tax and Northern Ireland rates

In terms of gross and disposable weekly household income, Table 5.1 illustrates that this actually increased between 2007 and 2008. This has been reported elsewhere[10], and is partly due to increases in social security benefits and reductions in taxes. Both then fell in 2009 and increased only slightly in 2010, despite the fact that the economy was no longer in recession. This is consistent with labour market statistics reported in the introduction, with unemployment increasing gradually during the recession and only falling gradually during the recovery. Continued limited wage growth is also likely to have contributed to the fall in disposable income in 2010.

Comparing COICOP expenditure in 2007 with 2010, statistically significant falls in expenditure are observed for clothing and footwear, health, recreation and culture, restaurants and hotels, and miscellaneous goods and services. Conversely, there have been significant[11] increases in education; and housing, fuel and power expenditure. It should again be noted that some of the changes in expenditure, although significant, were relatively small. For instance, average household clothing expenditure fell by only 43p in real terms. Figure 5.2 illustrates the change in COICOP expenditure between 2006 and 2010.

Figure 5.2   **COICOP item expenditure as a percentage of total COICOP expenditure time series**
United Kingdom

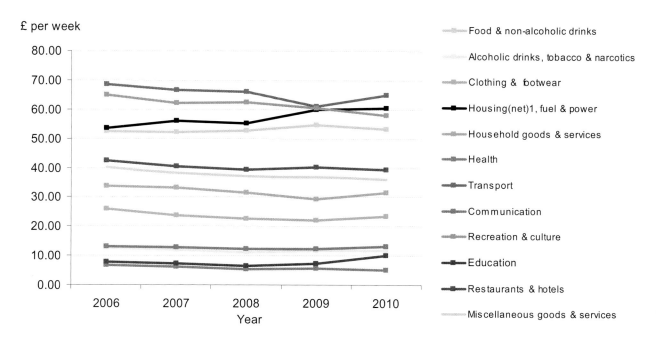

Considering the extent of the fall in GDP, we find little change in average household expenditure for COICOP items during and immediately following the recession. This is surprising, as we had expected to see much greater falls in discretionary item expenditure. However, this is consistent with evidence reported here and elsewhere that average household disposable income actually grew during the recession, and only began to fall later so that 'the pain was delayed but not avoided'[12].

OAC super group and group composition

Evidence elsewhere has suggested that the impact of the recession upon employment and disposable income was disproportionate[8]. The average household classification may therefore be masking many impacts. To investigate further, we classify households according to Output Area Classification (OAC) super group and group. *Family Spending* 2010 details the methodology for deriving Output Areas, of which there are over 223,000. To assist with developing a clearer understanding of these geographic areas, output areas were clustered into groups based upon a number of characteristics. This resulted in seven super groups, which can be further disaggregated into 21 groups. These super groups and groups share a series of characteristics, but may be distributed over different parts of the UK. The characteristics common to each super group are outlined in *Family Spending* 2010 (Table 5.3, page 106):
www.ons.gov.uk/ons/rel/family-spending/family-spending/2010-edition/family-spending-2010--living-costs-and-food-survey-2009-.pdf

Table 5.2   **OAC super groups and groups classification (percentages)**
United Kingdom

	2007	2010
1 Blue Collar Communities	**16**	**17**
1A Terraced Blue Collar	4	4
1B Younger Blue Collar	6	6
1C Older Blue Collar	6	7
2 City Living	**7**	**7**
2A Transient Communities	2	2
2B Settled in the City	5	5
3 Countryside	**11**	**13**
3A Village Life	5	5
3B Agricultural	3	3
3C Accessible Countryside	4	4
4 Prospering Suburbs	**23**	**22**
4A Prospering Younger Families	4	4
4B Prospering Older Families	7	6
4C Prospering Semis	7	7
4D Thriving Suburbs	5	5
5 Constrained by Circumstances	**12**	**12**
5A Senior Communities	2	2
5B Older Workers	7	8
5C Public Housing	2	3
6 Typical Traits	**20**	**20**
6A Settled Households	5	6
6B Least Divergent	5	6
6C Young Families in Terraced Homes	5	5
6D Aspiring Households	5	4
7 Multicultural	**11**	**9**
7A Asian Communities	7	5
7B Afro-Caribbean Communities	4	3

Totals may not add up due to the independent rounding of component categories

N.B. Super groups are shown in bold

Table 5.2 reports the proportion of Living Costs and Food Survey respondents in each OAC super group (in bold) and group over time. The percentage of respondents in each super group remained relatively stable between 2007 and 2010, with only a slight increase in the Countryside and Constrained by Circumstances groups, and a fall in the Multicultural group. When we disaggregate each super group, further differences in the proportion contributing to the sample are observed. For instance, within the Typical Traits category there has been an increase in Young Families in Terraced Homes and a fall in Aspiring Households. This illustrates that broad super group classification could be masking differences at the group level.

COICOP expenditure by OAC super group

Overall COICOP expenditure

Table 5.3 reports mean weekly household COICOP expenditure and income by OAC super group in 2010. A detailed discussion of the differences in expenditure by OAC group in 2009 was reported in *Family Spending* 2010. Similar trends are observed here. For instance, housing, fuel and power made up the greatest component of total COICOP expenditure for all super groups except Countryside and Prospering Suburbs for which transport costs make up the largest proportion.

Table 5.3   **Mean COICOP weekly household expenditure and income 2010, by OAC super group (weighted)**
United Kingdom

Commodity or service	Blue collar communities	City living	Countryside	Prospering suburbs	Constrained by circumstances	Typical Traits	Multicultural	All households
	Average weekly household expenditure (£)							
Food & non-alcoholic drinks	49.70	48.00	59.90	61.00	40.50	51.60	55.90	53.20
Alcoholic drinks, tobacco & narcotics	13.50	10.80	13.80	11.00	11.60	11.50	9.70	11.80
Clothing & footwear	20.80	26.60	23.70	26.50	16.00	24.50	25.10	23.40
Housing(net)[1], fuel & power	53.50	115.10	60.10	48.90	51.30	56.30	83.10	60.40
Household goods & services	22.40	39.00	39.30	41.00	19.40	30.50	25.60	31.40
Health	3.40	6.00	5.30	6.50	2.50	5.40	6.30	5.00
Transport	50.80	61.70	83.00	84.80	33.00	64.40	63.00	64.90
Communication	12.40	13.80	13.50	13.50	9.80	13.30	16.00	13.00
Recreation & culture	49.50	48.30	81.10	73.20	36.90	53.70	49.80	58.10
Education	1.50	37.50	15.30	8.00	1.60	10.70	12.60	10.00
Restaurants & hotels	30.80	52.10	40.70	48.10	24.40	38.80	42.70	39.20
Miscellaneous goods & services	25.70	42.00	45.60	44.60	20.70	37.20	32.80	35.90
Income (£)								
Disposable weekly household income	458.40	681.80	639.90	692.80	361.50	589.50	626.70	578.40
Gross weekly household income	537.60	861.50	776.30	851.90	416.00	714.70	757.50	700.50

1 Excluding mortgage interest payments, council tax and Northern Ireland rates

City Living reports the highest gross weekly household income (£861.50), whilst Prospering Suburbs (£692.80) report the highest disposable weekly household income (which takes account of national insurance contributions and income tax). The Constrained by Circumstances super group report both the lowest gross (£416.00) and disposable (£361.50) household incomes. The difference between the super groups with the highest and lowest average disposable weekly household income is therefore fairly large at £331.30.

Table 5.4   Mean COICOP weekly household expenditure and income 2007, by OAC super group (weighted)
2010 prices
United Kingdom

Commodity or service	Blue Collar Communities	City Living	Countryside	Prospering Suburbs	Constrained by Circumstances	Typical Traits	Multicultural	All households
	Average weekly household expenditure (£)							
Food & non-alcoholic drinks	49.60	45.50	57.80	58.90	40.70	52.00	49.90	52.10
Alcoholic drinks, tobacco & narcotics	13.80	11.30	14.30	11.30	13.00	11.70	8.80	12.10
Clothing & footwear	20.10	25.40	22.40	27.10	14.70	25.00	26.60	23.80
Housing(net)[1], fuel & power	47.90	101.90	56.50	48.10	46.80	52.50	73.90	56.10
Household goods & services	29.60	30.40	43.10	37.60	20.70	35.90	28.10	33.30
Health	3.30	9.10	9.00	9.20	3.90	4.90	4.70	6.20
Transport	50.00	62.40	88.20	90.40	35.60	71.30	47.80	66.70
Communication	12.00	13.40	13.10	12.90	10.00	13.20	14.40	12.90
Recreation & culture	50.40	58.60	79.30	79.40	39.10	65.40	43.20	62.10
Education	3.00	12.10	13.10	8.30	1.50	5.30	12.90	7.40
Restaurants & hotels	30.70	48.20	45.50	46.90	24.20	44.00	36.50	40.30
Miscellaneous goods & services	28.20	40.10	47.50	48.70	22.30	38.70	34.20	38.20
Income (£)								
Disposable weekly household income	490.70	593.00	679.60	697.50	353.30	609.70	551.70	577.80
Gross weekly household income	559.80	743.60	864.20	873.60	409.00	757.60	679.90	713.60

1 Excluding mortgage interest payments, council tax and Northern Ireland rates

In 2007, Prospering Suburbs reported the highest real gross (£873.60) and real disposable (£697.50) household income ('real' income refers to income after the affects of inflation have been accounted for). These figures are above the equivalent highest household incomes reported in 2010 (disposable income decreased by £4.70 between 2007 and 2010 for Prospering Suburbs in real terms). As in 2010, Constrained by Circumstances report both the lowest real gross (£409.00) and real disposable (£353.30) incomes. For this group therefore, real disposable income increased by £8.20 between 2007 and 2010. Real disposable income increased for three super groups between 2007 and 2010. However, the increases between super groups vary greatly. Real disposable income increased by £75.00 for Multicultural and £88.80 for the City Living super group. In contrast, it decreased £32.30 for Blue Collar Communities and £39.70 for the Countryside super group. At the average household level, Table 5.1 showed very little change in real average household disposable income when comparing 2007 with 2010. This highlights that examining data at the average household level masks many differences.

Expenditure is also presented as a percentage of total COICOP expenditure by OAC super group for 2010 (Figure 5.3) and 2007 (Figure 5.4).

Figure 5.3   **OAC mean COICOP expenditure graph by OAC super group (as a percentage of total COICOP expenditure), 2010**
United Kingdom

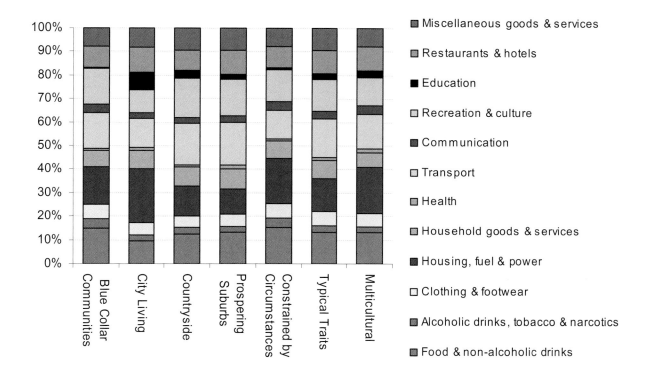

Figure 5.4   **Mean COICOP expenditure graph by OAC super group (as a percentage of total COICOP expenditure), 2007**
United Kingdom

In terms of comparing COICOP item expenditure in 2010 with 2007 by OAC super group, we consider that there are two potential impacts. Households may have changed their weekly expenditure on particular items, but they may not have changed the relative importance of the item in terms of their overall COICOP expenditure. This would account for households altering their total expenditure in 2010. The analysis supports this, as Table 5.1 illustrates a fall in expenditure for most COICOP categories between 2007 and 2010. We therefore examine if households have significantly changed the amount spent on each item, and also if they have changed the relative importance placed on a particular item in terms of their total COICOP expenditure.

Table 5.5 reports statistical significance levels of differences in mean real household expenditure in 2007 and 2010 by super group, following the calculation of t statistics. Bold figures indicate statistical significance above the 10 per cent significance levels, while red indicates an increase in expenditure, and blue a decrease. This highlights that:

- Blue Collar Communities, Prospering Suburbs and Typical Traits have two significant categories of reduced expenditure.

- The Countryside super group have one significant category of reduced expenditure.

- City Living and Typical Traits each have one significant category of increased expenditure.

- The Multicultural super group has two significant categories of increased expenditure.

Table 5.5 Statistical significance of mean differences in 2010 and 2007 expenditure
United Kingdom

	Blue Collar communities	City Living	Countryside	Prospering suburbs	Constrained by circumstances	Typical Traits	Multicultural
Food & non-alcoholic drinks	0.90	0.60	0.30	0.20	0.80	0.70	0.10
Alcoholic drinks, tobacco & narcotics	0.70	0.80	**0.70**	0.60	0.20	0.70	**0.20**
Clothing & footwear	0.90	0.80	0.90	0.40	1.00	0.80	1.00
Housing(net)[1], fuel & power	**0.00**	0.50	0.20	0.90	0.20	**0.10**	0.20
Household goods & services	**0.00**	0.20	0.50	0.50	0.30	0.20	0.80
Health	1.00	0.30	0.00	0.10	0.20	0.30	0.30
Transport	1.00	0.80	0.40	0.30	0.30	0.20	**0.00**
Recreation & culture	0.60	0.30	0.90	**0.00**	0.20	**0.00**	0.20
Education	0.30	**0.10**	0.60	1.00	**0.80**	0.20	0.90
Restaurants & hotels	0.80	0.70	**0.10**	0.90	0.50	**0.10**	**0.00**
Communication	1.00	0.70	0.80	0.70	**0.40**	0.80	0.20
Miscellaneous goods & services	**0.00**	0.80	0.60	**0.10**	0.20	0.90	0.80

1 Excluding mortgage interest payments, council tax and Northern Ireland rates

Bold font signifies significantly different expenditure in 2010
Blue increase in 2010
Red fall in 2010
Black font signifies no change

Table 5.6 then ranks each COICOP item in terms of amount spent in 2010 for each OAC super group. Red indicates a decrease and blue an increase in the importance of the item in terms of total expenditure in 2010 compared with 2007. For instance, the amount spent on housing, fuel and power by Blue Collar Communities increased in 2010 compared with 2007, with this group spending the most on this compared with any other COICOP item.

Table 5.6 Rank order percentage of total COICOP expenditure
United Kingdom

	Transport	Housing, fuel(net)¹ & power	Recreation & culture	Food & non-alcoholic drinks	Restaurants & hotels	Miscellaneous goods & services	Household goods & services	Clothing & footwear	Communication	Alcoholic drinks, tobacco & narcotics	Education	Health
ALL GROUPS COMBINED	Transport	Housing, fuel(net)¹ & power	Recreation & culture	Food & non-alcoholic drinks	Restaurants & hotels	Miscellaneous goods & services	Household goods & services	Clothing & footwear	Communication	Alcoholic drinks, tobacco & narcotics	Education	Health
Blue collar communities	Housing, fuel(net)¹ & power	Transport	Recreation & culture	Food & non-alcoholic drinks	Restaurants & hotels	Miscellaneous goods & services	Household goods & services	Clothing & footwear	Alcoholic drinks, tobacco & narcotics	Communication	Health	Education
City living	Housing, fuel(net)¹ & power	Transport	Restaurants & hotels	Recreation & culture	Food & non-alcoholic drinks	Miscellaneous goods & services	Household goods & services	Education	Clothing & footwear	Alcoholic drinks, tobacco & narcotics	Communication	Health
Countryside	Transport	Recreation & culture	Housing, fuel(net)¹ & power	Food & non-alcoholic drinks	Restaurants & hotels	Miscellaneous goods & services	Household goods & services	Clothing & footwear	Education	Communication	Alcoholic drinks, tobacco & narcotics	Health
Prospering suburbs	Transport	Recreation & culture	Food & non-alcoholic drinks	Housing, fuel(net)¹ & power	Restaurants & hotels	Miscellaneous goods & services	Household goods & services	Clothing & footwear	Communication	Alcoholic drinks, tobacco & narcotics	Education	Health
Constrained by circumstances	Housing, fuel(net)¹ & power	Food & non-alcoholic drinks	Recreation & culture	Transport	Restaurants & hotels	Miscellaneous goods & services	Household goods & services	Clothing & footwear	Alcoholic drinks, tobacco & narcotics	Communication	Health	Education
Typical Traits	Transport	Housing, fuel(net)¹ & power	Recreation & culture	Food & non-alcoholic drinks	Restaurants & hotels	Miscellaneous goods & services	Household goods & services	Clothing & footwear	Communication	Alcoholic drinks, tobacco & narcotics	Education	Health
Multicultural	Housing, fuel(net)¹ & power	Transport	Recreation & culture	Food & non-alcoholic drinks	Restaurants & hotels	Miscellaneous goods & services	Household goods & services	Clothing & footwear	Communication	Education	Alcoholic drinks, tobacco & narcotics	Health

1 Excluding mortgage interest payments, council tax and Northern Ireland rates

Bold font signifies significantly different expenditure in 2010

Blue increase in 2010

Red fall in 2010

Black font signifies no change

Housing, fuel and power expenditure

We first consider the COICOP expenditure items that make up the greatest proportion of expenditure for both years for all groups. Housing, fuel and power expenditure (Figure 5.5) increased for all groups in 2010, but only significantly so for Blue Collar Communities (from £47.90 to £53.50) and Typical Traits (from £56.30 to £52.50). In 2007 expenditure on this item made up 14.2 per cent of the total Blue Collar Communities super group expenditure, increasing to 16 per cent in 2010. For Typical Traits, the percentage of total expenditure spent increased from 12.5 per cent to 14.3 per cent. Households are also likely to be fairly limited in their ability to look for alternatives for these items following price increases, which is likely to have resulted in the increase observed.

Figure 5.5   **Housing, fuel and power expenditure – percentage of total COICOP expenditure**
United Kingdom

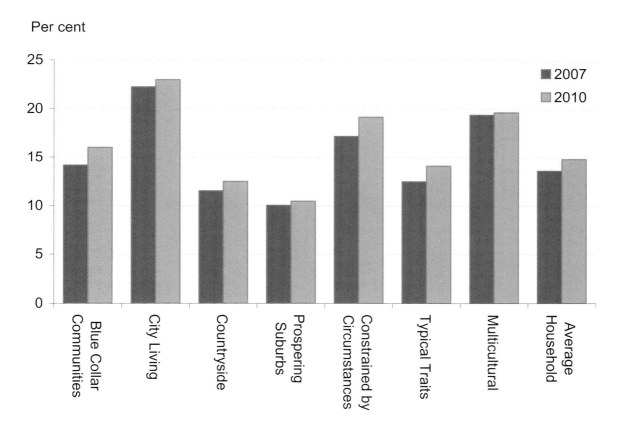

We disaggregate this COICOP item further. Figure 5.6 presents weekly expenditure on rentals for housing, which makes-up a substantial element of total housing, fuel and power expenditure, although it should be noted that the rental market in the UK is fairly small. Significant real increases in 2010 expenditure, compared with 2007[9] were observed for Blue Collar Communities, Countryside, Constrained by Circumstances and Typical Traits. This impact was fairly consistent for the majority of super groups. It should be noted that rental expenditure is smaller for the Prospering Suburbs and Countryside super groups due to the increased likelihood that households in this category do not live in a rental property. Mortgage payments are not considered in the

COICOP classification, as this is regarded as investment expenditure. Overall, this illustrates that some households renting property paid significantly more in 2010 compared with 2007 in real terms. Given the low interest rates during this period, those households with a variable rate mortgage are likely to have seen smaller increases in this key element of housing expenditure.

Figure 5.6   **Actual rentals for housing (subset of housing fuel and power) weekly household expenditure (£)**
2010 prices
United Kingdom

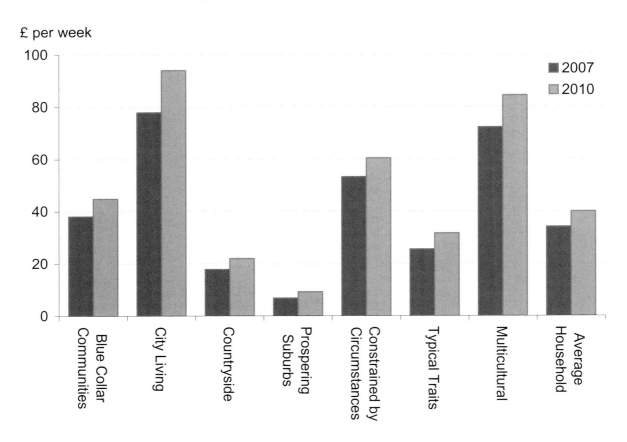

We also consider water and miscellaneous services relating to the dwelling expenditure (Figure 5.7). Fewer changes are observed, although there was a significant fall for City Living and a significant increase for Multicultural in real terms. The miscellaneous service element is likely to explain the fall for City Living and the relatively greater amount spent on this item by this super group. City Living households are more likely to live in a rental property (Table 5.3 Family Spending 2010), and the miscellaneous services category includes 'Other regular housing payments including service charge for rent'.

Figure 5.7   **Water and miscellaneous services relating to the dwelling (subset of housing fuel and power) weekly household expenditure (£)**
2010 prices
United Kingdom

£ per week

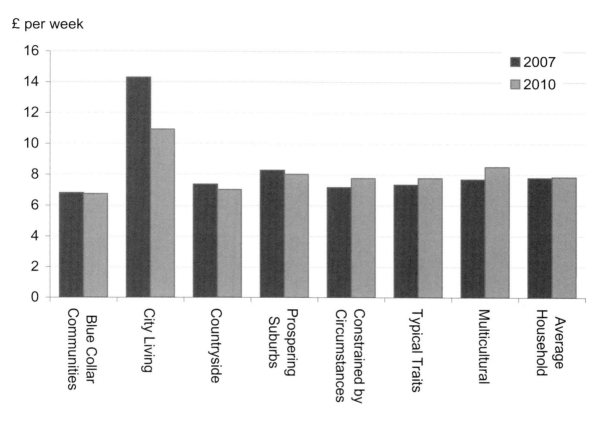

Figure 5.8 presents household expenditure on electricity, gas and other fuels. Similar to rental expenditure, Figure 5.8 illustrates that expenditure on electricity, gas and other fuels increased fairly uniformly across super groups in 2010. Increases were significant for all super groups except Constrained by Circumstances. This is particularly interesting as this super group also spend less on electricity, gas and other fuels than any other super group. All households pay the same proportion of fuel duty per unit of fuel purchased, regardless of their income. Because lower income households pay proportionally more of their income on fuel duty, we may therefore, expect Constrained by Circumstances households, who are more likely to have lower household income, to economise on this item, as suggested here.

Figure 5.8   **Electricity, gas and other fuels (subset of housing fuel and power) weekly household expenditure (£)**
2010 prices
United Kingdom

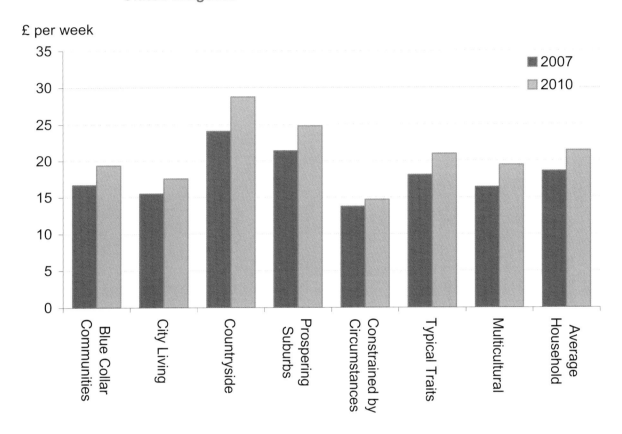

Overall, the rise in housing, fuel and power expenditure can mostly be explained by increases in rent and fuel costs. In terms of fuel payments, with the exception of the Constrained by Circumstances super group, there is a fairly uniform impact. This reflects the fact that this is essential expenditure, and households will have had limited opportunities to switch to cheaper alternatives when price increases above general inflation were observed across suppliers. For Constrained by Circumstances there appears to have been some economising on fuel in 2010, possibly as a response to fuel price increases over the period.

Transport expenditure

Transport costs expenditure (Figure 5.9), which also makes up a large percentage of total COICOP expenditure for many, did not change significantly at the average household level. However, it increased significantly in 2010 for the Multicultural super group, making up 12.5 per cent of this group's total expenditure in 2007 rising to 14.9 per cent in 2010. For many other super groups however, transport expenditure fell in 2010. We examine further to see if disaggregating transport expenditure reveals any other impacts.

Figure 5.9   **Transport expenditure percentage of total COICOP
expenditure**
United Kingdom

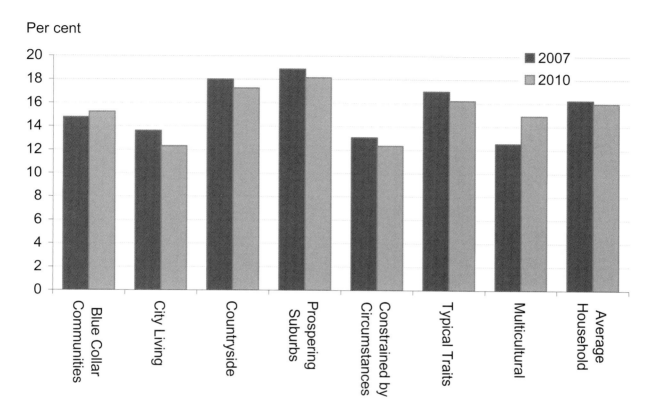

Per cent

Many households may be able to substitute one form of transport for another. We therefore
examine transport expenditure more closely. The Vehicle Scrappage Scheme, which subsidised
the purchase of new cars, ran from May 2009 to March 2010. This is likely to have resulted in an
increase in the purchase of new vehicles in the first quarter of 2010, but we would expect this to be
partially offset by a fall in purchases immediately after the scheme expired. We are unable to
examine this more closely because LCF data are annual. However, when interpreting these
results, the existence of this scheme should be considered.

Figure 5.10   **Purchase of vehicles (subset of transport) weekly household expenditure (£)**
2010 prices
United Kingdom

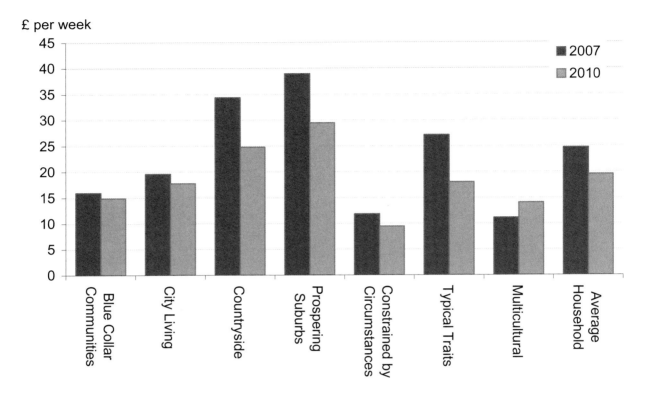

Many differences by OAC super group are highlighted. For instance, in 2010 there was a significant fall in the amount spent on Vehicle Purchase for Countryside, Typical Traits and Prospering Suburbs (Figure 5.10). Countryside and Prospering Suburbs super groups then significantly increased their expenditure on operation of personal transport, which includes fuel (Figure 5.11). This suggests that in the face of rising fuel prices, these super groups altered their expenditure so that the increase in fuel payments was partially offset by spending less on vehicle purchases.

Figure 5.11 **Operation of personal transport (subset of transport) weekly household expenditure (£)**
2010 prices
United Kingdom

£ per week

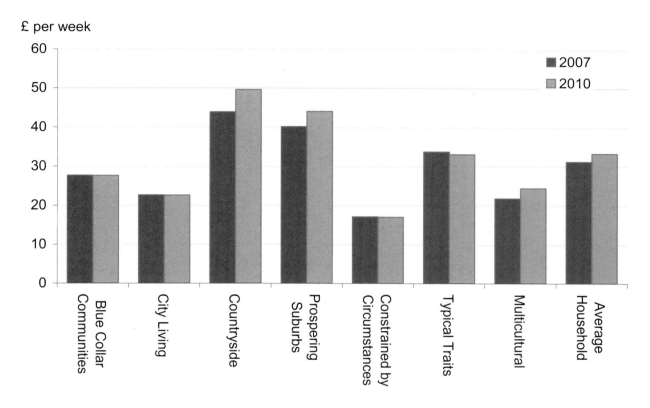

There were also significant increases in transport services expenditure, which includes public transport (Figure 5.12) for Typical Traits and Multicultural super groups. The increase in transport services expenditure by the Multicultural super group is particularly large, and closer examination reveals this increase to be largely down to an increase in international air fare expenditure. This is particularly interesting, as this element of expenditure could be considered essential by the Multicultural household super group. In the face of rising air fare prices above general inflation, visiting family and friends abroad is likely to be an essential component of the Multicultural super group outgoings. Other household super groups may be able to reduce this element of discretionary expenditure following price increases. Increases in air fuel duty will therefore have had a greater impact upon the Multicultural super group.

Figure 5.12 **Transport services (subset of transport) weekly household expenditure (£)**
2010 prices
United Kingdom

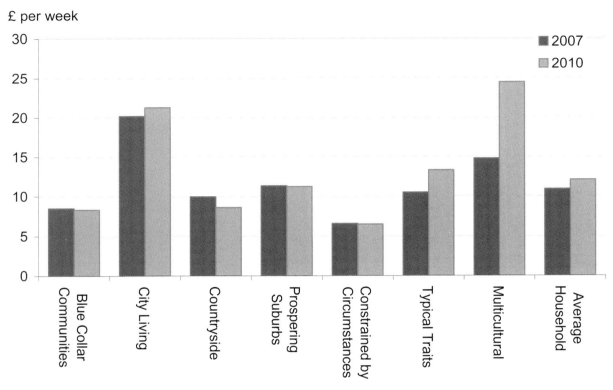

It is likely that changes in transport expenditure were observed partly as a result of changing fuel prices. Countryside and Prospering Suburbs appear to have continued to use personal transport, spending more on operation but spending less on purchasing vehicles. Conversely, Typical Traits and Multicultural reduced the amount spent on personal, private transport and increased their consumption of public transport services.

Overall, although there were few significant differences in overall transport expenditure by OAC super group, closer examination reveals there to be significant differences in the composition of this expenditure during the recession. While Countryside and Prospering Suburbs super groups have limited options and spent more on personal private transport, spending more on fuel but partially offsetting by spending less on vehicle purchases, Typical Traits and Multicultural super groups reduced private transport expenditure and spent more on public transport services. This supports the suggestion that although the average household may not have significantly altered their expenditure on main COICOP items, there may be differences in expenditure when we disaggregate. Households may reallocate their expenditure between general COICOP items. The ability for this type of reallocation depends upon household characteristics. The recession, while not appearing to have had a substantial impact upon overall COICOP expenditure for the average household, may have altered expenditure patterns in complex ways that are largely determined by specific household characteristics.

Recreation and culture expenditure

Recreation and culture expenditure makes up a similar proportion of total COICOP expenditure as some of the necessity items for the average household (£58.10 in 2010, which is very similar to the £53.20 spent on food). This supports the earlier suggestion that what may once have been considered discretionary items are now considered essential. Table 5.6 illustrates that this item makes up the second largest expenditure component in 2010 for Countryside and Prospering Suburbs. Interestingly, although recreation and culture expenditure remained the second largest expenditure item for Prospering Suburbs, Figure 5.13 shows the expenditure on recreation and culture as a proportion of total expenditure in 2007 and 2010. The mean amount spent on these items by this super group fell significantly in 2010 (from £79.40 to £73.20). There was also a statistically significant fall in the amount spent by Typical Traits (from £65.40 to £53.70). In terms of recreation and culture expenditure therefore, the amount spent fell in 2010 for many super groups. We may however, have expected a greater fall in expenditure for this traditionally non-essential item.

Figure 5.13 **Recreation and culture expenditure percentage of total COICOP expenditure**
United Kingdom

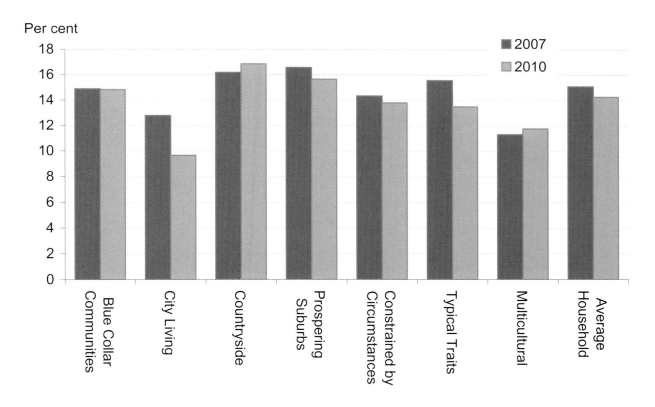

We investigate this possibility further by disaggregating recreation and culture expenditure to see if households made changes within the recreation and culture category. This reveals a significant fall in expenditure on package holidays abroad by the Prospering Suburbs and City Living super groups (Figure 5.14). In addition, for Prospering Suburbs, there was a significant fall for sports admissions, subscriptions, leisure class fees and equipment hire (Figure 5.15). In contrast, for the

City Living super group, there was a significant increase in this type of expenditure in 2010. This indicates that there is a degree of item selection within the recreation category, with the details of this differing by complex household circumstances and preferences.

Figure 5.14   **Package holidays abroad (subset of recreation and culture) weekly household expenditure (£)**
2010 prices
United Kingdom

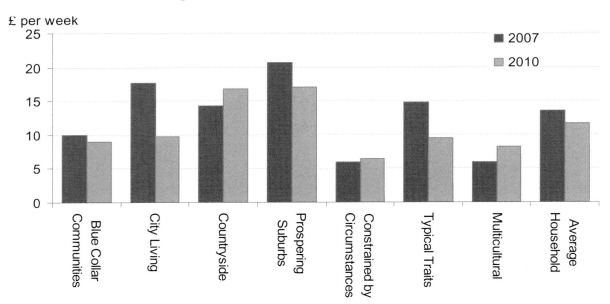

Figure 5.15   **Sports admissions, subscriptions, leisure class fees and equipment hire (subset of recreation and culture) weekly household expenditure**
2010 prices
United Kingdom

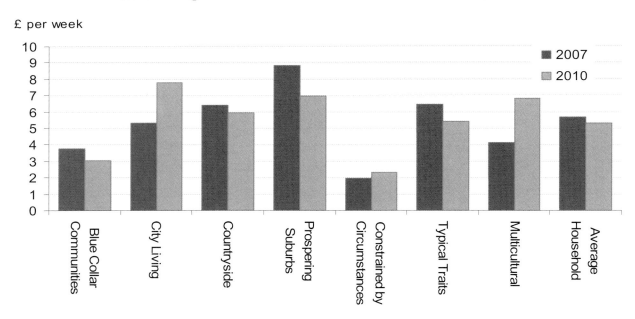

Restaurants and hotels expenditure

Some significant differences by OAC super group were also observed for restaurant and hotel expenditure (Figure 5.16). These traditionally non-essential items typically rank fifth in terms of amount spent. Significant falls in expenditure, comparing 2010 with 2007, were observed for Countryside (falling from £45.20 to £40.70), Multicultural (from £38.80 to £36.50), and Typical Traits (from £38.80 to £36.50). However, this item did not change the rank in terms of proportion spent for any of these groups, which is surprising. Again, it could be that this item is no longer considered discretionary by many groups.

Figure 5.16 **Restaurants and hotels expenditure percentage of total COICOP expenditure**
United Kingdom

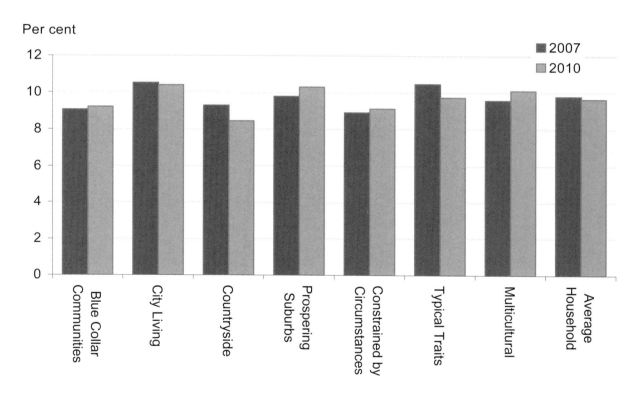

We investigate further to see if there are any within-group expenditure changes. When we do this, we do find further differences by OAC super group. For instance, the amount spent on takeaway meals eaten at home fell significantly for Constrained by Circumstances and Typical Traits (Figure 5.17). The amount spent on alcoholic drinks away from home also fell for Countryside, Constrained by Circumstances and Typical Traits (Figure 5.18), although it should be noted that such expenditure is prone to under-reporting.

Figure 5.17 **Takeaway meals eaten at home (subset of restaurants and hotels) weekly household expenditure (£)**
2010 prices
United Kingdom

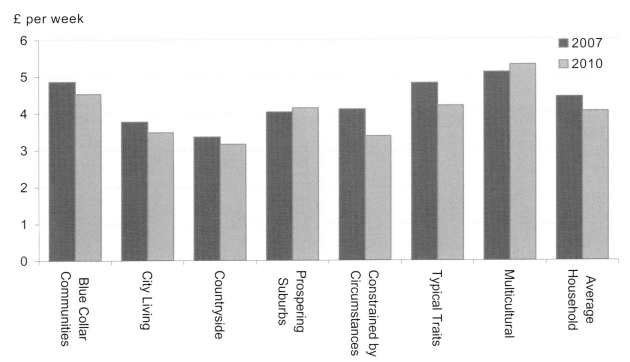

Figure 5.18 **Alcoholic drinks away from home (subset of restaurants and hotels) weekly household expenditure (£)**
2010 prices
United Kingdom

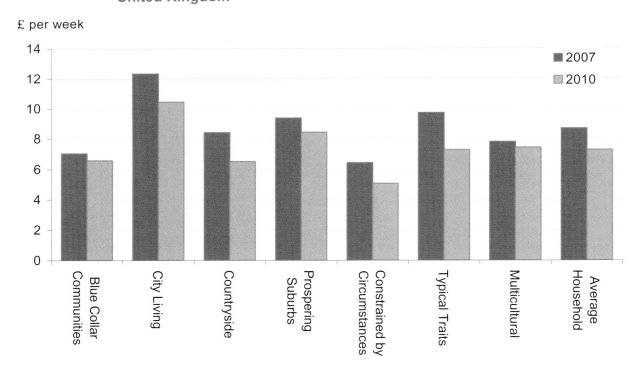

Overall, there are fewer changes than expected in restaurant and hotel expenditure. It appears that rather than significantly reduce the total amount spent, many households have instead chosen to reallocate expenditure between these types of items, perhaps to gain more value for their money in the face of rising prices and stagnant wage growth. This appears to be particularly true for the Constrained by Circumstances and Typical Traits super groups.

Household goods and services

Household goods and services expenditure ranks seventh in terms of relative amount spent for all groups in 2010. The only statistically significant change comparing 2010 with 2007 is a fall for Blue Collar Communities, where the amount spent fell from £29.60 to £22.40 (Figure 5.19). Further examination reveals that a large proportion of this expenditure fall for this group was due to a significant fall in household appliances expenditure (Figure 5.20). When we examine tools and equipment for house and garden expenditure, significant falls in 2010 expenditure are observed for Typical Traits and Constrained by Circumstances (Figure 5.21)

Figure 5.19 **Household goods and services expenditure percentage of total COICOP expenditure**
United Kingdom

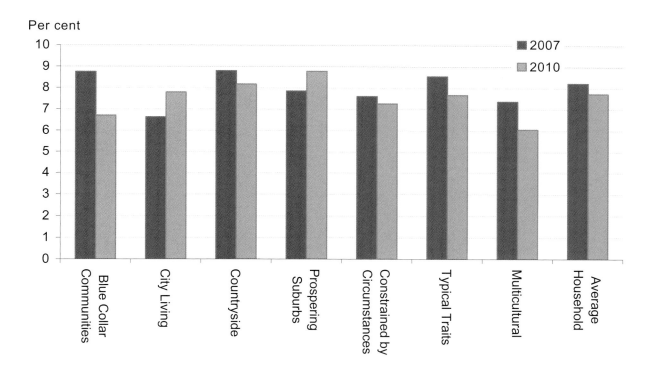

Figure 5.20   **Household appliances (subset of household goods and furnishings) weekly household expenditure (£)**
2010 prices
United Kingdom

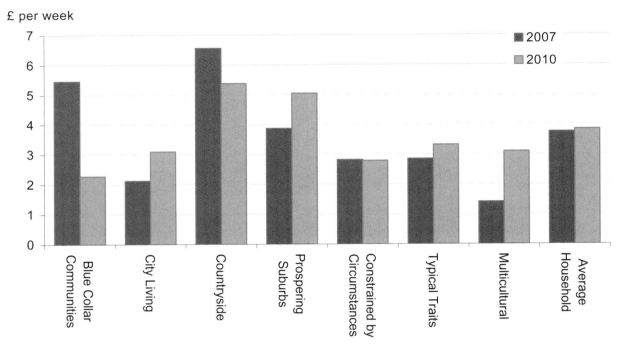

Figure 5.21   **Tools and equipment for house and garden (subset of household goods and services) weekly household expenditure (£)**
2010 prices
United Kingdom

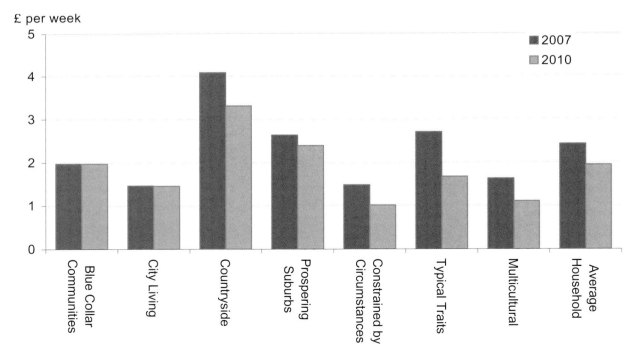

COICOP items with no significant change

For several COICOP expenditure items, there is no significant change in expenditure from 2007 to 2010 for any OAC super group. For instance, there were no significant changes in communication, health, and alcoholic drinks, tobacco and narcotics (although this is often under-reported) expenditure. These items could be considered non-essential. In addition, total expenditure for these items is relatively small for each group for both years examined, and they also make up a small proportion of total COICOP expenditure in 2010 as illustrated by Table 5.6. Households did not significantly change their expenditure on these items during the recession, which goes against expectations, as we would on average expect households to reduce non-essential item expenditure during uncertain economic conditions.

Figure 5.22 **Alcoholic drinks (subset of alcoholic drinks, tobacco and narcotics) weekly household expenditure (£)**
2010 prices
United Kingdom

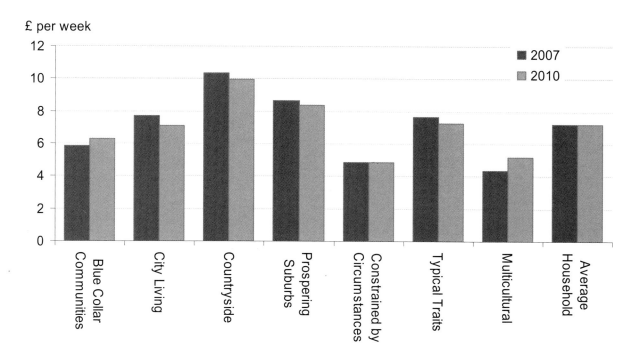

When we disaggregate expenditure on these COICOP items, we again find few differences in mean expenditure between 2007 and 2010. The only significant change is observed for alcohol expenditure (disaggregating the alcohol, tobacco and narcotics category), with a significant increase observed for the Multicultural super group in 2010, as shown in Figure 5.22. However, these results should be interpreted with caution, as there is thought to be under-reporting in this expenditure category.

Figure 5.23 **Clothing and footwear expenditure percentage of total COICOP expenditure**
United Kingdom

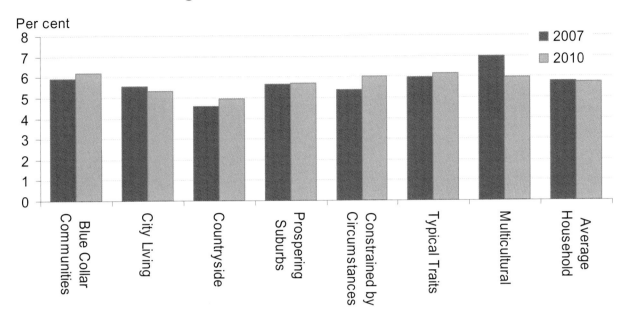

Figure 5.24 **Footwear (subset of clothing and footwear) weekly household expenditure (£)**
2010 prices
United Kingdom

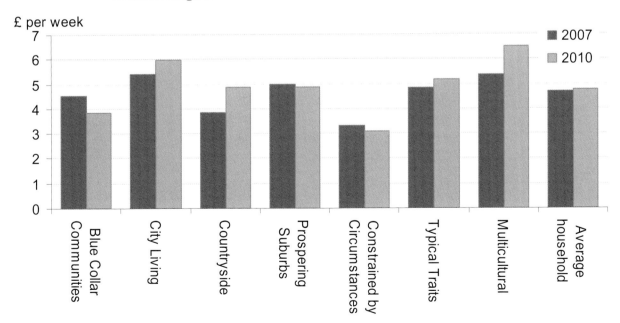

Household mean expenditure on food and non-alcoholic drinks, and clothing and footwear (Figure 5.23), also exhibit no significant change for any super group. There is also limited change in their order of total expenditure for each group, with food typically the third greatest expenditure item, and clothing typically the eighth largest. However, these are essential items, so we would not necessarily expect to see large changes in expenditure, even during a recession. Disaggregating clothing and footwear expenditure however, we do find a significant fall in 2010 footwear expenditure for Blue Collar Communities (Figure 5.24).

Overall, when we examine expenditure by OAC super group, we do find some significant changes in household expenditure comparing 2007 with 2010. This is consistent with evidence reported elsewhere that although the recession had little impact on disposable income for the average household, this impact is disproportionate. The impact upon a particular household will depend upon a combination of many factors, including working status, tenure, number of children and preferences for products. Therefore, when we disaggregate COICOP items, we find within-category differences by group. Although households' overall expenditure on particular items may have remained unchanged, this analysis suggests households reallocate expenditure within some COICOP categories, presumably to gain the most value in the face of changing economic conditions. Households' ability to reallocate expenditure will very much depend upon circumstances. For instance Countryside are limited in their ability to move from private to public transport, which is why we observe differences when we disaggregate transport expenditure by OAC super group.

COICOP expenditure by OAC group

Overall COICOP Expenditure

Our analysis so far has illustrated that our ability to assess the impact of the recession upon households appears to depend upon how specifically households are defined, with examination at the average household level masking complex differences. We therefore disaggregate the OAC super groups further to see if more differences are observed. The groups in each super group and the percentage of each category in the sample are listed in Table 5.2. Mean weekly household expenditure for each COICOP item and OAC group are reported in Table 5.7 for 2010 and Table 5.8 for 2007 (in 2010 prices).

Table 5.7   **Mean COICOP weekly household expenditure (£) by OAC group 2010**
United Kingdom

Commodity or service	Terraced Blue Collar	Younger Blue Collar	Older Blue Collar	Transient Communities	Settled in the City	Village Life	Agricultural	Accessible Countryside	Prospering Younger Families	Prospering Older Families	Prospering semis
	Average weekly household expenditure										
Food & non-alcoholic drinks	47.60	50.00	50.80	44.70	49.70	58.70	59.50	60.80	65.60	61.60	56.80
Alcoholic drinks, tobacco & narcotics	13.10	15.50	11.20	8.70	11.70	14.40	14.50	12.80	10.80	11.00	10.20
Clothing & footwear	21.50	18.10	20.90	26.30	25.40	20.50	17.50	23.80	31.00	29.00	23.00
Housing(net)[1], fuel & power	52.70	59.70	49.70	149.80	97.20	59.00	62.90	62.00	47.90	50.50	45.20
Household goods & services	27.70	16.60	24.10	18.40	50.40	38.50	34.60	43.60	41.70	41.30	33.60
Health	2.10	3.40	3.90	5.50	6.30	4.50	5.30	6.50	5.10	7.20	5.30
Transport	46.20	39.60	64.70	58.20	63.20	83.80	79.60	85.20	89.10	85.60	70.70
Communication	12.40	13.10	11.50	15.00	13.10	14.60	11.00	14.20	14.50	12.90	13.10
Recreation & culture	43.40	41.90	60.60	39.30	53.10	81.10	73.40	84.50	79.80	81.00	61.50
Education	2.60	0.90	1.50	36.30	30.10	10.20	20.30	19.40	5.50	4.20	7.00
Restaurants & hotels	26.30	26.80	34.80	51.80	51.20	37.40	38.80	44.00	48.00	49.80	40.70
Miscellaneous goods & services	20.80	22.90	31.50	40.50	42.90	46.30	40.30	48.80	50.50	46.80	34.50

Commodity or service	Thriving suburbs	Senior Communities	Older Workers	Public Housing	Settled Households	Least Divergent	Young Families in Terraced Homes	Aspiring Households	Asian Communities	Afro-Caribbean Communities
	Average weekly household expenditure									
Food & non-alcoholic drinks	61.90	32.40	42.70	39.90	53.00	54.80	45.10	54.70	60.80	47.90
Alcoholic drinks, tobacco & narcotics	12.30	11.30	10.50	14.50	10.40	13.30	10.90	11.00	10.30	8.90
Clothing & footwear	24.20	11.00	17.10	13.70	25.50	26.20	19.30	27.20	26.30	23.20
Housing(net)[1], fuel & power	53.40	55.30	51.70	47.80	46.10	56.50	60.30	66.00	78.00	91.10
Household goods & services	52.80	16.20	21.10	15.30	39.70	24.50	33.10	24.10	24.90	26.60
Health	8.50	2.70	2.90	1.20	6.50	7.10	3.30	4.20	6.10	6.50
Transport	102.70	19.30	37.30	31.60	63.60	74.90	48.20	73.20	69.90	51.80
Communication	13.80	8.20	9.50	11.70	13.20	12.90	12.70	14.90	15.00	17.50
Recreation & culture	77.50	20.80	40.10	39.40	64.10	56.20	39.20	57.20	48.50	52.00
Education	17.30	0.90	2.30	0.40	6.60	7.20	1.30	34.40	11.20	14.70
Restaurants & hotels	55.70	17.30	26.70	22.40	39.90	38.50	30.50	50.10	38.20	50.10
Miscellaneous goods & services	51.70	12.50	21.30	24.50	46.10	35.60	26.70	39.90	33.70	31.30

1 Excluding mortgage interest payments, council tax and Northern Ireland rates

Table 5.8   Mean COICOP weekly household expenditure (£) by OAC group 2007
2010 prices
United Kingdom

Commodity or service	Terraced Blue Collar	Younger Blue Collar	Older Blue Collar	Transient Communities	Settled in the City	Village Life	Agricultural	Accessible Countryside	Prospering Younger Families	Prospering Older Families	Prospering semis
	Average weekly household expenditure (£)										
Food & non-alcoholic drinks	50.50	48.80	49.80	38.30	49.00	54.80	59.70	60.00	62.50	59.60	55.30
Alcoholic drinks, tobacco & narcotics	13.80	14.90	12.70	11.00	11.50	14.30	15.80	13.40	13.60	11.30	9.70
Clothing & footwear	19.50	17.60	23.10	15.40	30.40	22.10	23.10	22.30	31.20	27.20	23.60
Housing(net)[1], fuel & power	48.80	47.40	48.00	106.30	99.80	57.20	61.00	52.90	47.70	45.70	40.10
Household goods & services	30.90	19.30	39.90	22.10	34.50	46.20	42.90	39.70	39.60	39.30	29.60
Health	2.40	2.40	4.70	5.00	11.10	6.50	14.30	8.70	13.70	6.90	6.70
Transport	42.20	48.30	57.10	57.60	64.80	78.10	97.80	94.00	98.40	91.60	70.60
Communication	13.20	11.70	11.50	11.90	14.20	12.90	13.10	13.50	15.80	12.80	11.90
Recreation & culture	53.10	45.80	53.50	43.20	66.30	73.50	102.60	71.70	85.50	88.10	74.00
Education	2.00	5.10	1.60	7.80	14.20	6.30	10.80	22.40	8.60	4.00	6.10
Restaurants & hotels	25.80	32.30	32.40	47.70	48.50	42.60	39.80	52.40	59.60	46.10	41.10
Miscellaneous goods & services	27.20	25.40	32.00	40.00	40.10	43.80	48.10	51.40	53.20	52.50	40.60

Commodity or service	Thriving suburbs	Senior Communities	Older Workers	Public Housing	Settled Households	Least Divergent	Young Families in Terraced Homes	Aspiring Households	Asian Communities	Afro-Caribbean Communities
	Average weekly household expenditure (£)									
Food & non-alcoholic drinks	60.70	34.40	40.80	45.40	53.40	52.90	45.20	56.00	50.70	48.50
Alcoholic drinks, tobacco & narcotics	11.90	9.00	13.00	15.90	11.20	10.30	12.20	13.40	9.50	7.60
Clothing & footwear	30.20	13.00	14.40	16.90	30.60	21.50	19.50	28.10	30.30	20.50
Housing(net)[1], fuel & power	62.90	41.50	47.50	48.90	44.50	52.60	56.80	57.20	65.00	88.60
Household goods & services	45.10	19.90	19.40	25.20	37.50	41.20	26.40	37.50	30.50	24.00
Health	12.10	8.60	3.30	2.00	4.30	4.60	4.60	6.30	4.20	5.50
Transport	110.90	23.10	36.90	41.20	70.30	67.60	54.90	92.90	52.40	40.10
Communication	13.40	8.20	10.60	9.40	13.90	12.40	11.90	14.70	15.00	13.50
Recreation & culture	75.40	31.20	38.40	47.00	70.40	66.80	51.60	71.80	46.90	37.00
Education	16.90	0.80	2.20	0.00	7.00	3.20	2.00	8.90	11.90	14.50
Restaurants & hotels	49.10	20.40	24.70	25.70	44.30	39.50	41.00	51.60	38.60	33.00
Miscellaneous goods & services	52.50	14.70	23.20	25.10	40.10	43.40	30.80	39.50	38.80	26.50

1 Excluding mortgage interest payments, council tax and Northern Ireland rates

COICOP items with no significant OAC super group change

We first consider the expenditure items where no significant change was observed by OAC super group. This was the case with food and non-alcoholic drinks expenditure. When we examine changes by OAC group, we find only a small but significant increase for Asian Communities. There are also no significant differences in clothing expenditure. This is consistent with expectations: substitution within these essential broad-based COICOP items is difficult for all households, regardless of their circumstances. We may observe substitution between brands during recession, but we are unable to examine this using the LCF Survey data.

When we consider traditionally non-essential COICOP items however, we do begin to see differences. For instance, alcohol and tobacco expenditure did not change significantly at the super group level, but when we disaggregate, there is a significant fall for Prospering Younger Families and a significant increase for Older Workers. For less essential items, we do begin to see changes in expenditure when we define household characteristics more closely. No significant changes in communication expenditure were observed for the average household, or at the super group level. However, when we examine at group level as illustrated by Figure 5.25, we observe significant increases in 2010 expenditure for Transient Communities and Afro-Caribbean Communities. Conversely, there is a significant fall in communication expenditure for Older Workers. The increases are likely to be as a result of technological change, with certain groups keen to increase their expenditure following communication innovations. However, this is not the case for all households, as illustrated by the fall in expenditure for Older Workers

Figure 5.25 Communication weekly household expenditure (£) by OAC group
2010 prices
United Kingdom

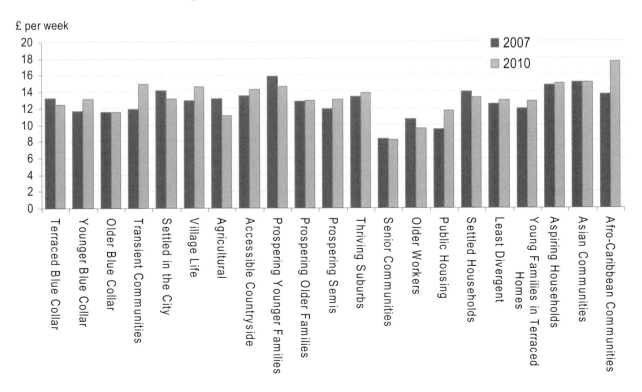

We also examine if there are differences by OAC group for recreation expenditure. Although we observed some significant falls in 2010 for some super groups, there were not as many as may have been expected for these traditionally non-essential items. Figure 5.26 illustrates there are lots of differences in 2010 expenditure, with significant falls for Prospering Semis and Young Families in Terraced Homes, and significant increase for Afro-Caribbean Communities. Many of the differences observed are insignificant, possibly due to the reduced sample size we have for each OAC group following this more detailed disaggregation. The fact that many more differences in recreation expenditure are observed when we define household characteristics more specifically however, illustrates how important it is to consider these to appreciate the full impact of the recession on household expenditure.

Figure 5.26   **Recreation and culture weekly household expenditure (£) by OAC group**
2010 prices
United Kingdom

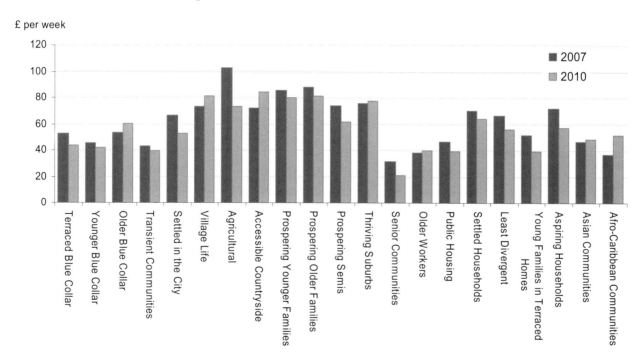

Similarly for restaurant and hotel expenditure, there are lots of differences at group level, but only a few are significant, again possibly due to the reduced sample size at this level of OAC disaggregation. Figure 5.27 illustrates there are significant falls for Accessible Countryside, Prospering Younger Families, and Younger Families in Terraced Homes. We also observe a significant increase for Afro-Caribbean Communities.

Figure 5.27 **Restaurant and hotel weekly household expenditure (£) by OAC group**
2010 prices
United Kingdom

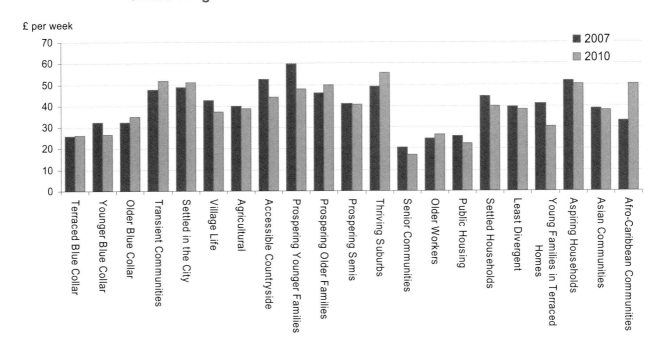

At the super group level, we observed a significant fall in 2010 furnishings, household equipment and carpets expenditure for Blue Collar Communities. When we disaggregate this, we find further significant differences, as illustrated by Figure 5.28. Significant falls are found for Older Blue Collar, Public Housing, Least Divergent, and Aspiring Households. This supports the idea that the average impact does not present us with the full effect upon expenditure for many households.

Figure 5.28 **Household goods and services weekly household expenditure (£) by OAC group**
2010 prices
United Kingdom

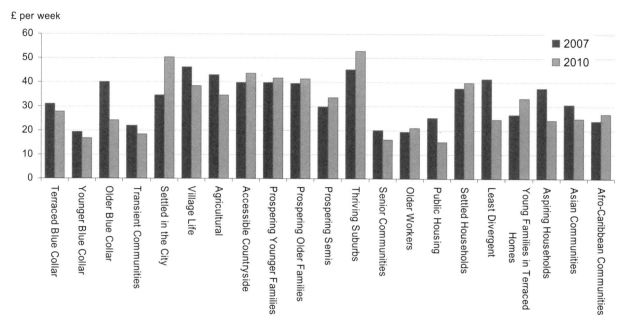

COICOP items with significant OAC super group changes

We also examine whether or not there are any changes at the super group level for those COICOP items where we observe the most differences comparing 2007 and 2010 expenditure at the group level. For instance we observed many differences in transport expenditure between groups. Figure 5.29 presents mean weekly household expenditure at the group level. This illustrates the large variation in transport expenditure between groups. In addition, significant increases in 2010 expenditure were observed for Asian and Afro-Caribbean Communities. This is consistent with earlier findings, where we reported a significant rise in transport expenditure for the Multicultural super group, mostly attributable to an increase in air fare expenditure. This further examination illustrates this rise was consistent across both Multicultural groups. In addition, a significant fall in 2010 transport expenditure was reported for Prospering Younger Families. Although falls are also observed for other Prospering Suburbs super groups, namely Prospering Older Families and Thriving Suburbs, they are not significant. Transport expenditure for the remaining Prospering Suburbs super group, Prospering Semis is relatively unchanged in 2010. Examining at the group level therefore, allows us to define more closely the type of household that reduced their transport expenditure over this period.

Figure 5.29 **Transport weekly household expenditure (£) by OAC group**
2010 prices
United Kingdom

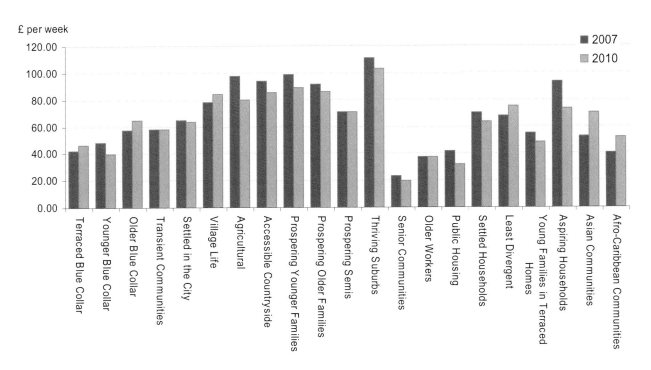

We also examine housing, fuel and power expenditure at the group level. We observed significant expenditure increases for this COICOP item for the Blue Collar Communities and Typical Traits super groups comparing 2007 with 2010 expenditure. Figure 5.30 illustrates changes at the OAC group level. Significant increases in 2010 expenditure are observed for Younger Blue Collar Workers but not for the other Blue Collar Communities super groups where small insignificant increases in expenditure are reported. For the groups within the Typical Traits super group category, small increases are observed but none are significant. We do, however, observe significant increases for groups where no overall super group significant change was observed. Within the City Living category, a significant increase is reported for the Transient Communities group. Significant increases are also observed for Accessible Countryside (within the Countryside super group) and Senior Communities (within the Constrained by Circumstances super group). Analysis at the group level was therefore masking these impacts.

Figure 5.30 **Housing, fuel and power weekly household expenditure (£) by OAC group**
2010 prices
United Kingdom

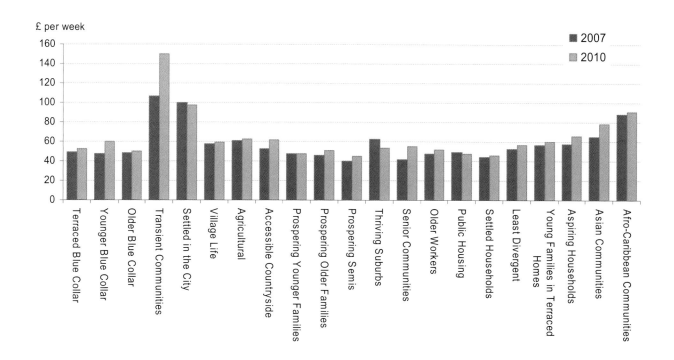

Disaggregating by OAC group illustrates that a consideration of characteristics are very important in considering the household impact of the recession. Whereas we find little difference in expenditure overall for the average household, when we examine first by super group, some differences are found. When we disaggregate groups even further, many more significant changes in expenditure are found. This suggests the impact of the recession upon household expenditure is very much linked to household group which reflects circumstances and preferences. The average and super group classification masks many of these changes.

Conclusion

Our analysis surprisingly showed that there was very little change in average household expenditure between 2007 and 2010. This is in contrast to the impact of previous recessions that were less severe, but saw broadly based impacts affecting swathes of communities. For instance, the 1980s recession resulted in significant unemployment. Similarly, there was a general widespread rise in unemployment in the 1990s. It may be that certain items that have traditionally been considered discretionary, such as restaurant and recreation expenditure, are no longer considered optional items by some groups. In this case, we may observe reallocation of expenditure within items as households alter item expenditure to account for changing prices and stagnant incomes, and we do find some evidence of this. We also begin to see differences in expenditure when we account for household characteristics in the form of OAC super groups disaggregated further into groups. It appears that in this recession, the combination of factors, including prices rises, low interest rates, limited wage growth and restricted access to borrowing, means that it is almost impossible to say what the impact of the recession is on a 'typical' household. The effect will depend very much upon household circumstances and consumption preferences. The finer our household classification, the more significant differences in expenditure we observe. Overall therefore, focus on one or other macroeconomic statistic is unlikely to tell us the household impact. To gain a fuller understanding, we need to take account of the complex combination of household circumstances and preferences.

1 Quarterly National Accounts q2 2011
(www.ons.gov.uk/ons/rel/naa2/quarterly-national-accounts/q2-2011/sbd-quarterly-national-accounts-q2-2011.pdf)
2 Chamberlain (2010) 'Economic Review', *Economic and Labour Market Review*, August 2010, pp. 6
3 ONS (2011) GDP Preliminary Estimates, q2 2011,
(www.ons.gov.uk/ons/rel/gva/gross-domestic-product--preliminary-estimate/q2-2011/gross-domestic-product---preliminary-estimate-q2-2011.pdf)
4 ONS (2009), The Impact of the Recession on the Labour Market, chapter 2
(www.ons.gov.uk/ons/rel/lmac/impact-of-the-recession-on-the-labour-market/impact-of-the-recession-on-the-labour-market/impact-of-the-recession-on-the-labour-market---impact-of-the-recession-on-the-labour-market.pdf)
5 Jenkins (2010) 'The Labour Market in the 1980s, 1990s and 2008/09 Recessions', *Economic and Labour Market Review*, August 2010, pp.33
6 'Household Expenditure by Output Area Classification', *Family Spending: A Report on the 2009 Living Costs and Food Survey*, 2010, pp. 99-105
7 Gregg and Wadsworth (2010) 'Unemployment and Inactivity in the 2008/09 Recession', *Economic and Labour Market Review,* August 2010, pp.44
8 Howell, Leaker and Barrett (2010) 'Impact of the Recession in Households', *Economic and Labour Market Review,* August 2010, pp.18
9 All statistics reported in this article have been deflated to reflect 2010 prices using All Items Retail Price Index data. We should be aware that the price of some goods will have increased at a greater rate relative to others, but this is not accounted for through the use of an All Items deflator.
10 ONS (2010) Impact of the Recession on Household Income, Expenditure and Saving
(www.ons.gov.uk/ons/rel/naa2/quarterly-national-acounts/impact-of-the-recession/the-impact-of-the-recession-on-household-income-and-expenditure.pdf)
11 Significant throughout this article means statistically significant above the 10 per cent significance level. This has been calculated following the calculation of t statistics in Stata.
12 Jenkins, S *et al.* (2011) 'The Great Recession and the Distribution of Household Income' Report for Fondazione Rodoldo Debenedetti
(www.frdb.org/upload/file/report_1_palermo.pdf)

Appendix A

Table A1 **Components of household expenditure, 2010**
United Kingdom

Commodity or service	Average weekly expenditure all house-holds (£)	Total weekly expenditure (£ million)	Recording house-holds in sample	Percentage standard error (full method)
Total number of households			5,260	
1 Food & non-alcoholic drinks	**53.20**	**1,400**	**5,230**	**0.9**
1.1 Food	48.90	1287	5,230	0.9
1.1.1 Bread, rice and cereals	5.00	131	5,120	1.0
1.1.1.1 Rice	0.40	10	1,310	4.5
1.1.1.2 Bread	2.50	67	4,950	1.1
1.1.1.3 Other breads and cereals	2.10	54	4,210	1.4
1.1.2 Pasta products	0.40	11	2,170	2.7
1.1.3 Buns, cakes, biscuits etc.	3.20	84	4,630	1.5
1.1.3.1 Buns, crispbread and biscuits	1.90	49	4,210	1.7
1.1.3.2 Cakes and puddings	1.30	35	3,360	2.0
1.1.4 Pastry (savoury)	0.70	19	1,920	2.5
1.1.5 Beef (fresh, chilled or frozen)	1.70	45	2,370	2.4
1.1.6 Pork (fresh, chilled or frozen)	0.60	17	1,290	3.5
1.1.7 Lamb (fresh, chilled or frozen)	0.70	18	910	5.7
1.1.8 Poultry (fresh, chilled or frozen)	2.00	52	2,710	2.1
1.1.9 Bacon and ham	1.00	26	2,440	2.3
1.1.10 Other meats and meat preparations	5.60	148	4,680	1.4
1.1.10.1 Sausages	0.80	21	2,430	2.4
1.1.10.2 Offal, pate etc.	0.10	3	770	4.7
1.1.10.3 Other preserved or processed meat and meat preparations	4.70	123	4,500	1.5
1.1.10.4 Other fresh, chilled or frozen edible meat	[0.00]	1	20	28.3
1.1.11 Fish and fish products	2.30	61	3,430	1.9
1.1.11.1 Fish (fresh, chilled or frozen)	0.70	20	1,270	3.5
1.1.11.2 Seafood, dried, smoked or salted fish	0.50	14	1,250	3.9
1.1.11.3 Other preserved or processed fish and seafood	1.00	27	2,640	2.2
1.1.12 Milk	2.60	67	4,820	1.4
1.1.12.1 Whole milk	0.60	15	1,430	3.0
1.1.12.2 Low fat milk	1.80	48	4,040	1.7
1.1.12.3 Preserved milk	0.20	5	400	7.5
1.1.13 Cheese and curd	1.80	48	3,780	2.4
1.1.14 Eggs	0.70	17	2,930	1.7
1.1.15 Other milk products	1.90	51	3,980	1.6
1.1.15.1 Other milk products	0.90	23	2,990	2.0
1.1.15.2 Yoghurt	1.00	27	2,820	2.1
1.1.16 Butter	0.40	10	1,590	2.7
1.1.17 Margarine, other vegetable fats and peanut butter	0.50	14	2,530	1.9
1.1.18 Cooking oils and fats	0.30	8	1,240	3.7
1.1.18.1 Olive oil	0.10	3	430	6.0
1.1.18.2 Edible oils and other edible animal fats	0.20	4	890	4.9
1.1.19 Fresh fruit	3.10	81	4,330	1.6
1.1.19.1 Citrus fruits (fresh)	0.50	13	2,170	2.6
1.1.19.2 Bananas (fresh)	0.50	13	3,210	1.7
1.1.19.3 Apples (fresh)	0.50	14	2,390	2.2
1.1.19.4 Pears (fresh)	0.20	4	1,060	3.5
1.1.19.5 Stone fruits (fresh)	0.40	10	1,460	3.5
1.1.19.6 Berries (fresh)	1.00	27	2,410	2.5
1.1.20 Other fresh, chilled or frozen fruits	0.40	9	1,450	3.8
1.1.21 Dried fruit and nuts	0.60	15	1,720	3.1
1.1.22 Preserved fruit and fruit based products	0.10	4	960	4.3
1.1.23 Fresh vegetables	4.00	105	4,720	1.4
1.1.23.1 Leaf and stem vegetables (fresh or chilled)	0.90	23	3,260	2.1
1.1.23.2 Cabbages (fresh or chilled)	0.40	10	2,460	2.3
1.1.23.3 Vegetables grown for their fruit (fresh, chilled or frozen)	1.40	37	3,900	1.6
1.1.23.4 Root crops, non-starchy bulbs and mushrooms (fresh, chilled or frozen)	1.30	35	4,200	1.6

Note: The commodity and service categories are not comparable with those in publications before 2001-02.
    The numbering is sequential, it does not use actual COICOP codes.
    Please see page ix for symbols and conventions used in this report.

Table A1 **Components of household expenditure, 2010 (cont.)**
United Kingdom

Commodity or service	Average weekly expenditure all house- holds (£)	Total weekly expenditure (£ million)	Recording house- holds in sample	Percentage standard error (full method)
1 Food & non-alcoholic drinks (continued)				
1.1.24 Dried vegetables	0.00	1	210	10.8
1.1.25 Other preserved or processed vegetables	1.30	34	3,880	1.8
1.1.26 Potatoes	0.90	22	3,510	1.6
1.1.27 Other tubers and products of tuber vegetables	1.40	36	3,630	1.6
1.1.28 Sugar and sugar products	0.30	9	1,960	3.0
1.1.28.1 Sugar	0.20	6	1,660	3.1
1.1.28.2 Other sugar products	0.10	3	520	6.5
1.1.29 Jams, marmalades	0.30	8	1,540	3.9
1.1.30 Chocolate	1.60	43	3,220	2.8
1.1.31 Confectionery products	0.60	16	2,430	2.5
1.1.32 Edible ices and ice cream	0.50	13	1,660	2.9
1.1.33 Other food products	2.40	62	4,270	2.0
1.1.33.1 Sauces, condiments	1.20	32	3,420	1.8
1.1.33.2 Baker's yeast, dessert preparations, soups	0.90	23	2,800	4.0
1.1.33.3 Salt, spices, culinary herbs and other food products	0.30	7	1,310	6.0
1.2 Non-alcoholic drinks	4.30	114	4,740	1.3
1.2.1 Coffee	0.60	16	1,630	2.9
1.2.2 Tea	0.50	13	1,780	2.5
1.2.3 Cocoa and powdered chocolate	0.10	3	530	5.7
1.2.4 Fruit and vegetable juices	1.10	29	3,000	2.1
1.2.5 Mineral or spring waters	0.20	6	1,030	4.8
1.2.6 Soft drinks (inc. fizzy and ready to drink fruit drinks)	1.80	46	3,340	1.8
2 Alcoholic drink, tobacco & narcotics	**11.80**	**311**	**3,330**	**2.3**
2.1 Alcoholic drinks	7.20	189	2,820	2.8
2.1.1 Spirits and liqueurs (brought home)	1.50	39	780	4.9
2.1.2 Wines, fortified wines (brought home)	3.70	98	1,910	3.9
2.1.2.1 Wine from grape or other fruit (brought home)	3.20	85	1,780	3.7
2.1.2.2 Fortified wine (brought home)	0.10	4	170	9.5
2.1.2.3 Champagne and sparkling wines (brought home)	0.40	10	200	15.2
2.1.3 Beer, lager, ciders and perry (brought home)	1.90	51	1,480	3.7
2.1.3.1 Beer and lager (brought home)	1.60	43	1,250	4.0
2.1.3.2 Ciders and perry (brought home)	0.30	8	450	7.3
2.1.4 Alcopops (brought home)	0.10	2	110	12.4
2.2 Tobacco and narcotics	4.60	121	1,210	3.9
2.2.1 Cigarettes	3.90	103	1,060	4.3
2.2.2 Cigars, other tobacco products and narcotics	0.70	19	400	6.9
2.2.2.1 Cigars	0.10	2	30	26.0
2.2.2.2 Other tobacco	0.60	17	370	7.0
2.2.2.3 Narcotics	[0.00]	[0]	-	96.8
3 Clothing & footwear	**23.40**	**615**	**3,510**	**2.7**
3.1 Clothing	18.60	490	3,310	2.8
3.1.1 Men's outer garments	4.80	126	1,070	4.8
3.1.2 Men's under garments	0.40	10	390	7.4
3.1.3 Women's outer garments	8.40	221	1,970	3.7
3.1.4 Women's under garments	1.10	29	960	4.8
3.1.5 Boys' outer garments (5-15)	0.80	21	370	9.2
3.1.6 Girls' outer garments (5-15)	0.90	24	430	6.8
3.1.7 Infants' outer garments (under 5)	0.60	17	400	6.5
3.1.8 Children's under garments (under 16)	0.40	11	470	6.8

Note: The commodity and service categories are not comparable with those in publications before 2001-02.
The numbering is sequential, it does not use actual COICOP codes.
Please see page ix for symbols and conventions used in this report.

Table A1 Components of household expenditure, 2010 (cont.)
United Kingdom

Commodity or service	Average weekly expenditure all house-holds (£)	Total weekly expenditure (£ million)	Recording house-holds in sample	Percentage standard error (full method)
3 Clothing & footwear (continued)				
3.1.9 Accessories	0.80	20	790	5.7
3.1.9.1 Men's accessories	0.30	7	250	10.0
3.1.9.2 Women's accessories	0.30	9	420	7.6
3.1.9.3 Children's accessories	0.10	3	220	9.4
3.1.9.4 Protective head gear (crash helmets)	[0.00]	[1]	20	35.3
3.1.10 Haberdashery, clothing materials and clothing hire	0.20	5	200	11.9
3.1.11 Dry cleaners, laundry and dyeing	0.30	7	180	12.5
3.1.11.1 Dry cleaners and dyeing	0.20	6	140	13.3
3.1.11.2 Laundry, launderettes	0.00	1	40	18.7
3.2 Footwear	4.70	125	1,400	4.2
3.2.1 Footwear for men	1.40	38	390	6.5
3.2.2 Footwear for women	2.40	64	840	6.3
3.2.3 Footwear for children (5 to 15 years) and infants (under 5)	0.80	21	410	6.5
3.2.4 Repair and hire of footwear	0.10	2	50	20.3
4 Housing(net)[1], fuel & power	**60.40**	**1,589**	**5,250**	**1.5**
4.1 Actual rentals for housing	39.90	1051	1,620	2.7
4.1.1 Gross rent	39.90	1050	1,620	2.7
4.1.2 less housing benefit, rebates and allowances received	15.50	408	1,100	3.6
4.1.3 Net rent2	24.40	643	1,260	3.3
4.1.4 Second dwelling - rent	[0.00]	[1]	..	58.4
4.2 Maintenance and repair of dwelling	6.70	176	2,270	4.7
4.2.1 Central heating repairs	1.30	35	1,410	6.3
4.2.2 House maintenance etc.	3.60	94	960	7.2
4.2.3 Paint, wallpaper, timber	1.00	27	380	8.6
4.2.4 Equipment hire, small materials	0.80	20	380	14.9
4.3 Water supply and miscellaneous services relating to the dwelling	7.80	206	4,680	1.5
4.3.1 Water charges	6.60	175	4,590	0.8
4.3.2 Other regular housing payments including service charge for rent	1.10	29	530	8.3
4.3.3 Refuse collection, including skip hire	[0.10]	[2]	..	43.3
4.4 Electricity, gas and other fuels	21.40	563	4,970	1.0
4.4.1 Electricity	9.90	261	4,850	1.0
4.4.2 Gas	9.60	254	4,070	1.4
4.4.3 Other fuels	1.80	48	380	2.0
4.4.3.1 Coal and coke	0.20	6	100	57.3
4.4.3.2 Oil for central heating	1.50	40	270	1.7
4.4.3.3 Paraffin, wood, peat, hot water etc.	0.10	3	40	26.5
5 Household goods & services	**31.40**	**827**	**4,800**	**4.2**
5.1 Furniture and furnishings, carpets and other floor coverings	16.60	437	1,880	6.1
5.1.1 Furniture and furnishings	13.40	354	1,590	6.3
5.1.1.1 Furniture	12.50	329	1,100	6.7
5.1.1.2 Fancy, decorative goods	0.80	21	620	7.9
5.1.1.3 Garden furniture	0.10	4	50	23.9
5.1.2 Floor coverings	3.20	83	640	9.3
5.1.2.1 Soft floor coverings	3.00	78	600	9.6
5.1.2.2 Hard floor coverings	0.20	6	40	36.6
5.2 Household textiles	1.80	48	880	9.8
5.2.1 Bedroom textiles, including duvets and pillows	0.70	18	370	8.6
5.2.2 Other household textiles, including cushions, towels, curtains	1.10	30	630	14.8

Note: The commodity and service categories are not comparable with those in publications before 2001-02.
The numbering is sequential, it does not use actual COICOP codes.
Please see page ix for symbols and conventions used in this report.

1 Excluding mortgage interest payments, council tax and NI rates.
2 The figure included in total expenditure is net rent as opposed to gross rent

Table A1   **Components of household expenditure, 2010 (cont.)**
United Kingdom

Commodity or service	Average weekly expenditure all house-holds (£)	Total weekly expenditure (£ million)	Recording house-holds in sample	Percentage standard error (full method)
5 **Household goods & services (continued)**				
5.3 Household appliances	3.80	101	440	13.8
5.3.1 Gas cookers	[0.20]	[5]	..	48.1
5.3.2 Electric cookers, combined gas/electric cookers	0.30	8	40	41.3
5.3.3 Clothes washing machines and drying machines	0.50	13	60	29.2
5.3.4 Refrigerators, freezers and fridge-freezers	0.60	16	60	26.7
5.3.5 Other major electrical appliances, dishwashers, micro-waves vacuum cleaners, heaters etc.	1.50	40	130	27.3
5.3.6 Fire extinguisher, water softener, safes etc	[0.00]	[0]	..	94.2
5.3.7 Small electric household appliances, excluding hairdryers	0.40	11	180	11.3
5.3.8 Repairs to gas and electrical appliances and spare parts	0.30	8	50	21.0
5.3.9 Rental/hire of major household appliances	[0.00]	[1]	..	78.5
5.4 Glassware, tableware and household utensils	1.40	38	1,500	5.1
5.4.1 Glassware, china, pottery, cutlery and silverware	0.50	14	620	9.6
5.4.2 Kitchen and domestic utensils	0.50	14	870	5.8
5.4.3 Repair of glassware, tableware and household utensils	-	-	0	-
5.4.4 Storage and other durable household articles	0.40	10	470	7.4
5.5 Tools and equipment for house and garden	1.90	51	1,570	6.1
5.5.1 Electrical tools	0.20	6	50	28.5
5.5.2 Garden tools, equipment and accessories e.g. lawn mowers etc.	0.30	8	270	10.9
5.5.3 Small tools	0.30	9	340	11.0
5.5.4 Door, electrical and other fittings	0.60	15	410	11.2
5.5.5 Electrical consumables	0.50	14	960	5.7
5.6 Goods and services for routine household maintenance	5.80	152	4,400	4.0
5.6.1 Cleaning materials	2.20	59	3,580	2.0
5.6.1.1 Detergents, washing-up liquid, washing powder	1.00	27	2,400	2.2
5.6.1.2 Disinfectants, polishes, other cleaning materials etc.	1.20	31	2,900	2.7
5.6.2 Household goods and hardware	1.30	33	3,250	2.8
5.6.2.1 Kitchen disposables	0.70	19	2,730	2.7
5.6.2.2 Household hardware and appliances, matches	0.20	6	630	6.7
5.6.2.3 Kitchen gloves, cloths etc.	0.10	3	860	4.6
5.6.2.4 Pins, needles, tape measures, nails, nuts and bolts etc.	0.20	4	370	10.4
5.6.3 Domestic services, carpet cleaning, hire/repair of furniture/furnishings	2.30	60	810	9.5
5.6.3.1 Domestic services, including cleaners, gardeners, au pairs	1.80	47	340	11.8
5.6.3.2 Carpet cleaning, ironing service, window cleaner	0.50	12	540	7.1
5.6.3.3 Hire/repair of household furniture and furnishings	[0.00]	[0]	..	85.7
6 **Health**	**5.00**	**132**	**2,630**	**6.0**
6.1 Medical products, appliances and equipment	3.10	83	2,480	6.3
6.1.1 Medicines, prescriptions and healthcare products	1.80	47	2,370	5.3
6.1.1.1 NHS prescription charges and payments	0.20	6	200	10.5
6.1.1.2 Medicines and medical goods (not NHS)	1.30	34	2,160	4.0
6.1.1.3 Other medical products (e.g. plasters, condoms, hot water bottle etc.)	0.10	3	300	8.8
6.1.1.4 Non-optical appliances and equipment (e.g. wheelchairs, batteries for hearing aids, shoe build-up)	0.20	4	40	44.9
6.1.2 Spectacles, lenses, accessories and repairs	1.40	36	270	12.5
6.1.2.1 Purchase of spectacles, lenses, prescription sunglasses	1.30	35	230	12.8
6.1.2.2 Accessories/repairs to spectacles/lenses	0.00	1	50	21.9
6.2 Hospital services	1.90	50	380	11.2
6.2.1 Out patient services	1.90	49	380	11.3
6.2.1.1 NHS medical, optical, dental and medical auxiliary services	0.70	18	190	9.0
6.2.1.2 Private medical, optical, dental and medical auxiliary services	1.20	32	190	16.1
6.2.1.3 Other services	-	-	-	-
6.2.2 In-patient hospital services	[0.00]	[0]	..	55.6

Note: The commodity and service categories are not comparable with those in publications before 2001-02.
      The numbering is sequential, it does not use actual COICOP codes.
      Please see page ix for symbols and conventions used in this report.

Table A1   **Components of household expenditure, 2010 (cont.)**
United Kingdom

		Average weekly expenditure all house-holds (£)	Total weekly expenditure (£ million)	Recording house-holds in sample	Percentage standard error (full method)
Commodity or service					
7	**Transport**	**64.90**	**1,707**	**4,510**	**2.3**
7.1	Purchase of vehicles	19.50	513	1,180	5.6
7.1.1	Purchase of new cars and vans	6.50	172	280	12.0
7.1.1.1	Outright purchases	4.20	111	110	17.0
7.1.1.2	Loan/Hire Purchase of new car/van	2.30	61	180	12.0
7.1.2	Purchase of second hand cars or vans	12.20	320	890	5.2
7.1.2.1	Outright purchases	8.30	218	520	6.2
7.1.2.2	Loan/Hire Purchase of second hand car/van	3.90	102	410	7.8
7.1.3	Purchase of motorcycles	0.80	22	80	22.2
7.1.3.1	Outright purchases of new or second hand motorcycles	0.30	8	30	25.0
7.1.3.2	Loan/Hire Purchase of new or second hand motorcycles	[0.10]	[2]	20	26.7
7.1.3.3	Purchase of bicycles and other vehicles	0.40	12	40	36.5
7.2	Operation of personal transport	33.30	877	3,880	1.9
7.2.1	Spares and accessories	2.10	55	420	8.4
7.2.1.1	Car/van accessories and fittings	0.30	8	120	19.1
7.2.1.2	Car/van spare parts	1.50	39	230	10.7
7.2.1.3	Motorcycle accessories and spare parts	0.10	3	20	33.6
7.2.1.4	Bicycle accessories, repairs and other costs	[0.20]	[5]	90	22.7
7.2.2	Petrol, diesel and other motor oils	21.60	569	3,430	1.8
7.2.2.1	Petrol	15.60	410	2,810	1.9
7.2.2.2	Diesel oil	5.90	157	1,040	3.9
7.2.2.3	Other motor oils	0.10	3	80	13.7
7.2.3	Repairs and servicing	7.00	185	1,610	4.4
7.2.3.1	Car or van repairs, servicing and other work	6.90	183	1,610	4.4
7.2.3.2	Motorcycle repairs and servicing	0.10	2	20	32.9
7.2.4	Other motoring costs	2.60	68	2,050	4.5
7.2.4.1	Motoring organisation subscription (e.g. AA and RAC)	0.40	11	870	6.2
7.2.4.2	Garage rent, other costs (excluding fines), car washing etc.	0.80	21	340	10.2
7.2.4.3	Parking fees, tolls, and permits (excluding motoring fines)	0.80	21	1,090	6.4
7.2.4.4	Driving lessons	0.40	11	60	15.5
7.2.4.5	Anti-freeze, battery water, cleaning materials	0.10	3	300	10.4
7.3	Transport services	12.10	318	2,200	5.8
7.3.1	Rail and tube fares	2.80	73	740	6.0
7.3.1.1	Season tickets	1.00	27	160	11.4
7.3.1.2	Other than season tickets	1.70	46	640	5.8
7.3.2	Bus and coach fares	1.50	39	910	5.7
7.3.2.1	Season tickets	0.50	13	160	11.0
7.3.2.2	Other than season tickets	1.00	26	820	5.4
7.3.3	Combined fares	1.20	32	210	11.7
7.3.3.1	Combined fares other than season tickets	0.30	9	120	12.7
7.3.3.2	Combined fares season tickets	0.90	24	110	13.2
7.3.4	Other travel and transport	6.60	174	1,220	9.6
7.3.4.1	Air fares (within UK)	[0.10]	[3]	10	32.2
7.3.4.2	Air fares (international)	2.60	67	60	23.3
7.3.4.3	School travel	[0.00]	[1]	20	35.2
7.3.4.4	Taxis and hired cars with drivers	1.20	32	680	6.4
7.3.4.5	Other personal travel and transport services	0.30	8	320	9.1
7.3.4.6	Hire of self-drive cars, vans, bicycles	0.40	11	30	31.1
7.3.4.7	Car leasing	1.80	47	200	7.5
7.3.4.8	Water travel, ferries and season tickets	0.20	4	60	27.0

Note: The commodity and service categories are not comparable with those in publications before 2001–02.
The numbering is sequential, it does not use actual COICOP codes.
Please see page ix for symbols and conventions used in this report.

Table A1   Components of household expenditure, 2010 (cont.)
United Kingdom

		Average weekly expenditure all house- holds (£)	Total weekly expenditure (£ million)	Recording house- holds in sample	Percentage standard error (full method)
Commodity or service					
8	**Communication**	**13.00**	**343**	**5,000**	**1.4**
8.1	Postal services	0.50	13	860	6.6
8.2	Telephone and telefax equipment	0.60	17	160	13.5
8.2.1	Telephone purchase	0.10	2	30	20.4
8.2.2	Mobile phone purchase	0.50	14	130	15.5
8.2.3	Answering machine, fax machine, modem purchase	[0.00]	[0]	..	59.7
8.3	Telephone and telefax services	11.60	304	4,970	1.2
8.3.1	Telephone account	5.90	156	4,530	1.4
8.3.2	Telephone coin and other payments	0.00	1	50	19.2
8.3.3	Mobile phone account	4.50	119	2,340	2.2
8.3.4	Mobile phone - other payments	1.10	29	710	4.9
8.4	Internet subscription fees	0.30	9	270	16.4
9	**Recreation & culture**	**58.10**	**1,529**	**5,220**	**2.2**
9.1	Audio-visual, photographic and information processing equipment	7.20	189	1,460	6.3
9.1.1	Audio equipment and accessories, CD players	1.40	38	580	9.5
9.1.1.1	Audio equipment, CD players including in car	0.70	17	100	16.9
9.1.1.2	Audio accessories e.g. tapes, headphones etc.	0.80	20	510	10.1
9.1.2	TV, video and computers	5.20	137	1,050	7.9
9.1.2.1	Purchase of TV and digital decoder	1.50	40	100	16.1
9.1.2.2	Satellite dish purchase and installation	[0.00]	[0]	..	101.2
9.1.2.3	Cable TV connection	[0.00]	[0]	..	88.4
9.1.2.4	Video recorder	[0.00]	[0]	..	80.7
9.1.2.5	DVD player/recorder	0.20	5	30	28.1
9.1.2.6	Blank, pre-recorded video cassettes, DVDs	0.80	21	640	5.4
9.1.2.7	Personal computers, printers and calculators	2.50	65	340	12.5
9.1.2.8	Spare parts for TV, video, audio	0.10	3	70	24.4
9.1.2.9	Repair of audio-visual, photographic and information processing	0.10	3	30	27.1
9.1.3	Photographic, cine and optical equipment	0.50	14	110	18.6
9.1.3.1	Photographic and cine equipment	0.50	13	90	20.0
9.1.3.2	Camera films	[0.00]	[0]	20	27.5
9.1.3.3	Optical instruments, binoculars, telescopes, microscopes	[0.00]	[1]	10	50.1
9.2	Other major durables for recreation and culture	3.20	85	120	20.3
9.2.1	Purchase of boats, trailers and horses	[0.50]	[14]	10	48.0
9.2.2	Purchase of caravans, mobile homes (including decoration)	1.80	47	30	30.8
9.2.3	Accessories for boats, horses, caravans and motor caravans	[0.10]	[2]	20	39.0
9.2.4	Musical instruments (purchase and hire)	0.20	5	40	28.7
9.2.5	Major durables for indoor recreation	[0.00]	[0]	..	89.3
9.2.6	Maintenance and repair of other major durables	0.30	9	30	31.3
9.2.7	Purchase of motor caravan (new and second-hand) - outright purchase	[0.10]	[3]	..	71.2
9.2.8	Purchase of motor caravan (new and second-hand) - loan/HP	[0.20]	[5]	..	74.8
9.3	Other recreational items and equipment, gardens and pets	11.40	300	3,580	4.0
9.3.1	Games, toys and hobbies	2.00	52	1,270	5.0
9.3.2	Computer software and games	1.50	39	350	8.9
9.3.2.1	Computer software and game cartridges	0.90	24	310	7.1
9.3.2.2	Computer games consoles	0.50	14	60	19.0
9.3.3	Equipment for sport, camping and open-air recreation	1.20	31	410	20.6
9.3.4	Horticultural goods, garden equipment and plants etc.	2.50	65	1,840	3.9
9.3.4.1	BBQ and swings	0.00	1	40	29.4
9.3.4.2	Plants, flowers, seeds, fertilisers, insecticides	2.30	61	1,790	3.9
9.3.4.3	Garden decorative	0.10	1	60	18.9
9.3.4.4	Artificial flowers, pot pourri	0.00	1	40	23.9
9.3.5	Pets and pet food	4.30	114	1,990	6.9
9.3.5.1	Pet food	2.10	55	1,870	3.9
9.3.5.2	Pet purchase and accessories	0.90	23	610	16.7
9.3.5.3	Veterinary and other services for pets identified separately	1.40	36	180	16.9

Note: The commodity and service categories are not comparable with those in publications before 2001–02.
The numbering is sequential, it does not use actual COICOP codes.
Please see page ix for symbols and conventions used in this report.

Table A1 **Components of household expenditure, 2010 (cont.)**
United Kingdom

Commodity or service			Average weekly expenditure all house-holds (£)	Total weekly expenditure (£ million)	Recording house-holds in sample	Percentage standard error (full method)
9	**Recreation & culture (continued)**					
9.4	Recreational and cultural services		17.80	467	5,000	2.1
	9.4.1	Sports admissions, subscriptions, leisure class fees and equipment hire	5.30	140	1,810	4.7
		9.4.1.1 Spectator sports: admission charges	0.50	13	140	15.3
		9.4.1.2 Participant sports (excluding subscriptions)	1.00	26	740	5.8
		9.4.1.3 Subscriptions to sports and social clubs	1.80	48	810	8.4
		9.4.1.4 Leisure class fees	1.90	50	770	6.6
		9.4.1.5 Hire of equipment for sport and open air recreation	0.10	3	40	36.3
	9.4.2	Cinema, theatre and museums etc.	2.40	62	840	5.5
		9.4.2.1 Cinemas	0.60	17	430	5.9
		9.4.2.2 Live entertainment: theatre, concerts, shows	1.20	31	260	9.3
		9.4.2.3 Museums, zoological gardens, theme parks, houses and gardens	0.50	14	240	11.1
	9.4.3	TV, video, satellite rental, cable subscriptions and TV licences	6.00	157	4,620	1.3
		9.4.3.1 TV licences	2.30	62	4,450	0.5
		9.4.3.2 Satellite subscriptions	2.60	67	1,600	2.8
		9.4.3.3 Rent for TV/Satellite/VCR	0.30	7	180	9.4
		9.4.3.4 Cable subscriptions	0.70	19	710	5.9
		9.4.3.5 TV slot meter payments	[0.00]	[1]	10	44.4
		9.4.3.6 Video, cassette and CD hire	0.10	2	90	14.9
	9.4.4	Miscellaneous entertainments	1.10	28	1,090	6.4
		9.4.4.1 Admissions to clubs, dances, discos, bingo	0.50	13	440	8.8
		9.4.4.2 Social events and gatherings	0.20	6	250	13.2
		9.4.4.3 Subscriptions for leisure activities and other subscriptions	0.40	9	540	11.4
	9.4.5	Development of film, deposit for film development, passport photos, holiday and school photos	0.40	9	220	21.0
	9.4.6	Gambling payments	2.70	71	2,390	4.1
		9.4.6.1 Football pools stakes	0.00	1	50	16.8
		9.4.6.2 Bingo stakes excluding admission	0.20	5	120	14.5
		9.4.6.3 Lottery	1.90	50	2,180	3.6
		9.4.6.4 Bookmaker, tote, other betting stakes	0.60	15	490	12.2
9.5	Newspapers, books and stationery		5.90	155	4,490	2.0
	9.5.1	Books	1.40	37	1,100	5.3
	9.5.2	Stationery, diaries, address books, art materials	0.70	18	1,260	4.9
	9.5.3	Cards, calendars, posters and other printed matter	1.20	31	2,290	3.1
	9.5.4	Newspapers	1.80	46	3,030	2.2
	9.5.5	Magazines and periodicals	0.90	24	2,310	3.0
9.6	Package holidays		12.60	332	720	5.4
	9.6.1	Package holidays - UK	1.00	28	160	11.3
	9.6.2	Package holidays - abroad	11.60	304	580	5.8
10	**Education**		**10.00**	**264**	**400**	**12.4**
10.1	Education fees		9.80	257	310	12.6
	10.1.1	Nursery and primary education	1.30	34	50	19.4
	10.1.2	Secondary education	2.30	61	50	17.9
	10.1.3	Sixth form college/college education	0.60	16	40	31.0
	10.1.4	University education	4.90	129	120	21.4
	10.1.5	Other education	0.60	17	80	26.8
10.2	Payments for school trips, other ad-hoc expenditure		0.30	7	110	15.0
	10.2.1	Nursery and primary education	0.10	2	50	20.5
	10.2.2	Secondary education	0.10	3	50	23.9
	10.2.3	Sixth form college/college education	[0.00]	[1]	..	40.2
	10.2.4	University education	[0.00]	[0]	..	41.7
	10.2.5	Other education	[0.00]	[0]	..	72.5

Note: The commodity and service categories are not comparable with those in publications before 2001–02.
The numbering is sequential, it does not use actual COICOP codes.
Please see page ix for symbols and conventions used in this report.

Table A1   **Components of household expenditure, 2010 (cont.)**
United Kingdom

Commodity or service		Average expenditure all house- holds (£)	Total weekly expenditure (£ million)	Recording house- holds in sample	Percentage standard error (full method)
11	**Restaurants & hotels**	**39.20**	**1033**	**4,590**	**2.1**
11.1	Catering services	31.90	841	4,560	2.0
11.1.1	Restaurant and café meals	14.00	369	3,630	2.5
11.1.2	Alcoholic drinks (away from home)	7.30	192	2,260	3.4
11.1.3	Take away meals eaten at home	4.10	107	2,200	2.4
11.1.4	Other take-away and snack food	4.30	113	2,910	2.3
11.1.4.1	Hot and cold food	3.00	79	2,550	2.5
11.1.4.2	Confectionery	0.30	9	1,370	3.5
11.1.4.3	Ice cream	0.10	4	480	6.5
11.1.4.4	Soft drinks	0.80	22	1,890	2.8
11.1.5	Contract catering (food)	0.50	13	40	45.8
11.1.6	Canteens	1.80	47	1,430	4.7
11.1.6.1	School meals	0.70	18	500	9.4
11.1.6.2	Meals bought and eaten at the workplace	1.10	29	1,100	4.8
11.2	Accommodation services	7.30	192	940	4.9
11.2.1	Holiday in the UK	3.10	83	660	5.4
11.2.2	Holiday abroad	4.10	107	360	7.9
11.2.3	Room hire	[0.10]	[2]	10	48.2
12	**Miscellaneous goods and services**	**35.90**	**945**	**5,130**	**2.3**
12.1	Personal care	10.60	278	4,540	2.1
12.1.1	Hairdressing, beauty treatment	3.30	87	1,220	3.9
12.1.2	Toilet paper	0.80	21	2,320	2.0
12.1.3	Toiletries and soap	2.20	59	3,600	2.6
12.1.3.1	Toiletries (disposable including tampons, lipsyl, toothpaste etc.)	1.30	35	3,030	3.3
12.1.3.2	Bar of soap, liquid soap, shower gel etc.	0.40	10	1,490	4.5
12.1.3.3	Toilet requisites (durable including razors, hairbrushes, toothbrushes etc.)	0.50	13	1,120	4.9
12.1.4	Baby toiletries and accessories (disposable)	0.60	17	870	4.6
12.1.5	Hair products, cosmetics and electrical appliances for personal care	3.60	95	2,810	3.4
12.1.5.1	Hair products	0.80	20	1,550	3.6
12.1.5.2	Cosmetics and related accessories	2.60	69	2,150	3.8
12.1.5.3	Electrical appliances for personal care, including hairdryers, shavers etc.	0.20	6	110	14.4
12.2	Personal effects	3.00	80	1,290	9.0
12.2.1	Jewellery, clocks and watches and other personal effects	1.90	51	940	13.0
12.2.2	Leather and travel goods (excluding baby items)	0.80	21	430	8.4
12.2.3	Sunglasses (non-prescription)	0.10	2	80	18.5
12.2.4	Baby equipment (excluding prams and pushchairs)	0.10	3	30	28.3
12.2.5	Prams, pram accessories and pushchairs	[0.10]	[2]	10	53.5
12.2.6	Repairs to personal goods	0.10	2	30	35.1
12.3	Social protection	3.30	86	250	10.3
12.3.1	Residential homes	[0.10]	[3]	..	63.6
12.3.2	Home help	0.20	6	30	31.9
12.3.3	Nursery, crèche, playschools	0.60	17	90	17.4
12.3.4	Child care payments	2.30	61	150	12.2

Note: The commodity and service categories are not comparable with those in publications before 2001–02.
The numbering is sequential, it does not use actual COICOP codes.
Please see page ix for symbols and conventions used in this report.

Table A1   **Components of household expenditure, 2010 (cont.)**
United Kingdom

		Average weekly expenditure all house- holds (£)	Total weekly expenditure (£ million)	Recording house- holds in sample	Percentage standard error (full method)
Commodity or service					
12	**Miscellaneous goods and services (continued)**				
12.4	Insurance	15.10	397	4,620	1.8
	12.4.1 Household insurances	5.00	132	4,110	1.7
	12.4.1.1 Structure insurance	2.50	65	3,300	2.1
	12.4.1.2 Contents insurance	2.50	65	3,980	1.8
	12.4.1.3 Insurance for household appliances	0.10	2	110	18.4
	12.4.2 Medical insurance premiums	1.80	48	640	7.0
	12.4.3 Vehicle insurance including boat insurance	8.00	211	3,870	1.8
	12.4.3.1 Vehicle insurance	8.00	210	3,870	1.8
	12.4.3.2 Boat insurance (not home)	[0.00]	[0]	..	61.6
	12.4.4 Non-package holiday, other travel insurance	0.20	5	40	27.2
12.5	Other services	4.00	104	1,700	12.0
	12.5.1 Moving house	1.90	49	310	9.6
	12.5.1.1 Moving and storage of furniture	0.30	7	140	14.8
	12.5.1.2 Property transaction - purchase and sale	0.70	19	60	17.0
	12.5.1.3 Property transaction - sale only	0.50	12	40	21.7
	12.5.1.4 Property transaction - purchase only	0.30	7	70	15.2
	12.5.1.5 Property transaction - other payments	0.20	4	80	19.5
	12.5.2 Bank, building society, post office, credit card charges	0.50	12	890	6.6
	12.5.2.1 Bank and building society charges	0.40	11	780	6.9
	12.5.2.2 Bank and Post Office counter charges	[0.00]	[0]	10	37.9
	12.5.2.3 Annual standing charge for credit cards	0.00	1	140	13.1
	12.5.2.4 Commission travellers' cheques and currency	[0.00]	[0]	..	125.8
	12.5.3 Other services and professional fees	1.60	43	820	32.6
	12.5.3.1 Other professional fees including court fines	0.40	11	50	29.2
	12.5.3.2 Legal fees	[0.10]	[2]	10	39.8
	12.5.3.3 Funeral expenses	[0.50]	[12]	..	91.9
	12.5.3.4 TU and professional organisations	0.60	15	660	8.5
	12.5.3.5 Other payments for services e.g. photocopying	0.10	2	140	25.5
1-12	**All expenditure groups**	**406.30**	**10,696**	**5,260**	**1.2**
13	**Other expenditure items**	**67.30**	**1,771**	**4,930**	**2.0**
13.1	Housing: mortgage interest payments, council tax etc.	46.10	1214	4,590	1.8
	13.1.1 Mortgage interest payments	23.80	626	1,820	2.8
	13.1.2 Mortgage protection premiums	1.40	37	810	4.8
	13.1.3 Council tax, domestic rates	19.40	510	4,540	0.9
	13.1.5 Council tax, mortgage (second dwelling)	1.60	41	80	19.8
13.2	Licences, fines and transfers	3.40	90	3,780	2.9
	13.2.1 Stamp duty, licences and fines (excluding motoring fines)	0.50	13	100	16.7
	13.2.2 Motoring fines	0.10	3	20	22.4
	13.2.3 Motor vehicle road taxation payments less refunds	2.80	74	3,770	1.5
13.3	Holiday spending	7.10	187	230	10.0
	13.3.1 Money spent abroad	7.10	187	220	10.0
	13.3.2 Duty free goods bought in UK	[0.00]	[0]	..	82.7

Note: The commodity and service categories are not comparable with those in publications before 2001–02.
The numbering is sequential, it does not use actual COICOP codes.
Please see page ix for symbols and conventions used in this report.

Table A1   **Components of household expenditure, 2010 (cont.)**
United Kingdom

Commodity or service	Average weekly expenditure all house- holds (£)	Total weekly expenditure (£ million)	Recording house- holds in sample	Percentage standard error (full method)
13 Other expenditure items (continued)				
13.4 Money transfers and credit	10.60	280	2,700	4.5
13.4.1 Money, cash gifts given to children	0.20	5	140	15.1
13.4.1.1 Money given to children for specific purposes	0.20	4	130	11.1
13.4.1.2 Cash gifts to children (no specific purpose)	[0.00]	[1]	..	71.4
13.4.2 Cash gifts and donations	8.60	225	2,200	5.3
13.4.2.1 Money/presents given to those outside the household	3.50	91	850	7.6
13.4.2.2 Charitable donations and subscriptions	2.40	63	1,500	7.2
13.4.2.3 Money sent abroad	1.10	29	260	20.9
13.4.2.4 Maintenance allowance expenditure	1.60	42	140	10.2
13.4.3 Club instalment payments (child) and interest on credit cards	1.90	50	840	7.4
13.4.3.1 Club instalment payment	0.00	[0]	..	104.1
13.4.3.2 Interest on credit cards	1.90	50	840	7.4
Total expenditure	**473.60**	**12,467**	**5,260**	**1.2**
14 Other items recorded				
14.1 Life assurance, contributions to pension funds	19.70	518	2,580	4.2
14.1.1 Life assurance premiums e.g. mortgage endowment policies	3.80	100	1,680	4.8
14.1.2 Contributions to pension and superannuation funds etc.	10.80	283	1,420	3.3
14.1.3 Personal pensions	5.10	135	560	11.8
14.2 Other insurance including friendly societies	1.40	36	1,170	4.5
14.3 Income tax, payments less refunds	89.20	2,349	4,040	2.9
14.3.1 Income tax paid by employees under PAYE	67.90	1,788	2,680	2.9
14.3.2 Income tax paid direct e.g. by retired or unoccupied persons	2.70	70	130	27.8
14.3.3 Income tax paid direct by self-employed	5.70	149	240	12.8
14.3.4 Income tax deducted at source from income under covenant from investments or from annuities and pensions	10.30	270	2,140	3.8
14.3.5 Income tax on bonus earnings	3.60	94	730	15.2
14.3.6 Income tax refunds under PAYE	0.10	3	30	45.5
14.3.7 Income tax refunds other than PAYE	0.70	19	250	10.6
14.4 National insurance contribution	27.00	712	2,620	1.7
14.4.1 NI contributions paid by employees	26.90	709	2,600	1.8
14.4.2 NI contributions paid by non-employees	0.10	2	40	34.1
14.5 Purchase or alteration of dwellings (contracted out), mortgages	43.00	1,133	2,030	5.2
14.5.1 Outright purchase of houses, flats etc. including deposits	[0.20]	[4]	10	33.2
14.5.2 Capital repayment of mortgage	19.70	519	1,410	2.9
14.5.3 Central heating installation	1.40	36	140	16.8
14.5.4 DIY improvements: Double glazing, kitchen units, sheds etc.	0.70	19	70	33.8
14.5.5 Home improvements - contracted out	16.80	443	760	9.6
14.5.6 Bathroom fittings	0.60	16	70	42.4
14.5.7 Purchase of materials for Capital Improvements	0.20	6	50	29.6
14.5.8 Purchase of second dwelling	3.40	88	40	43.6
14.6 Savings and investments	6.00	158	820	7.8
14.6.1 Savings, investments (excluding AVCs)	5.10	133	630	8.6
14.6.2 Additional Voluntary Contributions	0.70	19	70	18.9
14.6.3 Food stamps, other food related expenditure	0.20	5	180	15.0
14.7 Pay off loan to clear other debt	2.20	59	270	7.4
14.8 Windfall receipts from gambling etc.3	1.70	45	480	12.9

Note: The commodity and service categories are not comparable with those in publications before 2001–02.
The numbering is sequential, it does not use actual COICOP codes.
Please see page ix for symbols and conventions used in this report.

3 Expressed as an income figure as opposed to an expenditure figure.

Table A2 Expenditure on food and non-alcoholic drinks by place of purchase, 2010

United Kingdom

		Large supermarket chains[1]			Other outlets			Internet Expenditure[2]		
		Average weekly expenditure all house-holds (£)	Total weekly expenditure (£ million)	Recording house-holds in sample	Average weekly expenditure all house-holds (£)	Total weekly expenditure (£ million)	Recording house-holds in sample	Average weekly expenditure all house-holds (£)	Total weekly expenditure (£ million)	Recording house-holds in sample
1	**Food and non-alcoholic drinks**	**43.10**	**1,135**	**5,140**	**8.40**	**221**	**4,490**	**1.70**	**44**	**210**
1.1	Food	39.60	1,043	5,140	7.80	205	4,440	1.50	39	200
1.1.1	Bread, rice and cereals	4.10	108	4,930	0.70	19	1,980	0.10	4	170
1.1.2	Pasta products	0.40	10	1,990	0.00	1	200	0.00	0	80
1.1.3	Buns, cakes, biscuits etc.	2.70	71	4,380	0.40	11	1,410	0.10	2	140
1.1.4	Pastry (savoury)	0.70	17	1,800	0.00	1	130	0.00	1	60
1.1.5	Beef (fresh, chilled or frozen)	1.20	33	2,000	0.40	11	470	0.10	1	80
1.1.6	Pork (fresh, chilled or frozen)	0.50	12	1,040	0.20	4	260	0.00	0	20
1.1.7	Lamb (fresh, chilled or frozen)	0.40	11	700	0.30	7	220	[0.00]	[0]	20
1.1.8	Poultry (fresh, chilled or frozen)	1.60	42	2,420	0.30	8	400	0.10	2	100
1.1.9	Bacon and ham	0.80	22	2,110	0.20	4	420	0.00	1	70
1.1.10	Other meats and meat preparations	4.80	126	4,440	0.70	18	1,280	0.10	3	140
1.1.11	Fish and fish products	1.90	51	3,190	0.30	9	490	0.10	2	100
1.1.12	Milk	1.80	46	4,300	0.70	20	1,760	0.10	1	140
1.1.13	Cheese and curd	1.60	42	3,530	0.20	4	450	0.10	3	130
1.1.14	Eggs	0.50	13	2,350	0.10	4	700	0.00	0	90
1.1.15	Other milk products	1.70	46	3,790	0.10	3	460	0.10	2	140
1.1.16	Butter	0.30	9	1,460	0.00	1	140	0.00	0	50
1.1.17	Margarine, other vegetable fats and peanut butter	0.50	12	2,290	0.00	1	260	0.00	1	90
1.1.18	Cooking oils and fats	0.20	6	1,090	0.00	1	140	0.00	0	40
1.1.19	Fresh fruit	2.60	69	4,100	0.30	9	910	0.10	3	150
1.1.20	Other fresh, chilled or frozen fruits	0.30	8	1,290	0.00	1	170	0.00	0	50
1.1.21	Dried fruit and nuts	0.40	11	1,510	0.10	3	300	0.00	0	50
1.1.22	Preserved fruit and fruit based products	0.10	3	870	0.00	0	80	0.00	0	40
1.1.23	Fresh vegetables	3.40	89	4,510	0.50	12	1,200	0.20	4	170
1.1.24	Dried vegetables and other preserved and processed vegetables	0.70	18	3,000	0.60	17	2,420	0.00	1	120
1.1.25	Potatoes	0.70	18	3,110	0.10	4	620	0.00	1	110
1.1.26	Other tubers and products of tuber vegetables	1.20	31	3,350	0.10	4	740	0.00	1	100
1.1.27	Sugar and sugar products	0.30	7	1,650	0.10	1	330	0.00	0	70
1.1.28	Jams, marmalades	0.20	6	1,370	0.10	2	200	0.00	0	40
1.1.29	Chocolate	1.20	33	2,770	0.40	10	1,150	0.00	1	60
1.1.30	Confectionery products	0.40	10	1,900	0.20	5	1,010	0.00	0	30
1.1.31	Edible ices and ice cream	0.40	12	1,500	0.00	1	200	0.00	1	50
1.1.32	Other food products	1.90	51	4,010	0.40	9	850	0.10	2	150
1.2	Non-alcoholic drinks	3.50	93	4,430	0.60	16	1,670	0.20	5	160
1.2.1	Coffee	0.50	13	1,410	0.10	2	240	0.00	1	50
1.2.2	Tea	0.40	11	1,490	0.10	2	300	0.00	0	60
1.2.3	Cocoa and powdered chocolate	0.10	2	450	0.00	0	90	0.00	0	20
1.2.4	Fruit and vegetable juices (inc fruit squash)	1.00	25	2,760	0.10	3	400	[0.00]	1	100
1.2.5	Mineral or spring waters	0.20	5	870	0.00	1	200	0.00	0	20
1.2.6	Soft drinks	1.40	36	2,950	0.30	9	1,110	0.10	2	110

Note: The commodity and service categories are not comparable with those in publications before 2001–02.
The numbering is sequential; it does not use actual COICOP codes.
Please see page ix for symbols and conventions used in this report.

1 In 2010 the list of large supermarket chains was updated.
2 Includes internet expenditure from large supermarket chains.

136

Table A3 Expenditure on clothing and footwear by place of purchase, 2010

United Kingdom

	Large supermarket chains[1]			Clothing chains			Other outlets[2]		
	Average weekly expenditure all house-holds (£)	Total weekly expenditure (£ million)	Recording house-holds in sample	Average weekly expenditure all house-holds (£)	Total weekly expenditure (£ million)	Recording house-holds in sample	Average weekly expenditure all house-holds (£)	Total weekly expenditure (£ million)	Recording house-holds in sample
3 Clothing and footwear	**1.80**	**47**	**1,260**	**8.50**	**224**	**1,820**	**12.70**	**335**	**2,450**
3.1 Clothing	1.60	43	1,200	7.30	193	1,710	9.40	246	2,130
3.1.1 Men's outer garments	0.20	7	210	1.70	45	450	2.80	74	600
3.1.2 Men's under garments	0.10	2	110	0.20	5	180	0.10	3	130
3.1.3 Women's outer garments	0.60	15	430	3.70	97	1,020	4.20	109	1,060
3.1.4 Women's under garments	0.20	5	330	0.60	15	470	0.30	9	300
3.1.5 Boys' outer garments	0.10	3	120	0.20	6	110	0.50	13	200
3.1.6 Girls' outer garments	0.10	3	120	0.30	7	160	0.50	14	240
3.1.7 Infants' outer garments	0.10	3	130	0.30	7	160	0.20	6	170
3.1.8 Children's under garments	0.10	4	220	0.20	4	170	0.10	3	150
3.1.9 Accessories	0.10	1	170	0.30	7	250	0.40	10	450
3.1.9.1 Men's accessories	0.00	0	40	0.10	3	80	0.10	4	140
3.1.9.2 Women's accessories	0.00	1	80	0.10	4	150	0.20	5	220
3.1.9.3 Children's accessories	0.00	0	60	0.00	1	40	0.10	2	130
3.1.10 Haberdashery and clothing hire	0.00	0	30	[0.00]	[0]	10	0.20	4	160
3.2 Footwear	0.20	5	200	1.10	30	480	3.40	89	950
3.2.1 Men's	0.00	1	30	0.20	6	80	1.20	30	300
3.2.2 Women's	0.10	2	100	0.80	21	350	1.60	41	490
3.2.3 Children's	0.10	1	80	0.10	3	90	0.60	17	300

Note: The commodity and service categories are not comparable with those in publications before 2001-02.
The numbering system is sequential; it does not use actual COICOP codes.
Please see page ix for symbols and conventions used in this report.

1 In 2010 the list of large supermarket chains was updated.
2 Includes internet expenditure from large supermarket chains.

Table A4   **Household expenditure by gross income decile group, 2010**
United Kingdom

	Lowest ten per cent	Second decile group	Third decile group	Fourth decile group	Fifth decile group	Sixth decile group	Seventh decile group	Eighth decile group	Ninth decile group	Highest ten per cent	All house-holds
Lower boundary of group (£ per week)		160	238	315	413	522	651	801	1,015	1,368	
Weighted number of households (thousands)	2,630	2,640	2,620	2,630	2,630	2,640	2,630	2,640	2,630	2,630	26,320
Total number of households in sample	510	530	540	550	550	530	530	520	510	500	5,260
Total number of persons in sample	690	860	960	1,150	1,280	1,290	1,410	1,430	1,530	1,580	12,180
Total number of adults in sample	570	670	780	900	970	1,010	1,090	1,090	1,170	1,170	9,430
Weighted average number of persons per household	1.3	1.6	1.8	2.1	2.4	2.4	2.7	2.8	3.0	3.2	2.3
Commodity or service						*Average weekly household expenditure (£)*					
1 Food & non-alcoholic drinks	27.20	34.50	40.00	46.10	50.70	54.40	60.90	61.20	73.70	83.40	53.20
2 Alcoholic drinks, tobacco & narcotics	7.10	7.90	8.60	10.30	12.10	13.20	12.10	13.60	15.70	17.60	11.80
3 Clothing & footwear	6.80	8.30	11.70	13.50	18.60	21.90	25.40	32.30	40.50	54.90	23.40
4 Housing(net)¹, fuel & power	42.50	45.10	50.30	58.10	61.20	60.70	67.90	68.70	64.70	84.70	60.40
5 Household goods & services	14.40	16.20	17.90	25.40	22.50	31.40	38.60	31.10	37.30	79.60	31.40
6 Health	1.40	3.00	3.40	3.60	4.00	5.00	4.60	5.50	7.50	12.60	5.00
7 Transport	16.40	19.60	29.20	43.30	48.40	60.70	77.90	87.40	111.00	155.00	64.90
8 Communication	6.80	7.60	9.00	10.60	11.30	13.40	14.60	16.70	18.00	22.20	13.00
9 Recreation & culture	18.70	25.40	31.30	39.90	52.40	54.80	59.80	73.10	92.80	132.60	58.10
10 Education	[2.70]	[1.60]	[4.60]	5.70	8.50	3.90	9.30	7.80	19.30	37.00	10.00
11 Restaurants & hotels	10.50	12.40	16.60	23.30	30.30	35.80	44.80	54.10	67.60	97.10	39.20
12 Miscellaneous goods & services	12.20	14.10	19.60	23.50	29.00	38.40	37.20	48.30	54.20	82.60	35.90
1-12 All expenditure groups	166.90	195.60	242.20	303.10	349.00	393.40	453.00	499.90	602.20	859.20	406.30
13 Other expenditure items	18.70	19.10	30.50	41.30	51.00	65.40	78.20	90.90	118.50	159.30	67.30
Total expenditure	**185.60**	**214.80**	**272.70**	**344.40**	**400.00**	**458.80**	**531.30**	**590.80**	**720.70**	**1018.50**	**473.60**
Average weekly expenditure per person (£)											
Total expenditure	**141.00**	**130.90**	**154.50**	**166.50**	**169.90**	**188.30**	**194.80**	**211.10**	**237.80**	**320.30**	**203.10**

Note: The commodity and service categories are not comparable to those in publications before 2001-02.
Please see page ix for symbols and conventions used in this report.

1 Excluding mortgage interest payments, council tax and Northern Ireland rates

Table A5   Household expenditure as a percentage of total expenditure by gross income decile group, 2010
United Kingdom

	Lowest ten per cent	Second decile group	Third decile group	Fourth decile group	Fifth decile group	Sixth decile group	Seventh decile group	Eighth decile group	Ninth decile group	Highest ten per cent	All house-holds
Lower boundary of group (£ per week)		160	238	315	413	522	651	801	1,015	1,368	
Weighted number of households (thousands)	2,630	2,640	2,620	2,630	2,630	2,640	2,630	2,640	2,630	2,630	26,320
Total number of households in sample	510	530	540	550	550	530	530	520	510	500	5,260
Total number of persons in sample	690	860	960	1,150	1,280	1,290	1,410	1,430	1,530	1,580	12,180
Total number of adults in sample	570	670	780	900	970	1,010	1,090	1,090	1,170	1,170	9,430
Weighted average number of persons per household	1.3	1.6	1.8	2.1	2.4	2.4	2.7	2.8	3.0	3.2	2.3
Commodity or service						Percentage of total expenditure					
1 Food & non-alcoholic drinks	15	16	15	13	13	12	11	10	10	8	11
2 Alcoholic drinks, tobacco & narcotics	4	4	3	3	3	3	2	2	2	2	2
3 Clothing & footwear	4	4	4	4	5	5	5	5	6	5	5
4 Housing(net)[1], fuel & power	23	21	18	17	15	13	13	12	9	8	13
5 Household goods & services	8	8	7	7	6	7	7	5	5	8	7
6 Health	1	1	1	1	1	1	1	1	1	1	1
7 Transport	9	9	11	13	12	13	15	15	15	15	14
8 Communication	4	4	3	3	3	3	3	3	2	2	3
9 Recreation & culture	10	12	11	12	13	12	11	12	13	13	12
10 Education	[1]	[1]	[2]	2	2	2	2	1	3	4	2
11 Restaurants & hotels	6	6	6	7	8	8	8	9	9	10	8
12 Miscellaneous goods & services	7	7	7	7	7	8	7	8	8	8	8
1-12 All expenditure groups	90	91	89	88	87	86	85	85	84	84	86
13 Other expenditure items	10	9	11	12	13	14	15	15	16	16	14
Total expenditure	**100**	**100**	**100**	**100**	**100**	**100**	**100**	**100**	**100**	**100**	**100**

Note: The commodity and service categories are not comparable to those in publications before 2001-02.
Please see page ix for symbols and conventions used in this report.

1 Excluding mortgage interest payments, council tax and Northern Ireland rates.

Table A6 **Detailed household expenditure by gross income decile group, 2010**

United Kingdom

Commodity or service		Lowest ten per cent	Second decile group	Third decile group	Fourth decile group	Fifth decile group	Sixth decile group	Seventh decile group	Eighth decile group	Ninth decile group	Highest ten per cent	All households
Lower boundary of group (£ per week)			160	238	315	413	522	651	801	1,015	1,368	
Weighted number of households (thousands)		2,630	2,640	2,620	2,630	2,630	2,640	2,630	2,640	2,630	2,630	26,320
Total number of households in sample		510	530	540	550	550	530	530	520	510	500	5,260
Total number of persons in sample		690	860	960	1,150	1,280	1,290	1,410	1,430	1,530	1,580	12,180
Total number of adults in sample		570	670	780	900	970	1,010	1,090	1,090	1,170	1,170	9,430
Weighted average number of persons per household		1.3	1.6	1.8	2.1	2.4	2.4	2.7	2.8	3.0	3.2	2.3
		Average weekly household expenditure (£)										
1	**Food & non-alcoholic drinks**	**27.20**	**34.50**	**40.00**	**46.10**	**50.70**	**54.40**	**60.90**	**61.20**	**73.70**	**83.40**	**53.20**
1.1	Food	25.10	31.90	36.90	42.40	46.70	50.00	55.80	56.20	67.50	76.50	48.90
1.1.1	Bread, rice and cereals	2.60	3.50	3.70	4.20	4.80	5.10	5.90	5.80	7.00	7.30	5.00
1.1.2	Pasta products	0.20	0.30	0.20	0.30	0.40	0.40	0.50	0.50	0.70	0.80	0.40
1.1.3	Buns, cakes, biscuits etc.	1.70	2.20	2.60	2.80	3.20	3.50	3.80	3.40	4.10	4.70	3.20
1.1.4	Pastry (savoury)	0.40	0.40	0.40	0.60	0.60	0.70	0.80	1.00	1.10	1.30	0.70
1.1.5	Beef (fresh, chilled or frozen)	0.80	1.00	1.20	1.40	1.70	1.70	1.80	1.90	2.50	3.00	1.70
1.1.6	Pork (fresh, chilled or frozen)	0.30	0.50	0.60	0.70	0.70	0.60	0.70	0.70	1.00	0.70	0.60
1.1.7	Lamb (fresh, chilled or frozen)	0.30	0.40	0.60	0.70	1.10	0.70	0.60	0.50	0.80	1.40	0.70
1.1.8	Poultry (fresh, chilled or frozen)	0.90	1.00	1.20	1.60	1.90	2.00	2.50	2.50	3.10	3.20	2.00
1.1.9	Bacon and ham	0.50	0.60	0.90	1.00	1.00	1.00	1.10	1.20	1.40	1.40	1.00
1.1.10	Other meat and meat preparations	3.40	4.00	4.70	5.30	5.20	5.90	6.50	6.30	7.00	7.90	5.60
1.1.11	Fish and fish products	1.20	1.40	1.80	2.00	2.20	2.30	2.40	2.70	3.30	4.00	2.30
1.1.12	Milk	1.60	2.20	2.10	2.50	2.60	2.70	2.90	2.70	3.00	3.20	2.60
1.1.13	Cheese and curd	0.90	1.00	1.20	1.50	1.70	1.90	2.00	2.00	3.00	3.10	1.80
1.1.14	Eggs	0.40	0.50	0.60	0.60	0.60	0.70	0.80	0.70	0.80	1.00	0.70
1.1.15	Other milk products	0.90	1.30	1.40	1.50	1.70	2.00	2.30	2.30	2.70	3.20	1.90
1.1.16	Butter	0.20	0.20	0.40	0.40	0.40	0.40	0.40	0.40	0.50	0.50	0.40
1.1.17	Margarine, other vegetable fats and peanut butter	0.40	0.40	0.40	0.60	0.60	0.60	0.50	0.60	0.60	0.60	0.50
1.1.18	Cooking oils and fats	0.20	0.20	0.20	0.20	0.30	0.30	0.40	0.30	0.40	0.40	0.30
1.1.19	Fresh fruit	1.20	1.70	2.30	2.60	2.80	3.10	3.50	3.90	4.20	5.30	3.10
1.1.20	Other fresh, chilled or frozen fruits	0.10	0.20	0.30	0.30	0.30	0.30	0.40	0.50	0.50	0.70	0.40
1.1.21	Dried fruit and nuts	0.20	0.20	0.40	0.40	0.50	0.50	0.60	0.70	0.80	1.10	0.60
1.1.22	Preserved fruit and fruit based products	0.10	0.10	0.10	0.10	0.10	0.20	0.10	0.10	0.20	0.20	0.10
1.1.23	Fresh vegetables	1.80	2.30	2.70	3.20	3.40	3.90	4.50	5.10	6.00	7.20	4.00
1.1.24	Dried vegetables	[0.00]	[0.00]	0.00	[0.00]	[0.00]	0.00	0.10	0.00	0.10	0.10	0.00
1.1.25	Other preserved or processed vegetables	0.70	0.70	0.90	1.00	1.20	1.30	1.50	1.50	1.90	2.20	1.30
1.1.26	Potatoes	0.50	0.60	0.80	0.80	0.80	0.90	0.90	0.90	1.10	1.00	0.90
1.1.27	Other tubers and products of tuber vegetables	0.70	0.90	1.00	1.20	1.40	1.60	1.70	1.60	1.80	1.60	1.40
1.1.28	Sugar and sugar products	0.20	0.30	0.30	0.30	0.40	0.30	0.30	0.40	0.40	0.50	0.30
1.1.29	Jams, marmalades	0.20	0.20	0.30	0.30	0.30	0.20	0.30	0.30	0.40	0.40	0.30
1.1.30	Chocolate	0.70	1.10	1.10	1.30	1.70	1.50	2.10	1.90	2.30	2.60	1.60
1.1.31	Confectionery products	0.30	0.40	0.50	0.60	0.60	0.70	0.70	0.60	0.80	0.80	0.60
1.1.32	Edible ices and ice cream	0.20	0.30	0.40	0.50	0.50	0.60	0.60	0.50	0.70	0.70	0.50
1.1.33	Other food products	1.20	1.40	1.50	1.90	1.90	2.60	2.60	2.80	3.50	4.10	2.40
1.2	Non-alcoholic drinks	2.20	2.70	3.10	3.60	4.10	4.40	5.10	5.00	6.20	6.90	4.30
1.2.1	Coffee	0.40	0.40	0.50	0.50	0.60	0.60	0.60	0.70	0.80	1.00	0.60
1.2.2	Tea	0.30	0.40	0.40	0.50	0.50	0.50	0.50	0.50	0.70	0.60	0.50
1.2.3	Cocoa and powdered chocolate	0.10	0.10	0.10	0.10	0.10	0.10	0.10	0.10	0.10	0.20	0.10
1.2.4	Fruit and vegetable juices (inc. fruit squash)	0.50	0.60	0.80	0.90	0.90	1.10	1.30	1.30	1.60	2.20	1.10
1.2.5	Mineral or spring waters	0.10	0.10	0.10	0.20	0.10	0.20	0.40	0.30	0.40	0.40	0.20
1.2.6	Soft drinks (inc. fizzy and ready to drink fruit drinks)	0.80	1.00	1.20	1.50	1.70	1.90	2.20	2.00	2.50	2.50	1.80

Note: The commodity and service categories are not comparable to those in publications before 2001-02.

Please see page ix for symbols and conventions used in this report.

1 Excluding mortgage interest payments, council tax and Northern Ireland rates.

Table A6 **Detailed household expenditure by gross income decile group, 2010 (cont.)**
United Kingdom

Commodity or service		Lowest ten per cent	Second decile group	Third decile group	Fourth decile group	Fifth decile group	Sixth decile group	Seventh decile group	Eighth decile group	Ninth decile group	Highest ten per cent	All house-holds
						Average weekly household expenditure (£)						
2	**Alcoholic drink, tobacco & narcotics**	**7.10**	**7.90**	**8.60**	**10.30**	**12.10**	**13.20**	**12.10**	**13.60**	**15.70**	**17.60**	**11.80**
2.1	Alcoholic drinks	3.10	3.10	4.10	5.50	5.90	7.50	8.20	8.40	12.00	14.30	7.20
2.1.1	Spirits and liqueurs (brought home)	0.70	0.80	1.40	1.50	1.60	1.90	1.40	1.40	2.40	1.70	1.50
2.1.2	Wines, fortified wines (brought home)	1.20	1.10	1.70	2.20	2.70	3.30	4.40	4.60	6.40	9.50	3.70
2.1.3	Beer, lager, ciders and perry (brought home)	1.10	1.20	0.90	1.70	1.60	2.20	2.30	2.30	3.10	3.00	1.90
2.1.4	Alcopops (brought home)	[0.00]	[0.00]	[0.10]	[0.10]	[0.10]	[0.10]	[0.10]	[0.10]	[0.10]	[0.10]	0.10
2.2	Tobacco and narcotics	4.10	4.80	4.50	4.80	6.20	5.70	3.90	5.30	3.60	3.30	4.60
2.2.1	Cigarettes	2.90	4.00	3.70	4.30	5.00	4.90	3.30	4.70	3.30	2.90	3.90
2.2.2	Cigars, other tobacco products and narcotics	1.10	0.80	0.80	0.60	1.20	0.70	0.60	0.60	0.30	0.40	0.70
3	**Clothing & footwear**	**6.80**	**8.30**	**11.70**	**13.50**	**18.60**	**21.90**	**25.40**	**32.30**	**40.50**	**54.90**	**23.40**
3.1	Clothing	5.70	6.40	9.40	10.50	14.90	17.40	19.50	26.10	33.20	43.30	18.60
3.1.1	Men's outer garments	1.30	1.40	1.60	2.00	3.80	4.90	4.10	7.40	9.10	12.20	4.80
3.1.2	Men's under garments	0.20	0.20	0.10	0.30	0.40	0.30	0.30	0.80	0.80	0.60	0.40
3.1.3	Women's outer garments	2.60	2.90	4.50	5.20	6.10	7.30	9.80	11.50	14.40	19.90	8.40
3.1.4	Women's under garments	0.50	0.50	0.80	0.70	1.30	1.20	1.10	1.30	1.70	1.90	1.10
3.1.5	Boys' outer garments (5-15)	[0.20]	[0.10]	0.80	0.50	0.70	0.70	0.90	1.10	1.60	1.60	0.80
3.1.6	Girls' outer garments (5-15)	[0.30]	0.40	0.50	0.40	0.90	0.90	0.90	1.10	1.60	2.10	0.90
3.1.7	Infants' outer garments (under 5)	0.20	0.40	0.30	0.40	0.50	0.60	0.80	0.70	1.30	1.00	0.60
3.1.8	Children's under garments (under 16)	0.10	0.20	0.20	0.30	0.40	0.40	0.50	0.50	0.50	1.00	0.40
3.1.9	Accessories	0.20	0.10	0.40	0.40	0.50	0.70	0.70	1.10	1.60	1.90	0.80
3.1.10	Haberdashery and clothing hire	[0.00]	[0.10]	0.20	0.20	[0.10]	0.20	[0.20]	0.30	0.40	0.10	0.20
3.1.11	Dry cleaners, laundry and dyeing	[0.10]	[0.10]	[0.10]	[0.20]	[0.30]	[0.20]	[0.20]	0.30	[0.20]	1.00	0.30
3.2	Footwear	1.10	1.80	2.30	3.00	3.70	4.40	6.00	6.30	7.30	11.60	4.70
4	**Housing (net)¹, fuel & power**	**42.50**	**45.10**	**50.30**	**58.10**	**61.20**	**60.70**	**67.90**	**68.70**	**64.70**	**84.70**	**60.40**
4.1	Actual rentals for housing	71.80	58.10	46.80	43.70	39.00	28.90	31.80	29.50	19.80	30.00	39.90
4.1.1	Gross rent	71.80	58.10	46.80	43.70	38.80	28.90	31.70	29.50	19.80	29.80	39.90
4.1.2	less housing benefit, rebates & allowances rec'd	51.80	39.70	27.80	17.00	11.40	3.30	2.50	0.50	0.70	0.00	15.50
4.1.3	Net rent2	20.10	18.30	19.00	26.60	27.40	25.60	29.30	29.00	19.10	29.80	24.40
4.1.4	Second dwelling rent	-	[0.00]	-	[0.00]	[0.10]	-	[0.10]	-	-	[0.20]	[0.00]
4.2	Maintenance and repair of dwelling	2.10	2.80	5.10	4.70	5.60	5.80	8.20	7.80	11.30	13.60	6.70
4.3	Water supply and miscellaneous services relating to the dwelling	6.80	6.90	7.50	7.80	7.30	7.50	8.00	8.60	8.30	9.60	7.80
4.4	Electricity, gas and other fuels	13.50	17.10	18.70	19.00	20.70	21.80	22.30	23.30	26.00	31.50	21.40
4.4.1	Electricity	6.80	8.20	8.90	8.80	9.70	10.10	10.50	10.20	11.50	14.70	9.90
4.4.2	Gas	6.00	7.40	8.00	8.60	9.20	10.20	9.80	11.00	12.50	13.80	9.60
4.4.3	Other fuels	0.80	1.50	1.80	1.60	1.90	1.60	2.00	2.10	2.00	3.00	1.80
1.2.4	Fruit and vegetable juices (inc. fruit squash)	0.50	0.60	0.80	0.90	0.90	1.10	1.30	1.30	1.60	2.20	1.10
1.2.5	Mineral or spring waters	0.10	0.10	0.10	0.20	0.10	0.20	0.40	0.30	0.40	0.40	0.20
1.2.6	Soft drinks (inc. fizzy and ready to drink fruit drinks)	[0.80]	1.00	1.20	1.50	1.70	1.90	2.20	2.00	2.50	2.50	1.80

Note: The commodity and service categories are not comparable to those in publications before 2001-02.
The numbering system is sequential, it does not use actual COICOP codes.
Please see page ix for symbols and conventions used in this report.

1 Excluding mortgage interest payments, council tax and Northern Ireland rates.
2 The figure included in total expenditure is net rent as opposed to gross rent

Table A6 **Detailed household expenditure by gross income decile group, 2010 (cont.)**
United Kingdom

Commodity or service	Lowest ten per cent	Second decile group	Third decile group	Fourth decile group	Fifth decile group	Sixth decile group	Seventh decile group	Eighth decile group	Ninth decile group	Highest ten per cent	All households
					Average weekly household expenditure (£)						
5 Household goods & services	**14.40**	**16.20**	**17.90**	**25.40**	**22.50**	**31.40**	**38.60**	**31.10**	**37.30**	**79.60**	**31.40**
5.1 Furniture and furnishings, carpets and other floor coverings	8.80	9.00	8.40	11.90	10.10	18.20	21.60	15.90	18.90	43.40	16.60
5.1.1 Furniture and furnishings	8.10	7.00	6.90	9.00	7.40	14.00	17.60	12.80	16.10	35.50	13.40
5.1.2 Floor coverings	0.70	2.00	1.50	2.90	2.70	4.20	4.00	3.10	2.80	7.90	3.20
5.2 Household textiles	0.70	0.70	0.90	1.40	1.60	1.20	2.50	1.40	2.40	5.30	1.80
5.3 Household appliances	1.60	1.40	1.80	4.70	3.60	3.80	4.80	4.10	4.80	7.80	3.80
5.4 Glassware, tableware and household utensils	0.50	0.70	0.70	1.00	1.10	1.20	2.00	1.90	2.00	3.40	1.40
5.5 Tools and equipment for house and garden	0.70	0.60	0.90	1.40	1.30	1.90	2.50	2.10	2.60	5.40	1.90
5.6 Goods and services for routine household maintenance	2.20	3.80	5.20	5.00	4.80	5.00	5.20	5.70	6.60	14.20	5.80
5.6.1 Cleaning materials	1.10	1.60	1.70	2.00	2.10	2.20	2.60	2.50	3.00	3.60	2.20
5.6.2 Household goods and hardware	0.50	0.80	0.90	0.90	1.30	1.20	1.50	1.50	1.80	2.20	1.30
5.6.3 Domestic services, carpet cleaning, hire/repair of furniture/furnishings	0.70	1.50	2.60	2.10	1.40	1.60	1.10	1.70	1.70	8.40	2.30
6 Health	**1.40**	**3.00**	**3.40**	**3.60**	**4.00**	**5.00**	**4.60**	**5.50**	**7.50**	**12.60**	**5.00**
6.1 Medical products, appliances and equipment	0.80	2.30	2.20	2.50	2.90	4.10	3.20	2.80	4.60	6.10	3.10
6.1.1 Medicines, prescriptions, healthcare products etc.	0.70	0.90	1.30	1.30	1.70	2.20	1.90	1.90	2.90	3.00	1.80
6.1.2 Spectacles, lenses, accessories and repairs	[0.20]	[1.30]	[0.90]	1.20	1.20	1.90	1.30	0.90	1.60	3.10	1.40
6.2 Hospital services	[0.50]	[0.70]	1.20	1.10	1.10	0.90	1.40	2.70	2.90	6.40	1.90
7 Transport	**16.40**	**19.60**	**29.20**	**43.30**	**48.40**	**60.70**	**77.90**	**87.40**	**111.00**	**155.00**	**64.90**
7.1 Purchase of vehicles	3.80	3.10	6.20	13.00	12.30	17.90	26.90	25.90	34.50	51.50	19.50
7.1.1 Purchase of new cars and vans	[1.70]	[0.60]	[1.00]	4.20	[2.80]	8.50	11.00	5.80	13.00	16.80	6.50
7.1.2 Purchase of second hand cars or vans	1.90	2.50	5.10	8.50	9.10	8.80	15.00	18.80	20.40	31.70	12.20
7.1.3 Purchase of motorcycles and other vehicles	[0.10]	[0.10]	[0.10]	[0.30]	[0.40]	[0.60]	[1.00]	[1.40]	[1.10]	[3.10]	0.80
7.2 Operation of personal transport	9.60	12.20	15.80	22.60	27.50	34.00	38.40	43.80	58.50	70.80	33.30
7.2.1 Spares and accessories	1.30	1.10	0.80	1.40	1.20	1.50	2.30	2.40	5.00	3.90	2.10
7.2.2 Petrol, diesel and other motor oils	5.70	7.80	10.00	14.30	18.80	22.70	26.60	30.70	37.40	42.30	21.60
7.2.3 Repairs and servicing	1.90	2.40	3.90	4.70	5.50	7.20	7.10	7.40	11.70	18.50	7.00
7.2.4 Other motoring costs	0.60	1.00	1.00	2.20	2.00	2.50	2.50	3.30	4.40	6.10	2.60
7.3 Transport services	3.10	4.20	7.20	7.70	8.60	8.80	12.50	17.70	18.00	32.70	12.10
7.3.1 Rail and tube fares	0.50	0.70	1.00	0.60	1.00	1.80	2.20	3.50	6.20	10.20	2.80
7.3.2 Bus and coach fares	0.80	1.00	1.00	1.70	1.40	2.00	1.60	2.30	1.30	1.70	1.50
7.3.3 Combined fares	[0.10]	[0.40]	[0.60]	[0.60]	[1.00]	[0.40]	1.10	1.20	2.60	4.10	1.20
7.3.4 Other travel and transport	1.70	2.10	4.50	4.90	5.30	4.60	7.60	10.70	8.00	16.70	6.60
8 Communication	**6.80**	**7.60**	**9.00**	**10.60**	**11.30**	**13.40**	**14.60**	**16.70**	**18.00**	**22.20**	**13.00**
8.1 Postal services	0.20	0.30	0.40	0.50	0.30	0.50	0.40	0.70	0.70	1.10	0.50
8.2 Telephone and telefax equipment	0.20	0.20	0.20	0.40	0.30	0.80	0.60	0.60	1.00	2.10	0.60
8.3 Telephone and telefax services	[6.20]	[7.00]	[8.30]	[9.50]	[10.50]	11.90	[13.30]	15.00	[15.80]	18.10	11.60
8.4 Internet subscription fees	0.20	0.10	0.20	0.20	0.20	0.20	0.40	0.40	0.50	0.90	0.30

Note: The commodity and service categories are not comparable to those in publications before 2001–02.
The numbering system is sequential, it does not use actual COICOP codes.
Please see page ix for symbols and conventions used in this report.

Table A6   **Detailed household expenditure by gross income decile group, 2010 (cont.)**
United Kingdom

Commodity or service	Lowest ten per cent	Second decile group	Third decile group	Fourth decile group	Fifth decile group	Sixth decile group	Seventh decile group	Eighth decile group	Ninth decile group	Highest ten per cent	All households
					Average weekly household expenditure (£)						
9 Recreation & culture	**18.70**	**25.40**	**31.30**	**39.90**	**52.40**	**54.80**	**59.80**	**73.10**	**92.80**	**132.60**	**58.10**
9.1 Audio-visual, photographic and information processing equipment	1.40	3.40	3.60	5.40	5.20	6.20	7.40	6.60	13.30	19.30	7.20
9.1.1 Audio equipment and accessories, CD players	0.30	0.50	0.90	0.70	1.00	0.80	1.60	2.00	2.80	3.60	1.40
9.1.2 TV, video and computers	1.10	2.80	2.50	4.60	4.00	5.10	5.00	3.70	9.50	13.70	5.20
9.1.3 Photographic, cine and optical equipment	[0.00]	[0.10]	[0.20]	[0.10]	[0.10]	[0.30]	[0.70]	0.80	1.00	2.00	0.50
9.2 Other major durables for recreation and culture	[1.60]	[0.40]	[3.60]	[4.40]	[5.70]	[0.80]	[0.60]	[1.00]	[5.00]	[9.00]	3.20
9.3 Other recreational items and equipment, gardens and pets	3.90	5.00	6.10	8.00	8.60	11.00	11.40	13.90	20.80	25.20	11.40
9.3.1 Games, toys and hobbies	0.90	1.30	1.00	1.50	2.00	2.00	2.30	2.30	3.00	3.40	2.00
9.3.2 Computer software and games	[0.10]	[0.20]	[0.70]	1.20	1.10	1.60	2.30	2.30	2.40	2.70	1.50
9.3.3 Equipment for sport, camping and open-air recreation	0.30	0.10	0.20	0.40	0.40	0.50	1.50	1.20	1.40	5.60	1.20
9.3.4 Horticultural goods, garden equipment and plants	[0.60]	[1.10]	1.80	1.90	2.30	2.40	2.30	3.20	4.10	5.00	2.50
9.3.5 Pets and pet food	2.00	2.40	2.40	3.00	2.80	4.60	2.90	4.90	9.90	8.60	4.30
9.4 Recreational and cultural services	7.00	8.00	9.40	11.90	15.00	17.40	19.90	23.50	26.30	39.20	17.80
9.4.1 Sports admissions, subscriptions, leisure class fees and equipment hire	0.90	0.90	1.70	2.60	3.60	3.80	5.90	8.30	8.70	16.80	5.30
9.4.2 Cinema, theatre and museums etc.	0.50	0.70	0.40	0.80	2.00	2.10	2.70	3.10	3.80	7.40	2.40
9.4.3 TV, video, satellite rental, cable subscriptions, TV licences and the Internet	3.70	3.80	4.10	5.00	5.90	6.40	7.10	7.00	8.00	8.60	6.00
9.4.4 Miscellaneous entertainments	0.20	0.30	0.50	0.70	0.80	1.20	1.50	1.10	1.70	2.70	1.10
9.4.5 Development of film, deposit for film development, passport photos, holiday and school photos	[0.10]	[0.20]	[0.10]	[0.10]	0.20	[0.80]	0.20	0.40	0.40	1.10	0.40
9.4.6 Gambling payments	1.60	2.00	2.80	2.60	2.50	3.10	2.60	3.60	3.60	2.50	2.70
9.5 Newspapers, books and stationery	2.80	3.60	4.40	4.40	5.60	5.40	6.60	6.40	8.40	11.50	5.90
9.5.1 Books	0.50	0.60	0.60	0.80	1.10	1.00	1.50	1.50	2.30	4.00	1.40
9.5.2 Diaries, address books, cards etc.	0.70	0.90	1.00	1.20	1.70	1.70	2.40	2.30	3.00	3.70	1.80
9.5.3 Newspapers	1.10	1.60	2.00	1.70	1.80	1.90	1.80	1.50	1.90	2.30	1.80
9.5.4 Magazines and periodicals	0.40	0.60	0.70	0.70	0.90	0.80	1.00	1.10	1.20	1.60	0.90
9.6 Package holidays	2.00	5.00	4.00	5.80	12.30	14.00	[13.90]	[21.70]	[19.00]	[28.40]	12.60
9.6.1 Package holidays - UK	[0.30]	1.00	[0.80]	1.10	1.40	[0.90]	1.40	1.10	0.90	1.50	1.00
9.6.2 Package holidays - abroad	[1.70]	4.00	3.20	4.70	10.80	13.10	12.40	20.60	18.10	26.90	11.60
10 Education	**[2.70]**	**[1.60]**	**[4.60]**	**5.70**	**8.50**	**3.90**	**9.30**	**7.80**	**19.30**	**37.00**	**10.00**
10.1 Education fees	[2.50]	[1.60]	[4.50]	5.40	8.30	3.60	9.00	7.30	19.00	36.60	9.80
10.2 Payments for school trips, other ad-hoc expenditure	[0.20]	[0.00]	[0.20]	[0.30]	[0.20]	[0.20]	[0.30]	[0.50]	[0.30]	[0.40]	0.30
11 Restaurants & hotels	**10.50**	**12.40**	**16.60**	**23.30**	**30.30**	**35.80**	**44.80**	**54.10**	**67.60**	**97.10**	**39.20**
11.1 Catering services	10.10	10.70	14.40	19.20	26.90	30.20	38.00	45.10	52.30	72.70	31.90
11.1.1 Restaurant and café meals	4.30	5.20	6.90	8.90	12.00	12.20	16.60	18.70	21.60	34.20	14.00
11.1.2 Alcoholic drinks (away from home)	2.60	1.90	3.00	4.20	5.90	7.90	8.20	11.10	12.70	15.50	7.30
11.1.3 Take away meals eaten at home	1.60	2.30	2.20	2.80	4.00	4.00	5.00	5.40	5.90	7.30	4.10
11.1.4 Other take-away and snack food	1.20	1.10	1.80	2.40	3.70	4.10	5.80	6.50	7.00	9.30	4.30
11.1.5 Contract catering (food) and canteens	0.40	0.30	0.50	0.90	1.40	2.10	2.30	3.40	5.10	6.40	2.30
11.2 Accommodation services	[0.50]	1.70	2.20	4.00	3.50	5.60	6.80	9.00	15.30	24.40	7.30
11.2.1 Holiday in the UK	[0.40]	1.00	1.00	2.30	2.20	2.90	2.80	5.00	5.60	8.20	3.10
11.2.2 Holiday abroad	[0.00]	[0.70]	[1.10]	1.70	1.20	2.40	4.00	3.90	9.60	16.30	4.10
11.2.3 Room hire	[0.00]	-	[0.00]	[0.00]	[0.00]	[0.30]	[0.00]	[0.20]	[0.10]	-	[0.10]

Note: The commodity and service categories are not comparable to those in publications before 2001-02.
The numbering system is sequential, it does not use actual COICOP codes.
Please see page ix for symbols and conventions used in this report.

Table A6 Detailed household expenditure by gross income decile group, 2010 (cont.)
United Kingdom

Commodity or service	Lowest ten per cent	Second decile group	Third decile group	Fourth decile group	Fifth decile group	Sixth decile group	Seventh decile group	Eighth decile group	Ninth decile group	Highest ten per cent	All house-holds
					Average weekly household expenditure (£)						
12 Miscellaneous goods & services	**12.20**	**14.10**	**19.60**	**23.50**	**29.00**	**38.40**	**37.20**	**48.30**	**54.20**	**82.60**	**35.90**
12.1 Personal care	4.10	5.00	6.80	7.70	9.50	9.80	10.80	13.40	15.60	23.00	10.60
12.1.1 Hairdressing, beauty treatment	1.40	1.60	2.20	2.30	3.10	2.90	3.40	3.40	4.70	8.00	3.30
12.1.2 Toilet paper	0.40	0.50	0.70	0.70	0.80	0.80	0.90	0.90	1.10	1.10	0.80
12.1.3 Toiletries and soap	0.80	1.10	1.30	1.80	1.90	2.10	2.50	2.70	3.20	4.90	2.20
12.1.4 Baby toiletries and accessories (disposable)	0.30	0.60	0.30	0.60	0.40	0.70	0.80	0.90	0.80	0.80	0.60
12.1.5 Hair products, cosmetics and related electrical appliances	1.20	1.30	2.40	2.20	3.30	3.20	3.30	5.50	5.70	8.20	3.60
12.2 Personal effects	0.60	0.70	1.50	1.50	1.90	2.50	2.90	5.20	5.30	8.30	3.00
12.3 Social protection	[0.70]	[0.70]	[1.10]	1.60	1.80	2.90	2.60	5.40	4.80	11.10	3.30
12.4 Insurance	6.10	6.40	8.30	10.60	12.80	15.20	17.70	19.70	22.90	31.00	15.10
12.4.1 Household insurances - structural, contents and appliances	2.30	2.70	3.40	4.20	4.80	4.90	5.30	5.80	7.60	9.10	5.00
12.4.2 Medical insurance premiums	[0.90]	0.30	0.50	0.80	1.10	1.80	1.80	2.20	2.50	6.40	1.80
12.4.3 Vehicle insurance including boat insurance	2.50	3.30	4.10	5.50	6.60	8.40	10.60	11.40	12.70	15.00	8.00
12.4.4 Non-package holiday, other travel insurance	[0.50]	[0.00]	[0.30]	[0.00]	[0.30]	[0.00]	[0.00]	[0.30]	[0.10]	[0.50]	0.20
12.5 Other services	[0.80]	1.20	1.90	2.10	3.00	7.90	3.20	4.70	5.60	9.10	4.00
12.5.1 Moving house	[0.60]	0.70	1.20	1.40	1.60	1.20	1.50	1.70	3.60	5.30	1.90
12.5.2 Bank, building society, post office, credit card charges	0.10	0.20	0.20	0.40	0.30	0.40	0.60	0.60	0.70	1.10	0.50
12.5.3 Other services and professional fees	[0.10]	0.30	0.50	0.30	1.10	6.30	1.10	2.40	1.30	2.80	1.60
1-12 All expenditure groups	**166.90**	**195.60**	**242.20**	**303.10**	**349.00**	**393.40**	**453.00**	**499.90**	**602.20**	**859.20**	**406.30**
13 Other expenditure items	**18.70**	**19.10**	**30.50**	**41.30**	**51.00**	**65.40**	**78.20**	**90.90**	**118.50**	**159.30**	**67.30**
13.1 Housing: mortgage interest payments, council tax etc.	12.10	12.60	19.10	28.20	34.60	42.20	54.10	63.30	86.30	108.80	46.10
13.2 Licences, fines and transfers	1.10	1.10	2.00	2.40	2.90	3.30	3.90	4.90	5.20	7.40	3.40
13.3 Holiday spending	[1.80]	[1.40]	[2.30]	[2.40]	3.80	7.70	9.30	7.40	14.60	20.60	7.10
13.4 Money transfers and credit	3.70	4.10	7.20	8.30	9.80	12.20	11.00	15.20	12.40	22.50	10.60
13.4.1 Money, cash gifts given to children	[0.10]	[0.10]	[0.10]	0.30	[0.10]	[0.10]	[0.20]	[0.10]	[0.10]	0.60	0.20
13.4.2 Cash gifts and donations	3.10	3.40	5.90	5.90	8.10	9.70	8.80	12.10	9.70	18.80	8.60
13.4.3 Club instalment payments (child) and interest on credit cards	0.50	0.50	1.20	2.20	1.50	2.40	2.00	3.00	2.70	3.10	1.90
Total expenditure	**185.60**	**214.80**	**272.70**	**344.40**	**400.00**	**458.80**	**531.30**	**590.80**	**720.70**	**1018.50**	**473.60**
14 Other items recorded											
14.1 Life assurance and contributions to pension funds	2.20	1.60	2.70	6.10	7.50	10.70	16.60	30.60	38.30	80.90	19.70
14.2 Other insurance inc. friendly societies	0.30	0.40	0.40	0.60	1.10	1.30	1.70	2.00	2.50	3.40	1.40
14.3 Income tax, payments less refunds	7.40	4.90	11.80	22.60	38.20	59.90	79.10	111.30	160.50	397.50	89.20
14.4 National insurance contributions	0.50	0.90	2.50	6.20	12.30	21.50	31.10	48.00	63.30	84.40	27.00
14.5 Purchase or alteration of dwellings, mortgages	5.50	5.70	12.60	14.60	26.70	28.00	41.90	59.70	93.00	142.90	43.00
14.6 Savings and investments	[0.20]	0.80	1.10	1.70	2.00	3.10	5.60	7.70	12.90	24.80	6.00
14.7 Pay off loan to clear other debt	[0.40]	[0.60]	[0.70]	1.40	1.90	2.90	2.30	4.70	4.50	3.10	2.20
14.8 Windfall receipts from gambling etc[3]	0.90	1.60	2.10	1.10	1.70	1.80	1.40	2.40	2.60	1.50	1.70

Note: The commodity and service categories are not comparable to those in publications before 2001-02.
The numbering system is sequential, it does not use actual COICOP codes.
Please see page ix for symbols and conventions used in this report.

3 Expressed as an income figure as opposed to an expenditure figure.

Table A7 **Household expenditure by disposable income decile group, 2010**
United Kingdom

	Lowest ten per cent	Second decile group	Third decile group	Fourth decile group	Fifth decile group	Sixth decile group	Seventh decile group	Eighth decile group	Ninth decile group	Highest ten per cent	All house-holds
Lower boundary of group (£ per week)		157	226	290	369	454	549	670	814	1,087	
Weighted number of households (thousands)	2,630	2,630	2,640	2,630	2,640	2,630	2,630	2,630	2,630	2,630	26,320
Total number of households in sample	510	530	530	550	550	540	530	510	520	500	5,260
Total number of persons in sample	690	830	910	1,090	1,300	1,350	1,400	1,450	1,590	1,570	12,180
Total number of adults in sample	570	670	740	890	970	1,030	1,090	1,100	1,190	1,180	9,430
Weighted average number of persons per household	1.3	1.6	1.7	2.0	2.4	2.5	2.7	2.9	3.1	3.2	2.3
Commodity or service					Average weekly household expenditure (£)						
1 Food & non-alcoholic drinks	27.50	33.70	38.80	43.80	51.20	57.00	58.90	62.50	73.90	84.60	53.20
2 Alcoholic drinks, tobacco & narcotics	7.40	7.80	8.90	9.40	11.80	14.10	11.40	13.90	15.30	18.20	11.80
3 Clothing & footwear	7.20	7.70	11.60	13.00	18.60	21.20	26.50	33.40	39.70	55.00	23.40
4 Housing(net)[1], fuel & power	43.50	47.00	49.90	54.90	63.70	63.20	64.40	66.00	68.20	83.10	60.40
5 Household goods & services	16.50	14.80	17.70	21.50	26.40	31.50	30.20	37.40	40.30	77.90	31.40
6 Health	1.40	3.00	3.40	3.10	3.80	5.30	4.60	7.20	6.50	12.10	5.00
7 Transport	18.40	19.40	26.00	44.50	51.50	59.90	71.10	91.80	109.40	156.80	64.90
8 Communication	7.00	7.20	9.60	10.20	11.10	13.30	14.70	17.10	17.70	22.20	13.00
9 Recreation & culture	20.80	25.40	30.30	43.30	46.90	55.30	60.30	76.40	91.20	131.10	58.10
10 Education	[2.70]	[2.80]	[2.50]	5.00	3.70	10.40	5.40	10.90	21.20	35.60	10.00
11 Restaurants & hotels	11.60	12.10	17.60	21.50	29.50	36.10	44.00	53.60	70.90	95.50	39.20
12 Miscellaneous goods & services	12.70	13.70	19.80	23.60	28.10	37.80	36.60	49.00	54.70	83.00	35.90
1-12 All expenditure groups	176.70	194.60	236.10	293.80	346.40	405.20	428.10	519.30	608.90	855.00	406.30
13 Other expenditure items	20.90	20.40	32.40	40.60	51.80	63.20	75.50	96.40	117.50	154.10	67.30
Total expenditure	**197.60**	**215.00**	**268.50**	**334.40**	**398.20**	**468.40**	**503.70**	**615.70**	**726.40**	**1009.10**	**473.60**
Average weekly expenditure per person (£)											
Total expenditure	**149.30**	**137.10**	**157.00**	**167.90**	**167.70**	**184.50**	**188.90**	**214.60**	**235.10**	**316.70**	**203.10**

Note: The commodity and service categories are not comparable to those in publications before 2001-02.
    Please see page ix for symbols and conventions used in this report.

1 Excluding mortgage interest payments, council tax and Northern Ireland rates.

Table A8 **Household expenditure as a percentage of total expenditure by disposable income decile group, 2010**
United Kingdom

	Lowest ten per cent	Second decile group	Third decile group	Fourth decile group	Fifth decile group	Sixth decile group	Seventh decile group	Eighth decile group	Ninth decile group	Highest ten per cent	All households
Lower boundary of group (£ per week)		157	226	290	369	454	549	670	814	1,087	
Weighted number of households (thousands)	2,630	2,630	2,640	2,630	2,640	2,630	2,630	2,630	2,630	2,630	26,320
Total number of households in sample	510	530	530	550	550	540	530	510	520	500	5,260
Total number of persons in sample	690	830	910	1,090	1,300	1,350	1,400	1,450	1,590	1,570	12,180
Total number of adults in sample	570	670	740	890	970	1,030	1,090	1,100	1,190	1,180	9,430
Weighted average number of persons per household	1.3	1.6	1.7	2.0	2.4	2.5	2.7	2.9	3.1	3.2	2.3
Commodity or service					Percentage of total expenditure						
1 Food & non-alcoholic drinks	14	16	14	13	13	12	12	10	10	8	11
2 Alcoholic drinks, tobacco & narcotics	4	4	3	3	3	3	2	2	2	2	2
3 Clothing & footwear	4	4	4	4	5	5	5	5	5	5	5
4 Housing(net)1, fuel & power	22	22	19	16	16	13	13	11	9	8	13
5 Household goods & services	8	7	7	6	7	7	6	6	6	8	7
6 Health	1	1	1	1	1	1	1	1	1	1	1
7 Transport	9	9	10	13	13	13	14	15	15	16	14
8 Communication	4	3	4	3	3	3	3	3	2	2	3
9 Recreation & culture	11	12	11	13	12	12	12	12	13	13	12
10 Education	[1]	[1]	[1]	1	1	2	1	2	3	4	2
11 Restaurants & hotels	6	6	7	6	7	8	9	9	10	9	8
12 Miscellaneous goods & services	6	6	7	7	7	8	7	8	8	8	8
1-12 All expenditure groups	89	91	88	88	87	87	85	84	84	85	86
13 Other expenditure items	11	9	12	12	13	13	15	16	16	15	14
Total expenditure	100	100	100	100	100	100	100	100	100	100	100

Note: The commodity and service categories are not comparable to those in publications before 2001-02.
Please see page ix for symbols and conventions used in this report.

1 Excluding mortgage interest payments, council tax and Northern Ireland rates.

Table A9 **Household expenditure by age of household reference person, 2010**
United Kingdom

	Less than 30.00	30.00 to 49.00	50.00 to 64.00	65.00 to 74.00	75.00 or over	All house-holds
Weighted number of households (thousands)	2,810	9,540	7,020	3,420	3,530	26,320
Total number of households in sample	450	1,890	1,490	800	630	5,260
Total number of persons in sample	1,060	5,640	3,200	1,380	910	12,180
Total number of adults in sample	760	3,500	2,900	1,360	910	9,430
Weighted average number of persons per household	2.4	3.0	2.2	1.8	1.4	2.3
Commodity or service	Average weekly household expenditure (£)					
1　Food & non-alcoholic drinks	39.10	59.70	59.40	49.70	37.70	53.20
2　Alcoholic drinks, tobacco & narcotics	9.90	13.60	14.00	10.40	5.50	11.80
3　Clothing & footwear	21.50	31.20	25.40	15.10	7.70	23.40
4　Housing(net)1, fuel & power	91.70	68.40	52.00	47.50	42.90	60.40
5　Household goods & services	22.10	35.20	39.60	24.70	18.90	31.40
6　Health	1.70	4.30	6.50	6.40	5.20	5.00
7　Transport	57.00	80.50	76.90	49.00	20.20	64.90
8　Communication	13.70	15.70	13.70	9.70	7.00	13.00
9　Recreation & culture	40.60	65.50	72.00	54.90	27.50	58.10
10　Education	25.90	12.00	10.20	[0.40]	[1.00]	10.00
11　Restaurants & hotels	39.40	49.10	44.40	27.10	14.10	39.20
12　Miscellaneous goods & services	31.80	44.20	36.60	28.70	22.40	35.90
1-12　All expenditure groups	394.50	479.40	450.90	323.50	210.00	406.30
13　Other expenditure items	54.00	93.70	67.70	41.60	30.50	67.30
Total expenditure	**448.40**	**573.10**	**518.60**	**365.20**	**240.40**	**473.60**
Average weekly expenditure per person (£)						
Total expenditure	**190.10**	**193.70**	**233.10**	**208.20**	**173.40**	**203.10**

Note: The commodity and service categories are not comparable to those in publications before 2001-02.
　　　Please see page ix for symbols and conventions used in this report.

1 Excluding mortgage interest payments, council tax and Northern Ireland rates.

Table A10 **Household expenditure as a percentage of total expenditure by age of household reference person, 2010**

United Kingdom

	Less than 30.00	30.00 to 49.00	50.00 to 64.00	65.00 to 74.00	75.00 or over	All house- holds
Weighted number of households (thousands)	2,810	9,540	7,020	3,420	3,530	26,320
Total number of households in sample	450	1,890	1,490	800	630	5,260
Total number of persons in sample	1,060	5,640	3,200	1,380	910	12,180
Total number of adults in sample	760	3,500	2,900	1,360	910	9,430
Weighted average number of persons per household	2.4	3.0	2.2	1.8	1.4	2.3

Commodity or service	Percentage of total expenditure					
1 Food & non-alcoholic drinks	9	10	11	14	16	11
2 Alcoholic drinks, tobacco & narcotics	2	2	3	3	2	2
3 Clothing & footwear	5	5	5	4	3	5
4 Housing(net)1, fuel & power	20	12	10	13	18	13
5 Household goods & services	5	6	8	7	8	7
6 Health	0	1	1	2	2	1
7 Transport	13	14	15	13	8	14
8 Communication	3	3	3	3	3	3
9 Recreation & culture	9	11	14	15	11	12
10 Education	6	2	2	[0]	[0]	2
11 Restaurants & hotels	9	9	9	7	6	8
12 Miscellaneous goods & services	7	8	7	8	9	8
1-12 All expenditure groups	88	84	87	89	87	86
13 Other expenditure items	12	16	13	11	13	14
Total expenditure	**100**	**100**	**100**	**100**	**100**	**100**

Note: The commodity and service categories are not comparable to those in publications before 2001-02.

Please see page ix for symbols and conventions used in this report.

1 Excluding mortgage interest payments, council tax and Northern Ireland rates.

Table A11 **Detailed household expenditure by age of household reference person, 2010**
United Kingdom

	Less than 30	30 to 49	50 to 64	65 to 74	75 or over	All house-holds
Weighted number of households (thousands)	2,810	9,540	7,020	3,420	3,530	26,320
Total number of households in sample	450	1,890	1,490	800	630	5,260
Total number of persons in sample	1,060	5,640	3,200	1,380	910	12,180
Total number of adults in sample	760	3,500	2,900	1,360	910	9,430
Weighted average number of persons per household	2.4	3.0	2.2	1.8	1.4	2.3

Commodity or service	Average weekly household expenditure (£)					
1 Food & non-alcoholic drinks	**39.10**	**59.70**	**59.40**	**49.70**	**37.70**	**53.20**
1.1 Food	35.40	54.50	54.70	46.30	35.40	48.90
1.1.1 Bread, rice and cereals	4.10	6.00	5.20	4.30	3.10	5.00
1.1.2 Pasta products	0.50	0.60	0.40	0.20	0.10	0.40
1.1.3 Buns, cakes, biscuits etc.	2.10	3.50	3.50	3.20	2.70	3.20
1.1.4 Pastry (savoury)	0.80	1.00	0.70	0.40	0.30	0.70
1.1.5 Beef (fresh, chilled or frozen)	0.90	1.80	2.10	1.90	1.10	1.70
1.1.6 Pork (fresh, chilled or frozen)	0.30	0.60	0.90	0.80	0.60	0.60
1.1.7 Lamb (fresh, chilled or frozen)	0.30	0.60	0.90	0.80	0.60	0.70
1.1.8 Poultry (fresh, chilled or frozen)	1.50	2.40	2.40	1.60	0.80	2.00
1.1.9 Bacon and ham	0.50	1.00	1.30	1.10	0.70	1.00
1.1.10 Other meat and meat preparations	4.30	6.00	6.30	5.50	4.40	5.60
1.1.11 Fish and fish products	1.20	2.30	2.60	2.80	2.20	2.30
1.1.12 Milk	2.10	2.80	2.60	2.50	2.10	2.60
1.1.13 Cheese and curd	1.40	2.00	2.20	1.60	1.10	1.80
1.1.14 Eggs	0.50	0.70	0.70	0.60	0.50	0.70
1.1.15 Other milk products	1.40	2.20	2.00	1.80	1.50	1.90
1.1.16 Butter	0.20	0.30	0.40	0.50	0.50	0.40
1.1.17 Margarine, other vegetable fats and peanut butter	0.30	0.50	0.60	0.60	0.50	0.50
1.1.18 Cooking oils and fats	0.20	0.30	0.40	0.30	0.20	0.30
1.1.19 Fresh fruit	1.80	3.20	3.50	3.20	2.80	3.10
1.1.20 Other fresh, chilled or frozen fruits	0.30	0.40	0.40	0.30	0.20	0.40
1.1.21 Dried fruit and nuts	0.30	0.50	0.70	0.60	0.50	0.60
1.1.22 Preserved fruit and fruit based products	0.10	0.10	0.10	0.20	0.20	0.10
1.1.23 Fresh vegetables	2.70	4.50	4.60	3.80	2.70	4.00
1.1.24 Dried vegetables	[0.00]	0.00	0.00	0.00	0.00	0.00
1.1.25 Other preserved or processed vegetables	1.00	1.60	1.50	1.00	0.60	1.30
1.1.26 Potatoes	0.50	0.90	1.00	1.00	0.80	0.90
1.1.27 Other tubers and products of tuber vegetables	1.40	1.70	1.40	1.00	0.70	1.40
1.1.28 Sugar and sugar products	0.20	0.40	0.40	0.40	0.20	0.30
1.1.29 Jams, marmalades	0.10	0.20	0.30	0.40	0.40	0.30
1.1.30 Chocolate	1.20	1.90	1.90	1.30	1.20	1.60
1.1.31 Confectionery products	0.40	0.70	0.60	0.50	0.40	0.60
1.1.32 Edible ices and ice cream	0.40	0.60	0.50	0.50	0.30	0.50
1.1.33 Other food products	2.20	2.90	2.40	1.90	1.30	2.40
1.2 Non-alcoholic drinks	3.80	5.30	4.70	3.40	2.30	4.30
1.2.1 Coffee	0.30	0.60	0.80	0.70	0.40	0.60
1.2.2 Tea	0.20	0.50	0.60	0.60	0.50	0.50
1.2.3 Cocoa and powdered chocolate	0.10	0.10	0.10	0.10	0.10	0.10
1.2.4 Fruit and vegetable juices (inc. fruit squash)	1.10	1.40	1.20	0.80	0.60	1.10
1.2.5 Mineral or spring waters	0.10	0.30	0.30	0.10	0.10	0.20
1.2.6 Soft drinks (inc. fizzy and ready to drink fruit drinks)	1.90	2.30	1.80	1.10	0.70	1.80

Note: The commodity and service categories are not comparable to those in publications before 2001-02.
The numbering system is sequential, it does not use actual COICOP codes.
Please see page ix for symbols and conventions used in this report.

Table A11 **Detailed household expenditure by age of household reference person, 2010 (cont.)**
United Kingdom

		Less than 30	30 to 49	50 to 64	65 to 74	75 or over	All house-holds
Commodity or service		Average weekly household expenditure (£)					
2	**Alcoholic drink, tobacco & narcotics**	**9.90**	**13.60**	**14.00**	**10.40**	**5.50**	**11.80**
2.1	Alcoholic drinks	4.60	8.40	8.50	6.50	4.10	7.20
2.1.1	Spirits and liqueurs (brought home)	0.90	1.40	1.80	1.60	1.30	1.50
2.1.2	Wines, fortified wines (brought home)	1.80	4.30	4.60	3.40	2.20	3.70
2.1.3	Beer, lager, ciders and perry (brought home)	1.80	2.60	2.00	1.40	0.60	1.90
2.1.4	Alcopops (brought home)	[0.10]	0.10	0.10	[0.00]	-	0.10
2.2	Tobacco and narcotics	5.30	5.20	5.50	3.90	1.40	4.60
2.2.1	Cigarettes	4.40	4.40	4.80	3.10	1.20	3.90
2.2.2	Cigars, other tobacco products and narcotics	0.90	0.80	0.70	0.80	0.20	0.70
3	**Clothing & footwear**	**21.50**	**31.20**	**25.40**	**15.10**	**7.70**	**23.40**
3.1	Clothing	16.80	24.90	20.20	12.40	6.20	18.60
3.1.1	Men's outer garments	5.10	6.70	4.70	3.20	0.80	4.80
3.1.2	Men's under garments	[0.20]	0.40	0.50	0.30	0.30	0.40
3.1.3	Women's outer garments	7.40	9.70	10.60	6.30	3.50	8.40
3.1.4	Women's under garments	0.70	1.30	1.40	0.80	0.80	1.10
3.1.5	Boys' outer garments (5-15)	[0.30]	1.80	0.40	[0.30]	[0.00]	0.80
3.1.6	Girls' outer garments (5-15)	0.60	1.90	0.50	0.20	[0.00]	0.90
3.1.7	Infants' outer garments (under 5)	1.10	1.00	0.30	0.40	[0.10]	0.60
3.1.8	Children's under garments (under 16)	0.50	0.80	0.20	0.10	[0.10]	0.40
3.1.9	Accessories	0.70	1.00	0.90	0.50	0.20	0.80
3.1.10	Haberdashery, clothing materials and clothing hire	[0.10]	0.20	0.30	0.20	0.10	0.20
3.1.11	Dry cleaners, laundry and dyeing	[0.10]	0.30	0.30	0.20	0.20	0.30
3.2	Footwear	4.70	6.30	5.30	2.60	1.50	4.70
4	**Housing (net)[1], fuel & power**	**91.70**	**68.40**	**52.00**	**47.50**	**42.90**	**60.40**
4.1	Actual rentals for housing	91.00	48.70	23.60	25.40	22.30	39.90
4.1.1	Gross rent	91.00	48.60	23.50	25.40	22.30	39.90
4.1.2	less housing benefit, rebates & allowances rec'd	25.10	16.70	10.80	14.00	15.30	15.50
4.1.3	Net rent[2]	65.90	31.90	12.70	11.40	7.00	24.40
4.1.4	Second dwelling rent	-	[0.00]	[0.10]	-	-	0.00
4.2	Maintenance and repair of dwelling	2.50	6.90	7.70	7.10	7.00	6.70
4.3	Water supply and miscellaneous services relating to the dwelling	7.50	7.90	7.80	7.10	8.50	7.80
4.4	Electricity, gas and other fuels	15.80	21.60	23.60	22.00	20.40	21.40
4.4.1	Electricity	8.20	10.30	11.00	9.30	9.00	9.90
4.4.2	Gas	7.00	10.00	10.50	10.00	8.70	9.60
4.4.3	Other fuels	[0.60]	1.30	2.20	2.70	2.70	1.80

Note: The commodity and service categories are not comparable to those in publications before 2001-02.
The numbering system is sequential, it does not use actual COICOP codes.
Please see page ix for symbols and conventions used in this report.

1 Excluding mortgage interest payments, council tax and Northern Ireland rates.
2 The figure included in total expenditure is net rent as opposed to gross rent

Table A11 Detailed household expenditure by age of household reference person, 2010 (cont.)
United Kingdom

Commodity or service	Less than 30	30 to 49	50 to 64	65 to 74	75 or over	All house-holds
	Average weekly household expenditure (£)					
5 Household goods & services	**22.10**	**35.20**	**39.60**	**24.70**	**18.90**	**31.40**
5.1 Furniture and furnishings and floor coverings	14.70	20.50	19.90	10.40	7.30	16.60
5.1.1 Furniture and furnishings	13.10	16.30	16.40	7.80	5.60	13.40
5.1.2 Floor coverings	1.60	4.10	3.50	2.60	1.60	3.20
5.2 Household textiles	[1.40]	2.30	1.90	[1.80]	0.70	1.80
5.3 Household appliances	1.30	3.00	7.60	2.90	1.60	3.80
5.4 Glassware, tableware and household utensils	0.90	1.60	1.90	1.10	0.70	1.40
5.5 Tools and equipment for house and garden	1.30	2.40	2.50	1.30	0.70	1.90
5.6 Goods and services for routine household maintenance	2.40	5.40	5.90	7.10	7.90	5.80
5.6.1 Cleaning materials	1.40	2.60	2.40	2.20	1.50	2.20
5.6.2 Household goods and hardware	0.90	1.40	1.60	1.10	0.80	1.30
5.6.3 Domestic services, carpet cleaning, hire of furniture/furnishings	0.10	1.50	1.90	3.80	5.60	2.30
6 Health	**1.70**	**4.30**	**6.50**	**6.40**	**5.20**	**5.00**
6.1 Medical products, appliances and equipment	[1.10]	2.70	4.10	[4.30]	[3.10]	3.10
6.1.1 Medicines, prescriptions and healthcare products	0.90	1.60	2.40	1.90	1.40	1.80
6.1.2 Spectacles, lenses, accessories and repairs	0.20	1.00	1.60	2.40	[1.70]	1.40
6.2 Hospital services	0.60	1.70	2.50	2.10	[2.10]	1.90
7 Transport	**[57.00]**	**80.50**	**76.90**	**49.00**	**20.20**	**64.90**
7.1 Purchase of vehicles	15.80	26.60	20.20	16.20	5.20	19.50
7.1.1 Purchase of new cars and vans	2.20	8.00	6.80	8.60	3.30	6.50
7.1.2 Purchase of second hand cars or vans	13.00	17.20	12.50	7.30	1.90	12.20
7.1.3 Purchase of motorcycles and other vehicles	0.70	1.40	0.80	0.30	0.00	0.80
7.20 Operation of personal transport	26.40	40.40	40.90	26.40	11.30	33.30
7.2.1 Spares and accessories	1.30	2.70	2.80	1.40	0.30	2.10
7.2.2 Petrol, diesel and other motor oils	17.90	26.40	26.30	16.80	7.10	21.60
7.2.3 Repairs and servicing	5.20	8.30	8.60	5.80	3.00	7.00
7.2.4 Other motoring costs	1.90	3.00	3.20	2.40	0.90	2.60
7.3 Transport services	14.80	13.60	15.90	6.40	3.60	12.10
7.3.1 Rail and tube fares	3.30	3.30	3.50	1.10	0.80	2.80
7.3.2 Bus and coach fares	2.30	1.90	1.80	0.30	0.10	1.50
7.3.4 Combined fares	1.60	2.00	1.20	0.20	0.00	1.20
7.3.5 Other travel and transport	7.60	6.40	9.40	4.80	2.70	6.60
8 Communication	**13.70**	**15.70**	**13.70**	**9.70**	**7.00**	**13.00**
8.1 Postal services	0.40	0.50	0.50	0.50	0.60	0.50
8.2 Telephone and telefax equipment	0.60	0.90	0.80	0.30	0.10	0.60
8.3 Telephone and telefax services	12.20	13.90	12.20	8.80	6.20	11.60
8.4 Internet subscription fees	0.50	0.50	0.30	0.20	0.10	0.30

Note: The commodity and service categories are not comparable to those in publications before 2001-02.
The numbering system is sequential, it does not use actual COICOP codes.
Please see page ix for symbols and conventions used in this report.

Table A11 **Detailed household expenditure by age of household reference person, 2010 (cont.)**
United Kingdom

Commodity or service	Less than 30	30 to 49	50 to 64	65 to 74	75 or over	All house-holds
	Average weekly household expenditure (£)					
9 Recreation & culture	**40.60**	**65.50**	**72.00**	**54.90**	**27.50**	**58.10**
9.1 Audio-visual, photographic and information processing equipment	5.60	7.90	9.30	5.10	4.40	7.20
9.1.1 Audio equipment and accessories, CD players	0.80	2.00	1.70	1.10	0.30	1.40
9.1.2 TV, video and computers	4.70	5.10	7.00	3.60	4.00	5.20
9.1.3 Photographic, cine and optical equipment	[0.10]	0.80	0.60	[0.40]	[0.20]	0.50
9.2 Other major durables for recreation and culture	[0.50]	2.90	7.00	[1.00]	[0.90]	3.20
9.3 Other recreational items and equipment, gardens and pets	6.80	13.70	14.40	9.70	4.50	11.40
9.3.1 Games, toys and hobbies	1.80	3.10	1.70	1.10	0.40	2.00
9.3.2 Computer software and games	1.30	2.40	1.20	[0.80]	[0.20]	1.50
9.3.3 Equipment for sport, camping and open-air recreation	0.40	1.50	2.00	0.40	[0.00]	1.20
9.3.4 Horticultural goods, garden equipment and plants	1.00	2.10	3.20	3.20	2.40	2.50
9.3.5 Pets and pet food	2.40	4.60	6.30	4.20	1.50	4.30
9.4 Recreational and cultural services	13.90	22.10	19.90	15.80	6.70	17.80
9.4.1 Sports admissions, subscriptions, leisure class fees and equipment hire	4.30	8.10	5.20	3.10	1.10	5.30
9.4.2 Cinema, theatre and museums etc.	1.40	3.00	2.60	2.10	1.10	2.40
9.4.3 TV, video, satellite rental, cable subscriptions, TV licences	6.10	7.10	6.80	5.60	[1.50]	6.00
9.4.4 Miscellaneous entertainments	0.60	1.20	1.30	0.90	[0.80]	1.10
9.4.5 Development of film, deposit for film development, passport photos, holiday and school photos	[0.30]	0.50	0.20	0.50	[0.10]	0.40
9.4.6 Gambling payments	1.30	2.20	3.70	3.60	2.20	2.70
9.5 Newspapers, books and stationery	3.00	5.90	6.90	6.60	5.50	5.90
9.5.1 Books	1.10	1.80	1.50	1.00	0.80	1.40
9.5.2 Diaries, address books, cards etc.	1.00	2.10	2.30	1.70	1.20	1.80
9.5.3 Newspapers	0.30	1.00	2.10	3.00	2.90	1.80
9.5.4 Magazines and periodicals	0.60	1.00	1.00	1.00	0.60	0.90
9.6 Package holidays	10.90	12.90	14.40	16.70	5.40	12.60
9.6.1 Package holidays - UK	[0.20]	0.80	1.00	1.90	1.70	1.00
9.6.2 Package holidays - abroad	10.70	12.10	13.50	14.90	3.70	11.60
10 Education	**25.90**	**12.00**	**10.20**	**[0.40]**	**[1.00]**	**10.00**
10.1 Education fees	25.80	11.50	10.00	[0.40]	[1.00]	9.80
10.2 Payments for school trips, other ad-hoc expenditure	0.10	0.50	0.20	[0.00]	-	0.30
11.0 Restaurants & hotels	**39.40**	**49.10**	**44.40**	**27.10**	**14.10**	**39.20**
11.1 Catering services	35.10	39.90	35.10	21.90	11.30	31.90
11.1.1 Restaurant and café meals	11.10	16.10	16.30	12.80	7.50	14.00
11.1.2 Alcoholic drinks (away from home)	8.50	8.40	8.90	5.50	1.90	7.30
11.1.3 Take away meals eaten at home	5.40	5.50	3.90	1.90	1.40	4.10
11.1.4 Other take-away and snack food	4.90	6.70	4.00	1.50	0.50	4.30
11.1.5 Contract catering (food) and canteens	5.30	3.30	1.90	0.20	[0.10]	2.30
11.2 Accommodation services	4.20	9.20	9.30	5.10	2.80	7.30
11.2.1 Holiday in the UK	1.40	3.50	3.80	3.50	2.10	3.10
11.2.2 Holiday abroad	2.50	5.60	5.50	1.70	[0.70]	4.10
11.2.3 Room hire	[0.30]	[0.10]	[0.00]	-	[0.00]	[0.10]

Note: The commodity and service categories are not comparable to those in publications before 2001-02.
The numbering system is sequential, it does not use actual COICOP codes.
Please see page ix for symbols and conventions used in this report.

Table A11 Detailed household expenditure by age of household reference person, 2010 (cont.)
United Kingdom

Commodity or service	Less than 30	30 to 49	50 to 64	65 to 74	75 or over	All house-holds
	Average weekly household expenditure (£)					
12 Miscellaneous goods & services	**31.80**	**44.20**	**36.60**	**28.70**	**22.40**	**35.90**
12.1 Personal care	9.20	12.60	11.60	8.00	6.50	10.60
12.1.1 Hairdressing, beauty treatment	1.90	3.40	4.00	3.10	3.10	3.30
12.1.2 Toilet paper	0.60	0.90	0.90	0.70	0.60	0.80
12.1.3 Toiletries and soap	1.80	2.80	2.40	1.70	1.20	2.20
12.1.4 Baby toiletries and accessories (disposable)	1.50	1.10	0.20	0.10	0.10	0.60
12.1.5 Hair products, cosmetics and related electrical appliances	3.30	4.60	4.00	2.40	1.60	3.60
12.2 Personal effects	3.30	3.60	4.00	1.70	0.90	3.00
12.3 Social protection	4.40	6.70	[0.40]	[0.10]	1.80	3.30
12.4 Insurance	11.50	16.30	17.40	13.30	11.60	15.10
12.4.1 Household insurances - structural, contents and appliances	2.30	5.30	6.00	5.20	[4.40]	5.00
12.4.2 Medical insurance premiums	0.50	1.40	2.40	2.20	2.60	1.80
12.4.3 Vehicle insurance including boat insurance	8.70	9.50	8.90	5.60	3.90	8.00
12.4.4 Non-package holiday, other travel insurance	[0.00]	[0.20]	[0.10]	[0.30]	[0.60]	0.20
12.5 Other services	3.40	4.90	3.20	5.70	1.60	4.00
12.5.1 Moving house	1.60	2.70	1.60	1.70	[0.60]	1.90
12.5.2 Bank, building society, post office, credit card charges	0.50	0.70	0.40	0.30	0.10	0.50
12.5.3 Other services and professional fees	1.20	1.50	1.20	3.70	[0.90]	1.60
1-12 All expenditure groups	**394.50**	**479.40**	**450.90**	**323.50**	**210.00**	**406.30**
13.0 Other expenditure items	**54.00**	**93.70**	**67.70**	**41.60**	**30.50**	**67.30**
13.1 Housing: mortgage interest payments, council tax etc.	38.00	68.80	43.40	24.60	17.40	46.10
13.2 Licences, fines and transfers	2.80	3.90	4.10	3.00	1.70	3.40
13.3 Holiday spending	7.90	8.40	9.20	4.50	1.40	7.10
13.4 Money transfers and credit	5.30	12.60	10.90	9.50	9.90	10.60
13.4.1 Money, cash gifts given to children	[0.10]	0.30	[0.20]	[0.00]	[0.00]	0.20
13.4.2 Cash gifts and donations	3.20	9.60	8.70	8.90	9.60	8.60
13.4.3 Club instalment payments (child) and interest on credit cards	2.00	2.70	2.10	0.60	0.40	1.90
Total expenditure	**448.40**	**573.10**	**518.60**	**[365.20]**	**[240.40]**	**473.60**
14.0 Other items recorded						
14.1 Life assurance & contributions to pension funds	9.00	29.60	27.20	4.10	1.60	19.70
14.2 Other insurance inc. friendly societies	0.60	1.90	1.60	0.70	0.60	1.40
14.3 Income tax, payments less refunds	62.10	129.00	104.70	35.70	24.50	89.20
14.4 National insurance contributions	29.30	42.90	29.90	2.40	[0.50]	27.00
14.5 Purchase or alteration of dwellings, mortgages	20.50	62.50	52.70	22.10	9.20	43.00
14.6 Savings and investments	4.30	7.30	8.10	4.40	1.30	6.00
14.7 Pay off loan to clear other debt	3.10	3.40	2.00	[0.60]	[0.40]	2.20
14.8 Windfall receipts from gambling etc[3]	0.70	1.20	2.90	2.40	[0.90]	1.70

Note: The commodity and service categories are not comparable to those in publications before 2001-02.
The numbering system is sequential, it does not use actual COICOP codes.
Please see page ix for symbols and conventions used in this report.

3 Expressed as an income figure as opposed to an expenditure figure.

Table A12 Household expenditure as a percentage of total expenditure by age of household reference person, 2010
United Kingdom

	Less than 30	30 to 49	50 to 64	65 to 74	75 or over	All house-holds
Lower boundary of group (£ per week)[1]		238	413	651	1,015	
Average weighted number of households (thousands)	580	470	680	620	350	2,700
Total number of households in sample (over 3 years)	330	260	360	300	170	1,420
Total number of persons in sample (over 3 years)	740	650	900	760	410	3,450
Total number of adults in sample (over 3 years)	430	390	660	620	380	2,480
Weighted average number of persons per household	2.2	2.4	2.5	2.5	2.4	2.4
Commodity or service	Average weekly household expenditure (£)					
1 Food & non-alcoholic drinks	29.20	34.50	42.10	42.40	49.20	39.10
2 Alcoholic drinks, tobacco & narcotics	9.40	9.50	10.10	11.60	8.70	10.00
3 Clothing & footwear	11.00	15.50	20.00	26.50	37.20	21.10
4 Housing(net)2, fuel & power	57.30	87.00	95.70	95.50	114.40	88.40
5 Household goods & services	12.60	15.00	21.80	26.50	35.40	21.60
6 Health	0.60	1.70	2.50	3.80	6.10	2.70
7 Transport	15.10	33.40	51.00	80.00	113.50	55.10
8 Communication	6.70	10.20	14.40	16.70	18.10	13.10
9 Recreation & culture	17.30	29.10	39.50	54.00	68.80	40.10
10 Education	8.00	19.90	17.50	13.50	[26.30]	16.10
11 Restaurants & hotels	14.20	21.90	33.80	54.90	71.40	37.30
12 Miscellaneous goods & services	10.30	22.30	36.00	41.20	49.80	31.20
1-12 All expenditure groups	191.80	300.10	384.50	466.80	599.00	375.70
13 Other expenditure items	6.50	32.80	61.00	100.40	140.10	64.30
Total expenditure	**198.30**	**332.90**	**445.60**	**567.20**	**739.10**	**440.00**
Average weekly expenditure per person (£) **Total expenditure**	**90.50**	**136.00**	**180.90**	**231.20**	**302.30**	**183.60**

Note: The commodity and service categories are not comparable to those in publications before 2001-02.
    Please see page ix for symbols and conventions used in this report.

This table is based on a three year average.
1 Lower boundary of 2010 gross income quintile groups (£ per week).
2 Excluding mortgage interest payments, council tax and Northern Ireland rates.

Table A13 Household expenditure by gross income quintile group where the household reference person is aged 30 to 49, 2008-2010
United Kingdom

	Lowest twenty per cent	Second quintile group	Third quintile group	Highest twenty per cent	All house-holds
Lower boundary of group (£ per week)[1]		238	413	1,015	
Average weighted number of households (thousands)	1,060	1,300	1,860	2,900	9,650
Total number of households in sample (over 3 years)	680	880	1,230	1,820	6,220
Total number of persons in sample (over 3 years)	1,370	2,470	3,690	6,020	18,660
Total number of adults in sample (over 3 years)	840	1,340	2,210	3,950	11,610
Weighted average number of persons per household	2.0	2.8	2.9	3.3	2.9
Commodity or service	Average weekly household expenditure (£)				
1 Food & non-alcoholic drinks	33.20	46.60	51.90	75.10	58.50
2 Alcoholic drinks, tobacco & narcotics	9.40	11.20	12.50	15.60	12.90
3 Clothing & footwear	10.20	19.40	21.10	45.10	28.90
4 Housing(net)[2], fuel & power	49.60	65.30	60.50	69.80	63.50
5 Household goods & services	16.00	20.20	21.30	54.50	34.40
6 Health	1.10	2.00	4.00	7.00	4.30
7 Transport	22.20	41.30	53.90	127.50	77.10
8 Communication	8.20	11.90	13.80	18.70	14.90
9 Recreation & culture	21.10	33.50	50.00	109.90	66.90
10 Education	4.70	4.50	2.40	22.90	9.70
11 Restaurants & hotels	14.20	25.90	33.10	81.10	48.40
12 Miscellaneous goods & services	12.90	23.30	31.40	73.00	44.30
1-12 All expenditure groups	202.70	304.90	355.90	700.10	463.70
13 Other expenditure items	26.10	46.20	75.00	186.50	107.60
Total expenditure	**228.80**	**351.20**	**431.00**	**886.70**	**571.30**
Average weekly expenditure per person (£)					
Total expenditure	**117.30**	**127.50**	**148.70**	**271.60**	**194.10**

Note: The commodity and service categories are not comparable to those in publications before 2001-02.
Please see page lx for symbols and conventions used in this report.
This table is based on a three year average.

1 Lower boundary of 2010 gross income quintile groups (£ per week).
2 Excluding mortgage interest payments, council tax and Northern Ireland rates.

Table A14   **Household expenditure by gross income quintile group where the household reference person is aged 50 to 64, 2008-2010**

United Kingdom

	Lowest twenty per cent	Second quintile group	Third quintile group	Fourth quintile group	Highest twenty per cent	All house- holds
Lower boundary of group (£ per week)[1]		238	413	651	1,015	
Weighted number of households (thousands)	1,140	1,130	1,420	1,470	1,680	6,830
Total number of households in sample (over 3 years)	810	800	1,000	1,040	1,100	4,730
Total number of persons in sample (over 3 years)	1,060	1,360	2,100	2,540	3,200	10,270
Total number of adults in sample (over 3 years)	1,010	1,270	1,920	2,300	2,850	9,340
Weighted average number of persons per household	1.3	1.7	2.1	2.5	3.0	2.2
Commodity or service	Average weekly household expenditure (£)					
1 Food & non-alcoholic drinks	30.30	42.20	55.30	63.80	81.20	57.20
2 Alcoholic drinks, tobacco & narcotics	8.90	10.20	13.00	15.70	17.00	13.40
3 Clothing & footwear	7.30	12.20	19.60	25.60	46.00	24.10
4 Housing(net)[2], fuel & power	36.30	48.10	49.30	53.20	58.30	50.00
5 Household goods & services	14.80	23.10	29.90	36.30	58.40	34.70
6 Health	2.40	3.00	5.60	7.50	12.90	6.90
7 Transport	21.80	41.00	62.30	85.10	136.00	75.10
8 Communication	7.60	10.00	12.20	14.60	18.90	13.20
9 Recreation & culture	26.20	43.60	54.50	76.00	127.60	70.50
10 Education	[0.80]	4.10	3.60	7.90	21.50	8.50
11 Restaurants & hotels	12.20	20.80	34.80	46.90	85.90	43.80
12 Miscellaneous goods & services	14.00	21.40	30.20	40.50	65.10	36.80
1-12 All expenditure groups	182.70	279.70	370.40	473.20	728.80	434.30
13 Other expenditure items	21.20	41.60	56.20	79.30	125.80	70.10
Total expenditure	**203.90**	**321.40**	**426.60**	**552.40**	**854.60**	**504.40**
Average weekly expenditure per person (£)						
Total expenditure	**153.70**	**186.60**	**198.40**	**221.60**	**282.60**	**226.10**

Note: The commodity and service categories are not comparable to those in publications before 2001-02.
Please see page ix for symbols and conventions used in this report.
This table is based on a three year average.

1 Lower boundary of 2010 gross income quintile groups (£ per week).
2 Excluding mortgage interest payments, council tax and Northern Ireland rates.

Table A15   **Household expenditure by gross income quintile group where the household reference person is aged 65 to 74, 2008-2010**

United Kingdom

	Lowest twenty per cent	Second quintile group	Third quintile group	Fourth quintile group	Highest twenty per cent	All house-holds
Lower boundary of group (£ per week)[1]		238	413	651	1,015	
Average weighted number of households (thousands)	890	1,060	740	400	200	3,290
Total number of households in sample (over 3 years)	630	830	550	280	140	2,430
Total number of persons in sample (over 3 years)	770	1,410	1,070	650	340	4,230
Total number of adults in sample (over 3 years)	760	1,410	1,060	620	320	4,170
Weighted average number of persons per household	1.2	1.7	2.0	2.4	2.7	1.8
Commodity or service	Average weekly household expenditure (£)					
1 Food & non-alcoholic drinks	31.60	45.80	54.80	73.40	78.10	49.30
2 Alcoholic drinks, tobacco & narcotics	5.90	9.20	11.10	14.50	15.90	9.70
3 Clothing & footwear	5.60	9.90	16.60	24.90	32.00	13.50
4 Housing(net)[2], fuel & power	36.70	41.40	46.30	50.90	69.30	44.20
5 Household goods & services	11.40	20.90	27.60	44.60	57.90	24.80
6 Health	2.40	4.20	7.90	7.70	11.80	5.40
7 Transport	15.90	35.40	54.70	78.70	115.60	44.60
8 Communication	6.40	8.20	9.90	12.90	16.40	9.20
9 Recreation & culture	23.50	42.80	82.20	88.40	114.50	55.90
10 Education	[0.10]	0.90	0.50	2.30	5.30	1.00
11 Restaurants & hotels	9.70	19.70	30.70	47.00	74.60	26.00
12 Miscellaneous goods & services	12.00	20.80	32.40	37.30	66.90	25.90
1-12 All expenditure groups	161.30	259.20	374.70	482.80	658.40	309.60
13 Other expenditure items	17.30	40.50	45.50	83.70	95.60	43.90
Total expenditure	**178.60**	**299.70**	**420.20**	**566.50**	**754.00**	**353.50**
Average weekly expenditure per person (£) **Total expenditure**	**147.70**	**178.60**	**214.30**	**233.10**	**277.00**	**199.90**

Note: The commodity and service categories are not comparable to those in publications before 2001-02.

Please see page ix for symbols and conventions used in this report.

This table is based on a three year average.

1 Lower boundary of 2010 gross income quintile groups (£ per week).
2 Excluding mortgage interest payments, council tax and Northern Ireland rates.

Table A16   Household expenditure by gross income quintile group where the household
             reference person is aged 75 or over, 2008-2010

United Kingdom

	Lowest twenty per cent	Second quintile group	Third quintile group	Fourth quintile group	Highest twenty per cent	All house-holds
Lower boundary of group (£ per week)[1]		238	413	651	1,015	
Average weighted number of households (thousands)	1,540	1,230	500	180	70	3,520
Total number of households in sample (over 3 years)	850	780	330	110	50	2,130
Total number of persons in sample (over 3 years)	970	1,210	600	240	110	3,130
Total number of adults in sample (over 3 years)	970	1,200	600	240	110	3,120
Weighted average number of persons per household	1.1	1.5	1.8	2.2	2.5	1.4
Commodity or service	Average weekly household expenditure (£)					
1 Food & non-alcoholic drinks	27.60	37.50	49.20	59.20	78.00	36.80
2 Alcoholic drinks, tobacco & narcotics	3.10	4.90	8.30	8.90	14.80	5.00
3 Clothing & footwear	4.80	7.60	11.70	11.40	23.80	7.40
4 Housing (net)[2], fuel & power	33.80	41.50	42.50	55.20	87.10	39.90
5 Household goods & services	11.50	18.80	27.20	34.20	78.70	18.90
6 Health	3.90	5.00	9.50	9.50	15.60	5.70
7 Transport	7.70	19.60	32.40	38.90	77.50	18.40
8 Communication	5.40	6.30	8.10	10.20	14.50	6.50
9 Recreation & culture	15.90	25.10	54.30	66.70	87.70	28.40
10 Education	[0.00]	[0.70]	[1.30]	[2.50]	[0.70]	0.60
11 Restaurants & hotels	7.10	12.50	21.40	31.90	53.70	13.30
12 Miscellaneous goods & services	13.70	18.10	31.80	58.70	76.30	21.30
1-12 All expenditure groups	134.60	197.60	297.50	387.30	608.40	202.10
13 Other expenditure items	15.50	28.90	44.80	86.30	78.80	29.10
Total expenditure	**150.00**	**226.50**	**342.30**	**473.60**	**687.20**	**231.20**
Average weekly expenditure per person (£) **Total expenditure**	**135.90**	**153.50**	**194.10**	**218.60**	**273.20**	**163.90**

Note: The commodity and service categories are not comparable to those in publications before 2001-02.
Please see page ix for symbols and conventions used in this report.
This table is based on a three year average.

1 Lower boundary of 2010 gross income quintile groups (£ per week).
2 Excluding mortgage interest payments, council tax and Northern Ireland rates.

Table A17   **Household expenditure by economic activity status of the household reference person, 2010**

United Kingdom

Commodity or service	Employees			Self-employed	All in employment[1]	Unemployed	All economically active[1]	Economically inactive			All households
	Full-time	Part-time	All					Retired	Other	All	
Weighted number of households (thousands)	10,960	2,320	13,290	2,250	15,560	940	16,490	6,850	2,980	9,830	26,320
Total number of households in sample	2,130	480	2,610	460	3,070	180	3,250	1,400	610	2,020	5,260
Total number of persons in sample	5,690	1,150	6,830	1,250	8,090	440	8,530	2,170	1,480	3,650	12,180
Total number of adults in sample	4,240	840	5,070	920	6,000	270	6,270	2,150	1,010	3,160	9,430
Weighted average number of persons per household	2.7	2.5	2.6	2.7	2.7	2.4	2.6	1.5	2.5	1.8	2.3
Commodity or service					Average weekly household expenditure (£)						
1 Food & non-alcoholic drinks	60.00	51.30	58.50	68.20	59.80	40.30	58.70	41.80	48.80	43.90	53.20
2 Alcoholic drinks, tobacco & narcotics	13.90	10.90	13.30	13.50	13.40	12.00	13.30	7.60	13.40	9.40	11.80
3 Clothing & footwear	31.60	21.10	29.80	34.00	30.30	13.20	29.40	10.70	19.30	13.30	23.40
4 Housing (net)[2], fuel & power	70.50	72.40	70.90	63.10	69.70	54.50	68.80	44.20	50.60	46.20	60.40
5 Household goods & services	37.70	24.40	35.40	48.10	37.20	22.10	36.40	21.60	26.70	23.10	31.40
6 Health	5.60	4.40	5.40	5.70	5.40	0.90	5.10	5.60	3.00	4.80	5.00
7 Transport	90.60	55.10	84.40	88.00	84.80	47.40	82.70	31.30	43.30	34.90	64.90
8 Communication	16.20	13.40	15.70	15.10	15.60	11.10	15.30	8.20	11.40	9.10	13.00
9 Recreation & culture	72.10	54.20	69.00	76.10	69.90	30.30	67.70	38.80	49.20	42.00	58.10
10 Education	13.10	15.60	13.50	11.50	13.20	[3.80]	12.70	[0.50]	17.30	5.60	10.00
11 Restaurants & hotels	54.30	31.50	50.30	58.20	51.40	21.00	49.70	19.60	26.50	21.70	39.20
12 Miscellaneous goods & services	46.50	30.10	43.60	54.40	45.20	19.30	43.70	22.90	22.70	22.80	35.90
1-12 All expenditure groups	512.00	384.10	489.70	535.90	496.00	275.80	483.50	252.90	332.00	276.90	406.30
13 Other expenditure items	98.60	46.80	89.60	101.20	91.20	33.80	87.90	33.00	31.80	32.60	67.30
Total expenditure	**610.60**	**431.00**	**579.20**	**637.10**	**587.20**	**309.60**	**571.40**	**285.90**	**363.80**	**309.50**	**473.60**
Average weekly expenditure per person (£)											
Total expenditure	**227.60**	**175.00**	**219.00**	**232.20**	**220.90**	**129.30**	**216.20**	**188.30**	**147.00**	**171.10**	**203.10**

Note: The commodity and service categories are not comparable to those in publications before 2001-02.
Please see page ix for symbols and conventions used in this report.

1 Includes households where household reference person was on a Government supported training scheme.
2 Excluding mortgage interest payments, council tax and Northern Ireland rates.

Table A18   **Household expenditure by gross income: the household reference person is a full-time employee, 2010**

United Kingdom

	Lowest twenty per cent	Second quintile group	Third quintile group	Fourth quintile group	Highest twenty per cent	All house-holds
Lower boundary of group (£ per week)[1]		238	413	651	1,015	
Weighted number of households (thousands)	110	910	2,340	3,650	3,950	10,960
Total number of households in sample	20	180	460	710	760	2,130
Total number of persons in sample	30	280	1,060	1,950	2,360	5,690
Total number of adults in sample	30	240	790	1,450	1,740	4,240
Weighted average number of persons per household	1.3	1.6	2.3	2.8	3.1	2.7
Commodity or service	Average weekly household expenditure (£)					
1 Food & non-alcoholic drinks	27.40	34.30	46.80	57.10	77.30	60.00
2 Alcoholic drinks, tobacco & narcotics	[10.80]	10.40	12.50	12.60	16.70	13.90
3 Clothing & footwear	[3.70]	10.00	18.30	27.90	48.60	31.60
4 Housing (net)[2], fuel & power	66.00	70.40	68.80	67.60	74.50	70.50
5 Household goods & services	[14.00]	22.20	24.80	28.60	58.10	37.70
6 Health	[2.30]	1.00	2.60	4.30	9.60	5.60
7 Transport	[21.90]	39.70	54.90	79.20	135.80	90.60
8 Communication	[7.30]	10.20	12.20	16.30	20.10	16.20
9 Recreation & culture	15.80	27.70	47.80	63.20	106.40	72.10
10 Education	-	[6.90]	3.20	6.70	26.50	13.10
11 Restaurants & hotels	[8.00]	22.30	31.20	47.50	83.00	54.30
12 Miscellaneous goods & services	12.60	20.70	28.90	41.70	68.20	46.50
1-12 All expenditure groups	190.00	276.00	352.00	452.60	724.70	512.00
13 Other expenditure items	29.00	44.20	60.90	88.60	144.60	98.60
Total expenditure	**218.90**	**320.20**	**412.90**	**541.10**	**869.40**	**610.60**
Average weekly expenditure per person (£) **Total expenditure**	**163.50**	**201.40**	**178.10**	**195.80**	**279.30**	**227.60**

Note: The commodity and service categories are not comparable to those in publications before 2001-02.
       Please see page ix for symbols and conventions used in this report.

1 Lower boundary of 2010 gross income quintile groups (£ per week).
2 Excluding mortgage interest payments, council tax and Northern Ireland rates.

Table A19   **Household expenditure by gross income: the household reference person is self-employed, 2008-2010**
United Kingdom

	Lowest twenty per cent	Second quintile group	Third quintile group	Fourth quintile group	Highest twenty per cent	All house-holds
Lower boundary of group (gross income : £ per week)[1]		238	413	651	1,015	
Average weighted number of households (thousands)	170	320	490	490	590	2,070
Total number of households in sample (over 3 years)	120	220	340	340	370	1,390
Total number of persons in sample (over 3 years)	220	530	960	1,000	1,180	3,890
Total number of adults in sample (over 3 years)	160	380	670	750	880	2,840
Weighted average number of persons per household	1.8	2.4	2.8	3.0	3.2	2.8
Commodity or service	Average weekly household expenditure (£)					
1 Food & non-alcoholic drinks	39.00	49.40	63.20	70.00	79.00	65.10
2 Alcoholic drinks, tobacco & narcotics	6.60	9.20	12.40	14.50	17.30	13.30
3 Clothing & footwear	12.80	19.10	24.80	31.20	41.10	29.00
4 Housing(net)[2], fuel & power	59.70	62.10	56.10	65.50	68.70	63.10
5 Household goods & services	27.30	34.60	24.90	49.10	62.80	43.10
6 Health	2.20	2.40	6.40	4.70	6.70	5.10
7 Transport	40.60	67.30	66.80	88.10	130.00	88.30
8 Communication	10.00	14.20	14.50	17.20	21.10	16.70
9 Recreation & culture	29.00	39.50	55.30	76.70	115.90	73.40
10 Education	[10.50]	9.50	2.00	10.00	25.00	12.50
11 Restaurants & hotels	27.30	34.10	44.30	56.30	89.50	57.00
12 Miscellaneous goods & services	22.50	30.80	46.70	52.20	64.60	48.50
1-12 All expenditure groups	287.50	372.10	417.40	535.60	721.60	515.10
13 Other expenditure items	58.00	79.90	86.80	114.60	172.00	114.50
Total expenditure	**345.50**	**452.10**	**504.20**	**650.20**	**893.60**	**629.60**
Average weekly expenditure per person (£)						
Total expenditure	**192.20**	**184.80**	**182.10**	**220.00**	**282.60**	**224.90**

Note: The commodity and service categories are not comparable to those in publications before 2001-02.
    Please see page ix for symbols and conventions used in this report.
    This table is based on a three year average.

1 Lower boundary of 2010 gross income quintile groups (£ per week).
2 Excluding mortgage interest payments, council tax and Northern Ireland rates.

Table A20   **Household expenditure by number of persons working, 2010**

United Kingdom

	Number of persons working					All house-holds
	None	One	Two	Three	Four or more	
Weighted number of households (thousands)	9,520	7,190	7,880	1,380	360	26,320
Total number of households in sample	1,960	1,470	1,550	230	50	5,260
Total number of persons in sample	3,380	3,180	4,520	850	250	12,180
Total number of adults in sample	2,850	2,350	3,300	700	230	9,430
Weighted average number of persons per household	1.7	2.2	2.9	3.8	5.1	2.3
Weighted average age of head of household	65	47	42	49	49	52
Employment status of the household reference person[1]:						
- % working full-time or self-employed	0	66	88	89	92	49
- % working part-time	0	20	11	9	4	9
- % not working	100	14	2	2	4	42

Commodity or service	Average weekly household expenditure (£)					
1 Food & non-alcoholic drinks	40.20	49.00	65.00	85.40	98.80	53.20
2 Alcoholic drinks, tobacco & narcotics	8.80	11.40	14.60	18.10	16.40	11.80
3 Clothing & footwear	10.90	20.40	32.70	57.30	78.90	23.40
4 Housing(net)[2], fuel & power	44.60	66.20	69.30	72.30	120.50	60.40
5 Household goods & services	22.40	29.80	40.90	43.40	48.10	31.40
6 Health	4.30	4.90	5.50	7.90	7.10	5.00
7 Transport	29.70	64.90	92.10	122.50	175.90	64.90
8 Communication	8.30	12.70	16.70	22.30	30.30	13.00
9 Recreation & culture	36.60	54.40	79.50	93.40	93.30	58.10
10 Education	5.70	10.70	13.00	20.00	[5.40]	10.00
11 Restaurants & hotels	18.60	36.00	57.40	81.40	90.60	39.20
12 Miscellaneous goods & services	20.50	33.20	51.80	59.00	60.10	35.90
1-12 All expenditure groups	250.50	393.50	538.50	682.90	825.40	406.30
13 Other expenditure items	29.90	67.90	103.30	105.10	109.30	67.30
Total expenditure	**280.40**	**461.40**	**641.70**	**788.10**	**934.60**	**473.60**
Average weekly expenditure per person (£)						
Total expenditure	**165.70**	**214.50**	**222.20**	**208.70**	**183.90**	**203.10**

Note: The commodity and service categories are not comparable to those in publications before 2001-02.

Please see page ix for symbols and conventions used in this report.

1 Excludes households where the household reference person was on a Government-supported training scheme.
2 Excluding mortgage interest payments, council tax and Northern Ireland rates.

Table A21   **Household expenditure by age at which the household reference person completed continuous full-time education, 2010**
United Kingdom

	Aged 14 and under	Aged 15	Aged 16	Aged 17 and under 19	Aged 19 and under 22	Aged 22 or over
Weighted number of households (thousands)	390	3,280	6,900	4,330	2,740	3,620
Total number of households in sample	80	730	1,390	870	530	670
Total number of persons in sample	180	1,480	3,680	2,240	1,370	1,720
Total number of adults in sample	140	1,270	2,630	1,620	1,000	1,270
Weighted average number of persons per household	2.3	2.1	2.7	2.6	2.6	2.6
Weighted average age of head of household	50	57	46	44	43	41
Commodity or service			Average weekly household expenditure (£)			
1 Food & non-alcoholic drinks	42.10	49.70	54.50	57.10	61.80	61.70
2 Alcoholic drinks, tobacco & narcotics	13.90	13.80	14.50	12.80	12.10	11.20
3 Clothing & footwear	14.20	17.20	25.50	27.80	30.60	34.40
4 Housing(net)[1], fuel & power	50.30	50.70	56.70	62.30	71.00	88.60
5 Household goods & services	22.80	25.60	27.30	40.10	39.40	43.10
6 Health	2.20	3.80	4.00	4.70	7.80	6.20
7 Transport	28.10	48.20	61.80	86.20	96.00	94.40
8 Communication	10.50	10.70	13.90	15.50	16.10	15.90
9 Recreation & culture	40.90	50.20	56.50	69.40	79.40	75.90
10 Education	[0.00]	[1.60]	4.50	7.60	16.80	39.70
11 Restaurants & hotels	21.20	27.40	39.20	45.80	56.50	62.60
12 Miscellaneous goods & services	15.60	22.50	33.40	43.50	51.90	52.00
1-12 All expenditure groups	261.80	321.30	391.70	472.80	539.40	585.60
13 Other expenditure items	28.70	43.10	66.80	78.50	98.30	107.40
Total expenditure	**290.50**	**364.50**	**458.50**	**551.30**	**637.60**	**692.90**
Average weekly expenditure per person (£)						
Total expenditure	**125.80**	**174.00**	**171.00**	**212.10**	**248.70**	**271.70**

Note: The commodity and service categories are not comparable to those in publications before 2001-02.
     Please see page ix for symbols and conventions used in this report.

1 Excluding mortgage interest payments, council tax and Northern Ireland rates.

Table A22   Household expenditure by socio-economic classification of household reference person, 2010

United Kingdom

Commodity or service	Large employers & higher managerial	Higher professional	Lower managerial & professional	Intermediate	Small employers	Lower supervisory	Semi-routine	Routine	Long-term unemployed[1]	Students	Occupation not stated[2]	All households
Weighted number of households (thousands)	1,260	1,770	4,620	1,350	1,580	1,690	1,770	1,520	490	590	9,680	26,320
Total number of households in sample	260	350	910	270	330	320	350	290	100	100	1,990	5,260
Total number of persons in sample	710	930	2,430	640	870	860	870	770	270	260	3,560	12,180
Total number of adults in sample	520	680	1,790	480	650	650	650	590	150	180	3,090	9,430
Weighted average number of persons per household	2.8	2.6	2.7	2.4	2.7	2.7	2.6	2.7	2.9	2.7	1.8	2.3
	Average weekly household expenditure (£)											
1 Food & non-alcoholic drinks	70.80	67.40	62.30	50.20	64.00	57.90	48.10	53.10	38.30	48.50	43.80	53.20
2 Alcoholic drinks, tobacco & narcotics	16.60	14.40	14.10	10.90	11.60	14.90	12.30	12.80	10.10	7.80	9.30	11.80
3 Clothing & footwear	46.00	36.80	32.60	26.60	31.00	24.80	19.00	20.40	13.20	28.50	13.10	23.40
4 Housing (net)[3], fuel & power	74.50	77.00	66.70	64.50	58.70	65.50	73.40	69.40	44.40	109.30	45.30	60.40
5 Household goods & services	76.10	45.80	39.70	30.40	45.80	27.80	17.20	21.70	19.30	20.50	22.80	31.40
6 Health	6.40	8.10	6.90	3.60	5.30	3.50	2.60	3.00	0.60	3.90	4.90	5.00
7 Transport	120.90	106.90	100.80	66.90	77.40	73.60	49.20	58.70	19.40	75.50	34.30	64.90
8 Communication	17.00	14.30	17.10	14.50	14.20	17.10	14.20	13.90	8.80	14.80	9.00	13.00
9 Recreation & culture	104.70	91.00	77.50	61.00	68.60	61.30	45.30	48.30	16.60	38.90	41.10	58.10
10 Education	26.40	25.80	14.00	[9.00]	6.90	4.10	5.20	[4.10]	[1.70]	110.90	0.90	10.00
11 Restaurants & hotels	77.20	69.80	57.80	42.20	49.60	41.80	31.60	29.90	12.30	39.10	21.60	39.20
12 Miscellaneous goods & services	65.80	57.60	48.50	41.90	49.60	40.00	27.00	27.90	16.90	26.60	22.60	35.90
1-12 All expenditure groups	702.40	614.90	537.90	421.70	482.60	432.10	345.00	363.30	201.50	524.30	268.80	406.30
13 Other expenditure items	153.70	121.90	101.70	76.60	81.50	80.50	49.00	54.80	13.10	49.50	32.80	67.30
Total expenditure	856.10	736.80	639.60	498.30	564.20	512.60	394.00	418.10	214.60	573.90	301.50	473.60
Average weekly expenditure per person (£)												
Total expenditure	308.90	284.40	238.60	208.20	209.90	188.30	154.50	154.20	75.00	215.70	169.60	203.10

Note: The commodity and service categories are not comparable to those in publications before 2001-02.
Please see page ix for symbols and conventions used in this report.

1 Includes those who have never worked.
2 Includes those who are economically inactive.
3 Excludes mortgage interest payments, council tax and Northern Ireland rates.

Table A23   **Expenditure by household composition, 2010**
United Kingdom

| | Retired households | | | | Non-retired | | Retired and non-retired households | | | | | | |
| | State pension¹ | | Other retired | | | | One adult | | Two adults | | | Three or more adults | |
Commodity or service	One person	Two adults	One person	Two adults	One person	Two adults	with one child	with two or more children	with one child	with two children	with three or more children	without children	with children
Weighted number of households (thousands)	790	450	2,980	2,300	4,030	5,760	690	720	2,300	2,150	710	2,200	1,230
Total number of households in sample	150	110	570	530	800	1,170	160	170	450	450	160	370	200
Total number of persons in sample	150	220	570	1,050	800	2,340	310	590	1,340	1,820	830	1,220	950
Total number of adults in sample	150	220	570	1,050	800	2,340	160	170	900	910	310	1,220	650
Weighted average number of persons per household	1.0	2.0	1.0	2.0	1.0	2.0	2.0	3.5	3.0	4.0	5.3	3.4	4.8
						Average weekly household expenditure (£)							
1 Food & non-alcoholic drinks	28.10	47.80	29.70	56.70	26.20	54.30	36.90	56.10	58.90	77.10	85.70	78.60	96.20
2 Alcoholic drinks, tobacco & narcotics	4.70	8.70	5.10	10.60	7.60	15.40	9.50	11.00	13.40	12.80	14.20	16.80	20.20
3 Clothing & footwear	5.50	9.10	6.80	15.20	9.40	25.00	12.90	22.60	28.30	38.00	44.40	42.50	59.00
4 Housing(net)², fuel & power	37.00	41.20	41.60	46.10	55.30	71.40	57.40	63.40	68.20	62.40	68.70	70.80	77.90
5 Household goods & services	12.40	20.30	18.30	29.50	21.30	38.60	17.00	32.40	37.10	43.70	44.60	39.50	35.90
6 Health	2.30	4.60	4.50	8.00	2.40	5.20	2.50	2.30	5.10	5.50	5.20	6.90	9.20
7 Transport	9.50	27.80	19.50	50.60	38.20	90.10	22.10	36.60	68.90	94.90	83.10	116.60	97.20
8 Communication	5.60	7.80	6.70	9.80	8.70	14.50	10.10	11.00	16.50	17.50	15.40	19.40	23.70
9 Recreation & culture	16.50	37.00	25.90	61.50	31.20	74.10	29.90	39.30	67.30	80.80	83.60	83.80	86.10
10 Education	[2.60]	-	[0.10]	[0.40]	2.80	9.20	[5.20]	[2.50]	11.00	22.20	39.40	27.50	23.90
11 Restaurants & hotels	7.00	18.20	11.50	31.50	20.90	51.90	18.00	26.40	42.50	58.90	53.50	66.90	71.80
12 Miscellaneous goods & services	13.10	22.90	17.10	32.10	18.30	41.40	20.30	23.00	56.50	56.90	48.10	47.00	54.50
1-12 All expenditure groups	144.30	245.40	186.80	352.10	242.50	490.90	241.80	326.60	473.60	570.60	585.80	616.20	655.50
13 Other expenditure items	16.90	31.10	27.50	45.10	47.80	86.30	22.80	35.20	84.70	122.40	112.60	79.90	91.80
Total expenditure	**161.10**	**276.50**	**214.20**	**397.20**	**290.30**	**577.20**	**264.60**	**361.70**	**558.40**	**693.00**	**698.40**	**696.10**	**747.30**
Average weekly expenditure per person (£)													
Total expenditure	**161.10**	**138.30**	**214.20**	**198.60**	**290.30**	**288.60**	**132.30**	**103.80**	**186.10**	**173.30**	**130.80**	**206.40**	**155.30**

Note: The commodity and service categories are not comparable to those in publications before 2001-02.
    Please see page ix for symbols and conventions used in this report.

1 Mainly dependent on state pensions and not economically active - see definitions in Appendix B.
2 Excluding mortgage interest payments, council tax and Northern Ireland rates.

Table A24 **Expenditure of one person retired households mainly dependent on state pensions[1] by gross income quintile group, 2008-2010**
United Kingdom

	Lowest twenty per cent	Second quintile group	Third quintile group	Fourth quintile group	Highest twenty per cent	All house-holds
Lower boundary of group (gross income: £ per week)[2]		238.00	413.00	651.00	1015.00	
Average weighted number of households (thousands)	720	120	0	0	0	840
Total number of households in sample (over 3 years)	430	70	0	0	0	500
Total number of persons in sample (over 3 years)	430	70	0	0	0	500
Total number of adults in sample (over 3 years)	430	70	0	0	0	500
Weighted average number of persons per household	1.0	1.0	0.0	0.0	0.0	1.0
Commodity or service	Average weekly household expenditure (£)					
1 Food & non-alcoholic drinks	26.00	25.90	-	0.00	0.00	26.20
2 Alcoholic drinks, tobacco & narcotics	3.40	3.20	0.00	0.00	0.00	3.50
3 Clothing & footwear	4.90	7.50	0.00	0.00	0.00	5.20
4 Housing (net)[3], fuel & power	33.70	36.10	0.00	0.00	0.00	34.20
5 Household goods & services	10.50	19.70	0.00	0.00	0.00	12.20
6 Health	2.80	3.40	0.00	0.00	0.00	3.00
7 Transport	7.60	8.90	0.00	0.00	0.00	7.60
8 Communication	5.50	5.90	0.00	0.00	0.00	5.50
9 Recreation & culture	15.60	19.00	0.00	0.00	0.00	16.20
10 Education	[0.00]	[4.40]	[0.00]	0.00	0.00	0.90
11 Restaurants & hotels	6.80	9.70	0.00	0.00	0.00	7.10
12 Miscellaneous goods & services	12.70	13.00	0.00	0.00	0.00	12.80
1–12 All expenditure groups	129.60	156.90	-	0.00	0.00	134.50
13 Other expenditure items	16.50	22.50	0.00	0.00	0.00	17.40
Total expenditure	**146.10**	**179.40**	**0.00**	**0.00**	**0.00**	**151.90**
Average weekly expenditure per person (£) **Total expenditure**	**146.10**	**179.40**	**0.00**	**0.00**	**0.00**	**151.90**

Note: The commodity and service categories are not comparable to those in publications before 2001–02.
Please see page ix for symbols and conventions used in this report.

1 Includes those who have never worked.
2 Includes those who are economically inactive.
3 Excludes mortgage interest payments, council tax and Northern Ireland rates.

Table A25   Expenditure of one person retired households not mainly dependent on state pensions by gross income quintile group, 2008-2010

United Kingdom

	Lowest twenty per cent	Second quintile group	Third quintile group	Fourth quintile group	Highest twenty per cent	All house-holds
Lower boundary of group (gross income: £ per week)[1]		238	413	651	1,015	
Average weighted number of households (thousands)	1,510	950	270	60	..	2,800
Total number of households in sample (over 3 years)	910	590	170	40	..	1,720
Total number of persons in sample (over 3 years)	910	590	170	40	..	1,720
Total number of adults in sample (over 3 years)	910	590	170	40	..	1,720
Weighted average number of persons per household	1.0	1.0	1.0	1.0	1.0	1.0
Commodity or service	Average weekly household expenditure (£)					
1 Food & non-alcoholic drinks	26.80	29.80	34.70	37.80	[23.90]	28.80
2 Alcoholic drinks, tobacco & narcotics	3.60	4.80	7.60	12.00	[5.00]	4.50
3 Clothing & footwear	5.60	8.50	7.30	13.20	[2.00]	6.90
4 Housing(net)[2] fuel & power	35.40	41.60	43.40	69.50	[33.40]	39.10
5 Household goods & services	11.70	18.60	26.20	35.90	[49.20]	16.20
6 Health	3.80	3.90	6.70	15.30	-	4.30
7 Transport	10.00	20.70	33.60	31.30	[59.20]	16.70
8 Communication	5.70	6.80	8.40	8.20	[5.50]	6.40
9 Recreation & culture	17.80	27.40	43.90	67.40	[95.20]	25.10
10 Education	[0.00]	[0.70]	[1.40]	[1.30]	-	0.40
11 Restaurants & hotels	7.20	13.10	16.80	30.20	[28.20]	10.80
12 Miscellaneous goods & services	12.40	18.60	30.30	47.50	[37.70]	17.10
1-12 All expenditure groups	140.10	194.40	260.20	369.50	[339.30]	176.40
13 Other expenditure items	14.60	33.40	51.20	150.40	[60.50]	27.70
Total expenditure	**154.70**	**227.80**	**311.50**	**519.90**	**[399.80]**	**204.10**
Average weekly expenditure per person (£) **Total expenditure**	**154.70**	**227.80**	**311.50**	**519.90**	**599.60**	**204.10**

Note: The commodity and service categories are not comparable to those in publications before 2001-02.
Please see page ix for symbols and conventions used in this report.
This table is based on a three year average.

1 Lower boundary of 2010 gross income quintile groups (£ per week).
2 Excluding mortgage interest payments, council tax and Northern Ireland rates.

Table A26   **Expenditure of one adult non-retired households by gross income quintile group, 2008-2010**
United Kingdom

		Lowest twenty per cent	Second quintile group	Third quintile group	Fourth quintile group	Highest twenty per cent	All house-holds
Lower boundary of group (gross income: £ per week)[2]			238	413	651	1015	
Average weighted number of households (thousands)		1390	930	900	490	250	3960
Total number of households in sample (over 3 years)		890	610	570	300	140	2520
Total number of persons in sample (over 3 years)		890	610	570	300	140	2520
Total number of adults in sample (over 3 years)		890	610	570	300	140	2520
Weighted average number of persons per household		1.0	1.0	1.0	1	1	1.0
Commodity or service		Average weekly household expenditure (£)					
1	Food & non-alcoholic drinks	21.30	26.00	27.30	-	-	25.50
2	Alcoholic drinks, tobacco & narcotics	7.80	7.90	8.40	-	-	8.40
3	Clothing & footwear	4.30	8.00	12.60	-	-	9.20
4	Housing(net)[3], fuel & power	37.40	56.50	51.20	-	-	50.50
5	Household goods & services	9.70	14.60	20.80	-	-	19.30
6	Health	1.40	2.20	2.90	-	-	2.90
7	Transport	16.20	33.60	52.90	-	-	38.00
8	Communication	6.30	8.60	10.00	-	-	8.70
9	Recreation & culture	16.30	24.60	39.00	-	-	29.70
10	Education	2.50	1.90	1.40	-	-	[3.00]
11	Restaurants & hotels	10.30	17.40	26.90	-	-	21.90
12	Miscellaneous goods & services	9.30	17.90	22.00	-	-	18.90
1–12	All expenditure groups	142.80	219.10	275.40	-	-	235.80
13	Other expenditure items	18.80	46.50	71.60	-	-	58.90
Total expenditure		**161.60**	**265.60**	**347.00**	**-**	**-**	**294.70**
Average weekly expenditure per person (£)							
Total expenditure		**161.60**	**265.60**	**347.00**	**-**	**-**	**294.70**

Note: The commodity and service categories are not comparable to those in publications before 2001–02.
Please see page ix for symbols and conventions used in this report.
This table is based on a three year average.

1 Mainly dependent on state pensions and not economically active - see defintions in Appendix B.
2 Lower boundary of 2009 gross income quintile groups (£ per week).
3 Excluding mortgage interest payments, council tax and Northern Ireland rates.

Table A27 **Expenditure of one adult households with children by gross income quintile group 2008-2010**
United Kingdom

	Lowest twenty per cent	Second quintile group	Third quintile group	Fourth quintile group	Highest twenty per cent	All house-holds
Lower boundary of group (gross income: £ per week)[1]		238	413	651	1015	
Average weighted number of households (thousands)	570	490	240	70	30	1,390
Total number of households in sample (over 3 years)	410	370	180	60	20	1,040
Total number of persons in sample (over 3 years)	1,030	1,110	500	150	60	2,840
Total number of adults in sample (over 3 years)	410	370	180	60	20	1,040
Weighted average number of persons per household	2.5	2.9	2.7	2.6	2.8	2.7
Commodity or service	Average weekly household expenditure (£)					
1 Food & non-alcoholic drinks	37.00	49.00	49.50	53.50	70.60	44.90
2 Alcoholic drinks, tobacco & narcotics	9.10	8.30	10.10	9.50	[8.60]	9.10
3 Clothing & footwear	13.60	18.70	25.20	22.40	[40.00]	18.40
4 Housing(net)[2] fuel & power	51.70	60.00	62.90	65.70	72.20	57.90
5 Household goods & services	17.00	20.20	24.10	50.30	39.40	21.60
6 Health	0.80	2.00	2.30	[6.60]	[12.90]	2.10
7 Transport	16.80	28.70	44.10	49.20	98.20	29.10
8 Communication	7.50	10.80	15.50	14.40	18.70	10.60
9 Recreation & culture	23.50	35.80	50.90	62.20	[94.90]	36.00
10 Education	3.10	3.40	2.40	[10.60]	[36.00]	4.20
11 Restaurants & hotels	13.50	[20.90]	34.60	47.00	69.90	22.70
12 Miscellaneous goods & services	12.90	22.70	41.70	49.40	76.00	24.50
1–12 All expenditure groups	206.60	280.60	363.20	441.00	637.40	281.20
13 Other expenditure items	14.90	[35.30]	71.70	92.00	158.30	38.90
Total expenditure	**221.50**	**315.90**	**434.80**	**533.00**	**795.60**	**320.00**
Average weekly expenditure per person (£) **Total expenditure**	**89.00**	**107.20**	**162.10**	**208.00**	**288.80**	**118.90**

Note: The commodity and service categories are not comparable to those in publications before 2001-02.
   Please see page ix for symbols and conventions used in this report.
   This table is based on a three year average.

1 Lower boundary of 2010 gross income quintile groups (£ per week).
2 Excluding mortgage interest payments, council tax and Northern Ireland rates.

Table A28   **Expenditure of two adult households with children by gross income quintile group, 2008-2010**
United Kingdom

	Lowest twenty per cent	Second quintile group	Third quintile group	Fourth quintile group	Highest twenty per cent	All house-holds
Lower boundary of group (gross income £ per week)[1]		238	413	651	1015	
Average weighted number of households (thousands)	270	560	1,110	1,550	1,600	5,090
Total number of households in sample (over 3 years)	160	350	750	1,060	1,110	3,430
Total number of persons in sample (over 3 years)	560	1,380	2,900	4,020	4,230	13,090
Total number of adults in sample (over 3 years)	310	710	1,500	2,120	2,220	6,860
Weighted average number of persons per household	3.5	3.8	3.8	3.8	3.8	3.8
Commodity or service	Average weekly household expenditure (£)					
1 Food & non-alcoholic drinks	50.90	56.60	62.70	69.30	86.80	71.00
2 Alcoholic drinks, tobacco & narcotics	13.90	13.80	12.10	11.80	15.10	13.20
3 Clothing & footwear	16.00	23.70	22.80	30.70	45.70	32.20
4 Housing(net)[2], fuel & power	61.10	71.30	67.80	59.30	66.30	64.70
5 Household goods & services	15.30	25.30	22.20	36.50	63.40	39.40
6 Health	1.20	1.50	3.30	5.10	7.50	4.90
7 Transport	26.90	46.80	54.10	80.90	127.50	83.00
8 Communication	9.90	12.80	14.40	15.50	18.10	15.50
9 Recreation & culture	27.30	41.90	52.50	70.60	123.70	77.90
10 Education	8.50	6.50	3.70	4.80	40.00	16.00
11 Restaurants & hotels	21.60	30.10	32.90	46.80	78.60	50.60
12 Miscellaneous goods & services	19.70	28.20	36.80	50.20	87.00	54.80
1–12 All expenditure groups	272.30	358.50	385.30	481.60	759.70	523.30
13 Other expenditure items	24.60	49.80	69.20	110.10	197.10	117.30
Total expenditure	**296.90**	**408.30**	**454.50**	**591.70**	**956.80**	**640.60**
Average weekly expenditure per person (£) Total expenditure	**83.70**	**106.10**	**119.60**	**157.40**	**252.10**	**169.60**

Note: The commodity and service categories are not comparable to those in publications before 2001-02.
  Please see page ix for symbols and conventions used in this report.
  This table is based on a three year average.

1 Lower boundary of 2010 gross income quintile groups (£ per week).
2 Excluding mortgage interest payments, council tax and Northern Ireland rates.

Table A29   **Expenditure of two adult non-retired households by gross income quintile group, 2008-2010**
United Kingdom

	Lowest twenty per cent	Second quintile group	Third quintile group	Fourth quintile group	Highest twenty per cent	All house- holds
Lower boundary of group (gross income £ per week)[1]		238	413	651	1015	
Average weighted number of households (thousands)	320	700	1,320	1,700	1,660	5,700
Total number of households in sample (over 3 years)	200	480	890	1,090	1,040	3,700
Total number of persons in sample (over 3 years)	410	950	1,790	2,170	2,080	7,400
Total number of adults in sample (over 3 years)	410	950	1,790	2,170	2,080	7,400
Weighted average number of persons per household	2.0	2.0	2.0	2.0	2.0	2.0
Commodity or service	Average weekly household expenditure (£)					
1 Food & non-alcoholic drinks	42.80	47.20	51.40	51.90	57.90	52.40
2 Alcoholic drinks, tobacco & narcotics	12.90	11.70	12.90	13.80	14.70	13.60
3 Clothing & footwear	12.30	13.50	18.80	23.70	36.30	24.40
4 Housing(net)[2], fuel & power	55.80	60.00	62.30	63.90	69.20	64.10
5 Household goods & services	26.70	23.60	27.30	33.20	52.20	35.90
6 Health	3.10	3.40	5.50	5.20	9.20	6.10
7 Transport	35.50	48.80	58.50	82.10	123.60	82.20
8 Communication	9.50	10.80	12.30	14.30	16.20	13.70
9 Recreation & culture	37.70	47.30	52.50	69.60	106.40	71.80
10 Education	[5.20]	[10.80]	6.50	3.90	7.20	6.30
11 Restaurants & hotels	18.80	23.50	33.30	49.30	79.20	49.50
12 Miscellaneous goods & services	20.20	23.20	34.50	39.20	55.00	39.60
1–12 All expenditure groups	280.50	323.80	375.80	450.10	627.10	459.60
13 Other expenditure items	30.20	42.40	60.30	94.70	151.40	93.30
Total expenditure	**310.80**	**366.20**	**436.20**	**544.80**	**778.60**	**553.00**
Average weekly expenditure per person (£)						
Total expenditure	**155.40**	**183.10**	**218.10**	**272.40**	**389.30**	**276.50**

Note: The commodity and service categories are not comparable to those in publications before 2001–02.
Please see page ix for symbols and conventions used in this report.
This table is based on a three year average.

1 Lower boundary of 2010 gross income quintile groups (£ per week).
2 Excluding mortgage interest payments, council tax and Northern Ireland rates.

Table A30   **Expenditure of two adult retired households mainly dependent on state pensions[1]
by gross income quintile group, 2008-2010**
United Kingdom

	Lowest twenty per cent	Second quintile group	Third quintile group	Fourth quintile group	Highest twenty per cent	All house-holds
Lower boundary of group (£ per week)[2]		238	413	651	1015	
Average weighted number of households (thousands)	170	270	20	00	00	460
Total number of households in sample (over 3 years)	130	230	20	0	0	380
Total number of persons in sample (over 3 years)	270	450	30	0	0	750
Total number of adults in sample (over 3 years)	270	450	30	0	0	750
Weighted average number of persons per household	2.0	2.0	2.0	0.0	0.0	2.0
Commodity or service	Average weekly household expenditure (£)					
1 Food & non-alcoholic drinks	45.60	48.60	[46.70]	-	-	47.60
2 Alcoholic drinks, tobacco & narcotics	5.50	9.10	[7.70]	-	-	7.70
3 Clothing & footwear	6.40	10.10	[8.10]	-	-	8.60
4 Housing(net)[2], fuel & power	32.50	40.70	[53.80]	-	-	38.30
5 Household goods & services	14.30	21.80	[31.40]	-	-	19.60
6 Health	2.50	4.70	[5.90]	-	-	3.90
7 Transport	28.00	26.60	[46.40]	-	-	28.40
8 Communication	6.20	7.30	[7.00]	-	-	6.90
9 Recreation & culture	31.60	44.00	[54.20]	-	-	40.00
10 Education	-	0.30	-	-	-	[0.20]
11 Restaurants & hotels	11.70	16.00	[19.70]	-	-	14.80
12 Miscellaneous goods & services	17.70	20.60	[17.40]	-	-	19.50
1–12 All expenditure groups	201.80	249.70	[298.50]	-	-	235.50
13 Other expenditure items	27.80	33.20	[11.80]	-	-	30.10
Total expenditure	**229.60**	**282.80**	**[310.20]**	**-**	**-**	**265.60**
Average weekly expenditure per person (£) **Total expenditure**	**114.80**	**141.40**	**[155.10]**	**-**	**-**	**132.80**

Note: The commodity and service categories are not comparable to those in publications before 2001–02.
Please see page ix for symbols and conventions used in this report.
This table is based on a three year average.

1 Mainly dependent on the state pensions and not economically active -see defintions in Appendix B.
2 Lower boundary of 2010 gross income quintile groups (£ per week).
3 Excluding mortgage interest payments, council tax and Northern Ireland rates.

Table A31 **Expenditure of two adult retired households not mainly dependent on state pensions by gross income quintile group, 2008-2010**
United Kingdom

	Lowest twenty per cent	Second quintile group	Third quintile group	Fourth quintile group	Highest twenty per cent	All house-holds
Lower boundary of group (£ per week)[1]		238	413	651	1015	
Average weighted number of households (thousands)	180	950	730	270	110	2,260
Total number of households in sample (over 3 years)	120	730	550	210	90	1,690
Total number of persons in sample (over 3 years)	240	1,460	1,100	410	170	3,370
Total number of adults in sample (over 3 years)	240	1,460	1,100	410	170	3,370
Weighted average number of persons per household	2.0	2.0	2.0	2.0	2.0	2.0
Commodity or service	Average weekly household expenditure (£)					
1 Food & non-alcoholic drinks	46.60	51.20	56.20	63.50	68.80	54.90
2 Alcoholic drinks, tobacco & narcotics	10.60	8.90	10.00	12.00	19.40	10.30
3 Clothing & footwear	6.10	9.90	16.40	20.60	26.90	13.80
4 Housing(net)[2], fuel & power	37.70	41.10	41.70	52.70	76.70	44.40
5 Household goods & services	15.40	21.30	30.70	52.50	81.70	30.30
6 Health	2.70	5.40	9.60	6.60	13.60	7.10
7 Transport	21.00	33.70	47.90	69.10	92.40	44.70
8 Communication	9.00	7.90	9.00	10.20	14.20	9.00
9 Recreation & culture	30.60	39.60	86.00	95.80	120.50	64.60
10 Education	[0.20]	[0.20]	[0.50]	[1.40]	[1.70]	0.50
11 Restaurants & hotels	13.80	18.70	29.90	52.70	74.10	29.00
12 Miscellaneous goods & services	16.40	20.30	28.30	54.30	84.50	29.90
1–12 All expenditure groups	210.10	258.10	366.30	491.50	674.60	338.40
13 Other expenditure items	17.30	33.90	45.70	88.00	95.70	46.00
Total expenditure	**227.40**	**292.00**	**412.00**	**579.40**	**770.30**	**384.40**
Average weekly expenditure per person (£) **Total expenditure**	**113.70**	**146.00**	**206.00**	**289.70**	**385.20**	**192.20**

Note: The commodity and service categories are not comparable to those in publications before 2001–02.
Please see page ix for symbols and conventions used in this report.
This table is based on a three year average.

1 Lower boundary of 2009 gross income quintile groups (£ per week).
2 Excluding mortgage interest payments, council tax and Northern Ireland rates.

Table A32 **Household expenditure by tenure, 2010**
United Kingdom

	Owners			Social rented from			Private rented[4]				All tenures
	Owned outright	Buying with a mortgage[1]	All	Council[2]	Registered Social Landlord[3]	All	Rent free	Rent paid, unfurnished[5]	Rent paid, furnished	All	
Weighted number of households (thousands)	8,340	9,100	17,430	2,590	2,150	4,750	370	2,890	880	4,150	26,320
Total number of households in sample	1,770	1,830	3,600	500	430	920	70	540	130	740	5,260
Total number of persons in sample	3,400	5,080	8,480	1,070	920	1,990	130	1,280	290	1,710	12,180
Total number of adults in sample	3,120	3,660	6,780	750	640	1,390	110	910	240	1,260	9,430
Weighted average number of persons per household	1.9	2.8	2.4	2.2	2.2	2.2	1.8	2.4	2.3	2.3	2.3
Commodity or service					Average weekly household expenditure (£)						
1 Food & non-alcoholic drinks	53.80	64.40	59.30	38.70	40.50	39.50	37.30	44.70	40.30	43.10	53.20
2 Alcoholic drinks, tobacco & narcotics	10.30	13.40	12.00	11.70	11.00	11.40	12.90	12.80	7.50	11.70	11.80
3 Clothing & footwear	19.10	33.80	26.70	12.30	13.50	12.80	14.60	18.70	32.30	21.30	23.40
4 Housing(net)[6], fuel & power	42.70	42.50	42.60	57.90	63.90	60.60	30.70	135.50	176.70	134.90	60.40
5 Household goods & services	35.40	41.90	38.80	14.00	13.40	13.70	25.10	23.30	10.00	20.60	31.40
6 Health	7.10	5.80	6.40	1.70	2.90	2.20	3.40	2.40	2.60	2.50	5.00
7 Transport	60.80	93.60	77.90	26.00	26.10	26.00	45.20	54.90	56.90	54.40	64.90
8 Communication	10.70	16.30	13.60	10.30	10.70	10.50	11.20	13.00	15.70	13.40	13.00
9 Recreation & culture	65.20	73.10	69.40	27.60	34.10	30.50	41.10	46.30	29.10	42.10	58.10
10 Education	7.60	11.90	9.80	[0.70]	[3.00]	1.70	[0.60]	5.30	[77.60]	20.30	10.00
11 Restaurants & hotels	35.20	56.60	46.40	14.50	17.30	15.80	26.50	34.40	45.30	36.00	39.20
12 Miscellaneous goods & services	37.10	49.90	43.80	13.30	15.70	14.40	24.10	29.10	23.30	27.40	35.90
1-12 All expenditure groups	385.00	503.20	446.70	228.60	252.10	239.30	272.60	420.40	517.30	427.90	406.30
13 Other expenditure items	45.80	127.10	88.20	15.10	19.70	17.20	30.70	36.60	39.50	36.70	67.30
Total expenditure	**430.80**	**630.30**	**534.90**	**243.70**	**271.70**	**256.40**	**303.30**	**457.00**	**556.90**	**464.60**	**473.60**
Average weekly expenditure per person (£)											
Total expenditure	**223.80**	**225.60**	**224.90**	**111.60**	**124.40**	**117.40**	**169.80**	**192.50**	**242.70**	**201.60**	**203.10**

Note: The commodity and service categories are not comparable to those in publications before 2001-02.
Please see page ix for symbols and conventions used in this report.

1 Including shared owners (who own part of the equity and pay mortgage, part rent).
2 "Council" includes local authorities, New Towns and Scottish Homes, but see note 3 below.
3 Formerly Housing Associations.
4 All tenants whose accommodation goes with the job of someone in the household are allocated to "rented privately", even if the landlord is a local authority or housing association or Housing Action Trust, or if the accommodation is rent free. Squatters are also included in this category.
5 "Unfurnished" includes the answers: "partly furnished".
6 Excluding mortgage interest payments, council tax and Northern Ireland rates.

Table A33   Household expenditure by UK countries and regions, 2008-2010

	North East	North West	Yorks & the Humber	East Midlands	West Midlands	East	London	South East	South West	England	Wales	Scotland	Northern Ireland	United Kingdom
Average weighted number of households (thousands)	1,280	3,130	2,170	1,980	2,200	2,270	3,030	3,080	2,510	21,640	1,280	2,380	690	26,000
Total number of households in sample (over 3 years)	730	1,770	1,460	1,210	1,470	1,550	1,410	2,190	1,520	13,300	800	1,510	1,320	16,930
Total number of persons in sample (over 3 years)	1,620	4,100	3,400	2,820	3,550	3,610	3,470	5,190	3,440	31,200	1,840	3,330	3,390	39,750
Total number of adults in sample (over 3 years)	1,280	3,160	2,610	2,190	2,730	2,790	2,590	4,020	2,700	24,080	1,450	2,670	2,520	30,720
Weighted average number of persons per household	2.3	2.4	2.3	2.3	2.4	2.3	2.5	2.3	2.3	2.4	2.3	2.2	2.6	2.3
Commodity or service								Average weekly household expenditure (£)						
1 Food & non-alcoholic drinks	44.70	49.20	46.40	52.10	50.40	53.70	55.80	55.50	54.70	52.00	51.80	50.90	57.20	52.00
2 Alcoholic drinks, tobacco & narcotics	10.40	12.70	10.70	11.10	11.20	10.20	10.40	11.10	10.90	11.00	10.80	12.50	15.20	11.30
3 Clothing & footwear	19.40	22.20	19.10	19.50	20.80	22.30	27.00	21.20	19.50	21.60	19.90	21.70	39.40	22.00
4 Housing(net)[1], fuel & power	44.60	49.90	48.80	53.10	50.30	56.30	87.00	60.40	58.40	58.10	52.80	50.50	47.60	56.90
5 Household goods & services	26.40	27.50	29.10	27.00	27.30	33.70	34.10	33.10	33.00	30.50	22.70	27.60	29.10	29.80
6 Health	4.40	4.30	4.00	5.70	4.20	5.90	7.30	6.40	5.70	5.40	3.00	4.00	4.50	5.10
7 Transport	49.10	53.90	52.90	64.00	61.40	67.90	66.00	73.50	68.10	62.80	51.80	63.00	62.40	62.20
8 Communication	10.90	11.80	10.60	12.30	11.60	13.20	15.40	13.00	12.80	12.60	10.30	11.70	13.60	12.40
9 Recreation & culture	47.40	54.30	55.60	57.50	56.30	61.40	61.50	66.20	62.20	58.90	53.70	58.00	56.30	58.50
10 Education	2.30	5.40	3.80	9.10	4.90	5.80	16.60	9.80	8.30	7.90	9.70	5.20	8.10	7.70
11 Restaurants & hotels	30.50	36.90	34.10	36.60	36.30	38.60	51.40	41.00	36.90	38.90	30.30	36.50	46.30	38.40
12 Miscellaneous goods & services	27.70	33.70	29.30	35.30	33.20	39.10	42.90	41.40	36.70	36.30	26.30	33.10	37.00	35.50
1-12 All expenditure groups	317.90	361.70	344.50	383.20	367.80	408.00	475.60	432.60	407.30	396.00	343.10	374.60	416.70	391.90
13 Other expenditure items	54.70	68.80	61.00	66.20	62.20	85.30	102.20	90.50	75.30	76.50	50.90	72.60	66.10	74.60
Total expenditure	372.70	430.50	405.50	449.40	430.10	493.40	577.80	523.20	482.60	472.40	394.00	447.20	482.80	466.50
Average weekly expenditure per person (£)														
Total expenditure	163.50	181.70	179.20	191.70	178.20	217.40	228.70	223.80	207.80	200.40	171.50	205.90	187.80	199.10

Note: The commodity and service categories are not comparable to those in publications before 2001-02.
Please see page xiii for symbols and conventions used in this report.
This table is based on a three year average.
1 Excluding mortgage interest payments, council tax and Northern Ireland rates.

Table A34   Household expenditure as a percentage of total expenditure by UK countries and regions, 2008-2010

	North East	North West	Yorks & the Humber	East Midlands	West Midlands	East	London	South East	South West	England	Wales	Scotland	Northern Ireland	United Kingdom
Average weighted number of households (thousands)	1,280	3,130	2,170	1,980	2,200	2,270	3,030	3,080	2,510	21,640	1,280	2,380	690	26,000
Total number of households in sample (over 3 years)	730	1,770	1,460	1,210	1,470	1,550	1,410	2,190	1,520	13,300	800	1,510	1,320	16,930
Total number of persons in sample (over 3 years)	1,620	4,100	3,400	2,820	3,550	3,610	3,470	5,190	3,440	31,200	1,840	3,330	3,390	39,750
Total number of adults in sample (over 3 years)	1,280	3,160	2,610	2,190	2,730	2,790	2,590	4,020	2,700	24,080	1,450	2,670	2,520	30,720
Weighted average number of persons per household	2.3	2.4	2.3	2.3	2.4	2.3	2.5	2.3	2.3	2.4	2.3	2.2	2.6	2.3
Commodity or service								Percentage of total expenditure						
1 Food & non-alcoholic drinks	12	11	11	12	12	11	10	11	11	11	13	11	12	11
2 Alcoholic drinks, tobacco & narcotics	3	3	3	2	3	2	2	2	2	2	3	3	3	2
3 Clothing & footwear	5	5	5	4	5	5	5	4	4	5	5	5	8	5
4 Housing(net)¹, fuel & power	12	12	12	12	12	11	15	12	12	12	13	11	10	12
5 Household goods & services	7	6	7	6	6	7	6	6	7	6	6	6	6	6
6 Health	1	1	1	1	1	1	1	1	1	1	1	1	1	1
7 Transport	13	13	13	14	14	14	11	14	14	13	13	14	13	13
8 Communication	3	3	3	3	3	3	3	2	3	3	3	3	3	3
9 Recreation & culture	13	13	14	13	13	12	11	13	13	12	14	13	12	13
10 Education	1	1	1	2	1	1	3	2	2	2	2	1	2	2
11 Restaurants & hotels	8	9	8	8	8	8	9	8	8	8	8	8	10	8
12 Miscellaneous goods & services	7	8	7	8	8	8	7	8	8	8	7	7	8	8
1-12 All expenditure groups	85	84	85	85	86	83	82	83	84	84	87	84	86	84
13 Other expenditure items	15	16	15	15	14	17	18	17	16	16	13	16	14	16
Total expenditure	100	100	100	100	100	100	100	100	100	100	100	100	100	100

Note: The commodity and service categories are not comparable to those in publications before 2001-02.
    Please see page ix for symbols and conventions used in this report.
    This table is based on a three year average.

1 Excluding mortgage interest payments, council tax and Northern Ireland rates.

Table A35 Detailed household expenditure by UK countries and regions, 2008-2010

	North East	North West	Yorks & the Humber	East Midlands	West Midlands	East	London	South East	South West	England	Wales	Scotland	Northern Ireland	United Kingdom
Average weighted number of households (thousands)	1,280	3,130	2,170	1,980	2,200	2,270	3,030	3,080	2,510	21,640	1,280	2,380	690	26,000
Total number of households in sample (over 3 years)	730	1,770	1,460	1,210	1,470	1,550	1,410	2,190	1,520	13,300	800	1,510	1,320	16,930
Total number of persons in sample (over 3 years)	1,620	4,100	3,400	2,820	3,550	3,610	3,470	5,190	3,440	31,200	1,840	3,330	3,390	39,750
Total number of adults in sample (over 3 years)	1,280	3,160	2,610	2,190	2,730	2,790	2,590	4,020	2,700	24,080	1,450	2,670	2,520	30,720
Weighted average number of persons per household	2.3	2.4	2.3	2.3	2.4	2.3	2.5	2.3	2.3	2.4	2.3	2.2	2.6	2.3
Commodity or service						Average weekly household expenditure (£)								
1 Food & non-alcoholic drinks	**44.70**	**49.20**	**46.40**	**52.10**	**50.40**	**53.70**	**55.80**	**55.50**	**54.70**	**52.00**	**51.80**	**50.90**	**57.20**	**52.00**
1.1 Food	41.20	45.30	42.90	47.90	46.50	49.50	51.20	51.10	50.60	47.90	47.90	46.40	52.30	47.90
1.1.1 Bread, rice and cereals	4.50	4.70	4.50	5.00	5.00	4.90	5.20	4.90	5.00	4.90	4.90	5.00	6.20	4.90
1.1.2 Pasta products	0.40	0.40	0.40	0.40	0.30	0.40	0.50	0.40	0.40	0.40	0.40	0.50	0.40	0.40
1.1.3 Buns, cakes, biscuits etc.	2.80	3.10	3.00	3.10	3.10	3.30	2.90	3.50	3.60	3.20	3.20	3.20	4.00	3.20
1.1.4 Pastry (savoury)	0.70	0.70	0.70	0.80	0.60	0.80	0.60	0.80	0.70	0.70	0.70	0.70	0.70	0.70
1.1.5 Beef (fresh, chilled or frozen)	1.50	1.50	1.60	1.50	1.50	1.80	1.50	1.80	1.70	1.60	1.70	1.80	2.90	1.70
1.1.6 Pork (fresh, chilled or frozen)	0.50	0.60	0.60	0.70	0.80	0.70	0.60	0.70	0.70	0.70	0.70	0.50	0.70	0.60
1.1.7 Lamb (fresh, chilled or frozen)	0.50	0.70	0.50	0.60	0.70	0.70	1.10	0.80	0.70	0.70	0.70	0.30	0.40	0.70
1.1.8 Poultry (fresh, chilled or frozen)	1.60	1.80	1.70	1.80	2.00	2.10	2.40	2.00	2.00	2.00	1.80	1.80	2.40	2.00
1.1.9 Bacon and ham	0.80	1.10	0.90	1.00	1.00	0.90	0.70	1.00	1.00	1.00	1.10	0.90	1.40	1.00
1.1.10 Other meat and meat preparations	5.20	5.50	5.00	5.30	5.20	5.40	4.70	5.70	5.40	5.30	5.60	5.90	6.40	5.40
1.1.11 Fish and fish products	2.00	2.10	2.00	2.30	2.20	2.30	3.20	2.50	2.20	2.40	2.10	2.10	1.80	2.30
1.1.12 Milk	2.50	2.80	2.60	2.80	2.70	2.60	2.40	2.60	2.80	2.70	2.80	2.50	2.90	2.70
1.1.13 Cheese and curd	1.30	1.60	1.50	1.90	1.60	1.90	1.70	2.10	2.20	1.80	1.70	1.60	1.50	1.80
1.1.14 Eggs	0.60	0.60	0.60	0.60	0.60	0.60	0.70	0.70	0.70	0.60	0.60	0.60	0.60	0.60
1.1.15 Other milk products	1.60	1.80	1.70	1.90	1.90	2.10	2.00	2.10	2.10	1.90	1.80	1.80	2.00	1.90
1.1.16 Butter	0.40	0.40	0.40	0.40	0.30	0.30	0.40	0.40	0.40	0.40	0.40	0.40	0.40	0.40
1.1.17 Margarine, other vegetable fats and peanut butter	0.40	0.50	0.50	0.60	0.60	0.60	0.50	0.50	0.50	0.50	0.50	0.40	0.50	0.50
1.1.18 Cooking oils and fats	0.20	0.30	0.20	0.30	0.30	0.30	0.50	0.30	0.30	0.30	0.30	0.30	0.20	0.30
1.1.19 Fresh fruit	2.20	2.50	2.40	2.90	2.60	3.20	3.90	3.30	3.30	3.00	3.00	2.80	2.90	3.00
1.1.20 Other fresh, chilled or frozen fruits	0.20	0.30	0.30	0.30	0.30	0.30	0.50	0.40	0.30	0.30	0.30	0.40	0.30	0.30
1.1.21 Dried fruit and nuts	0.30	0.40	0.40	0.50	0.40	0.60	0.80	0.60	0.60	0.50	0.50	0.40	0.30	0.50
1.1.22 Preserved fruit and fruit based products	0.10	0.10	0.10	0.20	0.10	0.20	0.20	0.10	0.10	0.10	0.10	0.10	0.10	0.10
1.1.23 Fresh vegetables	3.00	3.20	3.30	3.90	3.60	4.10	5.10	4.40	4.20	4.00	3.60	3.10	3.10	3.80
1.1.24 Dried vegetables	0.00	0.00	0.00	0.00	0.00	0.00	0.10	0.00	0.00	0.00	0.00	0.10	0.00	0.00
1.1.25 Other preserved or processed vegetables	1.00	1.10	1.10	1.30	1.20	1.20	1.50	1.30	1.30	1.20	1.20	1.10	1.30	1.20
1.1.26 Potatoes	0.80	0.80	0.80	0.90	0.90	0.90	0.80	0.90	0.90	0.90	0.90	0.80	1.30	0.90
1.1.27 Other tubers and products of tuber vegetables	1.20	1.30	1.30	1.30	1.50	1.30	1.00	1.30	1.30	1.30	1.30	1.50	1.60	1.30

Note: The commodity and service categories are not comparable to those in publications before 2001-02.
The numbering system is sequential, it does not use actual COICOP codes.
Please see page ix for symbols and conventions used in this report.
This table is based on a three year average.

Table A35   Detailed household expenditure by UK countries and regions, 2008-2010 (cont.)

Average weekly household expenditure (£)

Commodity or service		North East	North West	Yorks & the Humber	East Midlands	West Midlands	East	London	South East	South West	England	Wales	Scotland	Northern Ireland	United Kingdom
1.1.28	Sugar and sugar products	0.30	0.30	0.30	0.30	0.40	0.30	0.30	0.40	0.40	0.30	0.40	0.30	0.20	0.30
1.1.29	Jams, marmalades	0.20	0.30	0.30	0.30	0.40	0.30	0.30	0.30	0.30	0.30	0.30	0.30	0.30	0.30
1.1.30	Chocolate	1.40	1.60	1.50	1.60	1.50	1.70	1.40	1.60	1.70	1.60	1.50	1.60	1.60	1.60
1.1.31	Confectionery products	0.60	0.60	0.60	0.60	0.60	0.60	0.50	0.60	0.60	0.60	0.60	0.60	0.70	0.60
1.1.32	Edible ices and ice cream	0.40	0.40	0.40	0.50	0.50	0.60	0.50	0.60	0.60	0.50	0.50	0.50	0.50	0.50
1.1.33	Other food products	2.00	2.10	2.00	2.40	2.20	2.40	2.70	2.50	2.40	2.30	2.30	2.40	2.50	2.40
1.2	Non-alcoholic drinks	3.50	4.00	3.50	4.10	3.90	4.20	4.70	4.40	4.10	4.10	3.90	4.50	4.90	4.20
1.2.1	Coffee	0.60	0.60	0.60	0.70	0.60	0.60	0.60	0.60	0.70	0.60	0.50	0.60	0.50	0.60
1.2.2	Tea	0.40	0.40	0.40	0.50	0.50	0.50	0.50	0.50	0.50	0.50	0.50	0.40	0.50	0.50
1.2.3	Cocoa and powdered chocolate	0.10	0.10	0.10	0.10	0.10	0.10	0.10	0.10	0.10	0.10	0.10	0.10	0.10	0.10
1.2.4	Fruit and vegetable juices (inc. fruit squash)	0.80	1.00	0.90	1.10	1.00	1.20	1.40	1.20	1.20	1.10	1.00	1.10	1.20	1.10
1.2.5	Mineral or spring waters	0.10	0.20	0.10	0.20	0.20	0.20	0.40	0.20	0.20	0.20	0.20	0.20	0.30	0.20
1.2.6	Soft drinks (inc. fizzy and ready to drink fruit drinks)	1.50	1.70	1.40	1.60	1.60	1.60	1.70	1.80	1.50	1.60	1.70	2.10	2.40	1.70
2	**Alcoholic drink, tobacco & narcotics**	**10.40**	**12.70**	**10.70**	**11.10**	**11.20**	**10.20**	**10.40**	**11.10**	**10.90**	**11.00**	**10.80**	**12.50**	**15.20**	**11.30**
2.1	Alcoholic drinks	6.30	7.40	6.10	6.90	6.10	6.50	6.20	7.30	7.20	6.70	6.20	7.10	6.30	6.70
2.1.1	Spirits and liqueurs (brought home)	1.00	1.60	1.10	1.20	1.40	1.30	1.00	1.20	1.20	1.20	1.20	2.00	1.70	1.30
2.1.2	Wines, fortified wines (brought home)	2.80	3.70	2.80	3.50	2.80	3.50	3.80	4.30	4.20	3.60	2.90	3.30	2.70	3.50
2.1.3	Beer, lager, ciders and perry (brought home)	2.40	2.10	2.20	2.10	1.90	1.70	1.40	1.70	1.70	1.90	2.00	1.80	1.60	1.90
2.1.4	Alcopops (brought home)	[0.10]	0.10	0.00	0.10	0.10	0.00	[0.00]	0.00	[0.00]	0.10	[0.00]	0.10	0.20	0.10
2.2	Tobacco and narcotics	4.10	5.30	4.60	4.20	5.10	3.70	4.10	3.80	3.70	4.30	4.50	5.40	9.00	4.50
2.2.1	Cigarettes	3.50	4.60	4.00	3.40	4.20	3.10	3.80	3.20	2.60	3.60	3.70	4.80	8.50	3.90
2.2.2	Cigars, other tobacco products and narcotics	0.60	0.60	0.60	0.80	0.90	0.60	0.30	0.60	1.10	0.70	0.80	0.60	0.50	0.70
3	**Clothing & footwear**	**19.40**	**22.20**	**19.10**	**19.50**	**20.80**	**22.30**	**27.00**	**21.20**	**19.50**	**21.60**	**19.90**	**21.70**	**39.40**	**22.00**
3.1	Clothing	15.70	17.60	15.60	15.70	16.50	18.20	21.40	17.30	16.00	17.40	15.80	17.90	31.40	17.70
3.1.1	Men's outer garments	4.20	4.30	3.90	4.20	4.30	4.30	5.00	4.30	3.90	4.30	4.30	5.00	7.70	4.40
3.1.2	Men's under garments	0.40	0.40	0.30	0.40	0.30	0.30	0.40	0.40	0.40	0.40	0.40	0.40	0.60	0.40
3.1.3	Women's outer garments	7.20	7.60	7.00	6.80	7.30	8.60	10.20	7.80	7.10	7.80	7.10	7.70	15.00	8.00
3.1.4	Women's under garments	0.90	1.00	0.80	1.00	1.10	1.00	1.10	1.10	1.10	1.00	0.90	1.20	1.60	1.10
3.1.5	Boys' outer garments (5-15)	0.80	0.90	0.80	0.70	0.80	0.80	0.80	0.60	0.60	0.70	0.60	0.70	1.60	0.70
3.1.6	Girls' outer garments (5-15)	0.60	1.10	0.90	0.80	0.80	1.00	1.00	0.80	0.90	0.90	1.10	0.80	1.80	0.90
3.1.7	Infants' outer garments (under 5)	0.50	0.80	0.70	0.50	0.70	0.60	0.70	0.60	0.50	0.60	0.50	0.70	1.10	0.60
3.1.8	Children's under garments (under 16)	0.40	0.40	0.30	0.30	0.30	0.40	0.50	0.30	0.30	0.40	0.30	0.30	0.60	0.40
3.1.9	Accessories	0.50	0.70	0.70	0.70	0.60	0.70	1.00	0.80	0.70	0.70	0.50	0.70	1.00	0.70
3.1.10	Haberdashery, clothing materials and clothing hire	0.10	0.30	0.10	0.10	0.30	0.10	0.20	0.20	0.30	0.20	0.10	0.30	0.20	0.20
3.1.11	Dry cleaners, laundry and dyeing	[0.10]	0.20	0.20	0.20	0.10	0.40	0.60	0.30	0.20	0.30	[0.10]	0.20	0.20	0.30
3.2	Footwear	3.70	4.60	3.50	3.80	4.20	4.10	5.60	3.90	3.60	4.20	4.10	3.80	8.00	4.30

1 Excluding mortgage interest payments, council tax and Northern Ireland rates.
2 The figure included in total expenditure is net rent as opposed to gross rent
3 Expressed as an income figure as opposed to an expenditure figure.

Table A35　　Detailed household expenditure by UK countries and regions, 2008-2010 (cont.)

Commodity or service	North East	North West	Yorks & the Humber	East Midlands	West Midlands	East	London	South East	South West	England	Wales	Scotland	Northern Ireland	United Kingdom
					Average weekly household expenditure (£)									
4　Housing (net)[1], fuel & power	**44.60**	**49.90**	**48.80**	**53.10**	**50.30**	**56.30**	**87.00**	**60.40**	**58.40**	**58.10**	**52.80**	**50.50**	**47.60**	**56.90**
4.1　Actual rentals for housing	31.30	30.20	27.90	28.30	30.00	31.20	75.40	36.80	32.20	37.50	26.50	27.40	25.20	35.70
4.1.1　Gross rent	31.30	30.00	27.90	28.20	30.00	31.20	75.30	36.70	32.20	37.40	26.50	27.40	25.10	35.60
4.1.2　less housing benefit, rebates & allowances rec'd	17.10	13.50	11.30	9.60	13.80	11.20	25.60	12.80	11.80	14.40	10.50	11.10	10.60	13.80
4.1.3　Net rent[2]	14.20	16.50	16.60	18.60	16.20	20.00	49.70	23.80	20.40	23.00	16.00	16.30	14.50	21.80
4.1.4　Second dwelling rent	-	[0.20]	[0.00]	[0.10]	-	-	[0.10]	[0.10]	[0.00]	[0.10]	-	[0.00]	[0.10]	[0.10]
4.2　Maintenance and repair of dwelling	4.80	5.40	5.70	7.00	6.70	7.20	8.50	7.90	8.40	7.00	7.60	5.80	5.20	6.90
4.3　Water supply and miscellaneous services relating to the dwelling	6.60	8.20	7.20	6.80	6.80	8.10	9.30	8.00	8.60	7.90	7.90	7.00	0.30	7.60
4.4　Electricity, gas and other fuels	19.10	19.60	19.30	20.60	20.70	20.90	19.50	20.60	21.00	20.20	21.30	21.40	27.40	20.50
4.4.1　Electricity	9.30	9.00	9.10	9.70	9.50	10.00	9.20	9.90	10.70	9.60	10.10	10.40	10.60	9.70
4.4.2　Gas	9.10	9.80	9.60	9.30	10.00	9.20	10.20	9.60	7.60	9.40	8.30	9.40	1.90	9.20
4.4.3　Other fuels	0.60	0.80	0.60	1.60	1.10	1.70	[0.10]	1.10	2.70	1.10	2.90	1.70	14.90	1.60
5　Household goods & services	**26.40**	**27.50**	**29.10**	**27.00**	**27.30**	**33.70**	**34.10**	**33.10**	**33.00**	**30.50**	**22.70**	**27.60**	**29.10**	**29.80**
5.1　Furniture and furnishings, carpets and other floor coverings	15.80	16.00	14.60	14.20	14.10	17.40	19.30	17.00	16.10	16.20	10.20	15.80	15.50	15.90
5.1.1　Furniture and furnishings	12.80	12.80	11.30	10.80	11.00	13.80	15.90	12.90	13.00	12.80	8.00	12.40	11.50	12.50
5.1.2　Floor coverings	3.00	3.20	3.30	3.40	3.10	3.60	3.40	4.10	3.20	3.40	2.20	3.50	4.00	3.40
5.2　Household textiles	1.60	1.30	1.50	1.40	1.40	2.00	1.80	1.90	1.60	1.60	1.20	1.80	1.70	1.60
5.3　Household appliances	1.20	2.80	5.60	2.40	2.60	4.60	3.50	4.10	5.20	3.70	2.60	2.10	3.00	3.50
5.4　Glassware, tableware and household utensils	1.00	1.30	1.30	1.20	1.20	1.50	1.70	1.60	1.50	1.40	1.40	1.10	1.40	1.40
5.5　Tools and equipment for house and garden	1.90	1.50	1.80	2.10	2.30	2.90	1.80	2.50	2.50	2.10	2.30	1.60	1.80	2.10
5.6　Goods and services for routine household maintenance	5.00	4.60	4.20	5.70	5.70	5.50	6.00	6.00	6.10	5.40	5.10	5.10	5.70	5.40
5.6.1　Cleaning materials	1.90	2.00	2.00	2.30	2.30	2.30	2.30	2.40	2.30	2.20	2.30	2.10	2.40	2.20
5.6.2　Household goods and hardware	1.00	1.10	1.10	1.30	1.10	1.20	1.20	1.40	1.50	1.20	1.20	1.10	1.40	1.20
5.6.3　Domestic services, carpet cleaning and hire/repair of furniture/furnishings	2.10	1.50	1.20	2.10	2.20	1.90	2.50	2.30	2.30	2.00	1.50	1.90	2.00	2.00
6　Health	**4.40**	**4.30**	**4.00**	**5.70**	**4.20**	**5.90**	**7.30**	**6.40**	**5.70**	**5.40**	**3.00**	**4.00**	**4.50**	**5.10**
6.1　Medical products, appliances and equipment	3.10	3.20	3.10	3.50	2.70	3.60	4.00	3.60	3.00	3.30	2.00	2.40	3.00	3.20
6.1.1　Medicines, prescriptions, healthcare products and equipment	1.70	2.20	1.50	2.30	1.40	1.90	2.20	1.80	1.90	1.90	1.20	1.40	1.90	1.80
6.1.2　Spectacles, lenses, accessories and repairs	1.40	1.10	1.60	1.10	1.30	1.70	1.80	1.80	1.10	1.40	0.80	1.10	1.10	1.40
6.2　Hospital services	1.30	1.00	0.90	2.30	1.50	2.30	3.30	2.80	2.70	2.10	1.00	1.60	1.50	2.00
7　Transport	**49.10**	**53.90**	**52.90**	**64.00**	**61.40**	**67.90**	**66.00**	**73.50**	**68.10**	**62.80**	**51.80**	**63.00**	**62.40**	**62.20**
7.1　Purchase of vehicles	15.00	18.00	16.70	24.20	23.40	19.40	17.90	23.00	23.50	20.30	13.90	23.10	13.30	20.00
7.1.1　Purchase of new cars and vans	3.80	5.70	4.50	8.20	7.10	6.40	7.10	7.40	7.80	6.60	5.10	7.40	5.40	6.60
7.1.2　Purchase of second hand cars or vans	10.50	12.00	11.80	15.10	15.00	11.60	10.20	15.00	14.50	12.90	8.20	15.00	7.50	12.70
7.1.3　Purchase of motorcycles and other vehicles	[0.70]	0.30	[0.40]	0.90	1.30	[1.40]	[0.50]	0.60	1.20	0.80	[0.50]	0.70	[0.40]	0.80

Table A35 Detailed household expenditure by UK countries and regions, 2008-2010 (cont.)

Commodity or service		North East	North West	Yorks & the Humber	East Midlands	West Midlands	East	London	South East	South West	England	Wales	Scotland	Northern Ireland	United Kingdom
								Average weekly household expenditure (£)							
7.2	Operation of personal transport	27.10	27.70	27.80	32.40	30.90	36.40	26.70	38.00	36.20	31.60	31.60	27.80	38.30	31.50
7.2.1	Spares and accessories	3.00	1.30	2.10	2.10	2.10	2.90	1.60	2.80	2.30	2.20	1.80	2.00	2.20	2.10
7.2.2	Petrol, diesel and other motor oils	18.20	19.30	18.70	21.20	20.80	23.20	15.50	23.20	23.30	20.40	22.90	19.20	28.40	20.60
7.2.3	Repairs and servicing	4.30	5.00	5.10	6.60	5.40	7.50	7.30	9.30	7.60	6.60	5.10	5.10	5.90	6.40
7.2.4	Other motoring costs	1.60	2.10	1.80	2.40	2.60	2.80	2.30	2.80	3.00	2.40	1.80	1.50	1.70	2.30
7.3	Transport services	7.10	8.20	8.40	7.40	7.10	12.10	21.50	12.40	8.50	10.80	6.30	12.10	10.80	10.70
7.3.1	Rail and tube fares	0.90	1.40	1.60	1.70	1.20	5.30	4.20	4.40	2.20	2.70	1.00	2.20	0.90	2.50
7.3.2	Bus and coach fares	1.70	1.70	1.70	1.30	1.60	0.90	1.40	1.00	1.10	1.40	1.00	1.70	1.10	1.40
7.3.3	Combined fares	[0.20]	0.10	0.40	[0.10]	[0.10]	1.10	7.40	1.00	[0.10]	1.40	[0.10]	[0.30]	[0.00]	1.20
7.3.4	Other travel and transport	4.30	5.00	4.80	4.30	4.20	4.70	8.40	6.00	5.00	5.40	4.30	7.90	8.80	5.60
8	Communication	10.90	11.80	10.60	12.30	11.60	13.20	15.40	13.00	12.80	12.60	10.30	11.70	13.60	12.40
8.1	Postal services	0.30	0.40	0.30	0.50	0.40	0.60	0.50	0.60	0.70	0.50	0.30	0.50	0.40	0.50
8.2	Telephone and telefax equipment	0.50	0.60	0.50	0.40	0.50	0.60	0.50	0.40	0.50	0.50	[0.60]	0.70	0.70	0.50
8.3	Telephone and telefax services	9.90	10.60	9.50	10.90	10.50	11.70	14.10	11.70	11.20	11.30	9.10	10.20	12.30	11.10
8.4	Internet subscription fees	0.20	0.20	0.30	0.50	0.20	0.30	0.40	0.30	0.40	0.30	0.40	0.40	0.10	0.30
9	Recreation & culture	47.40	54.30	55.60	57.50	56.30	61.40	61.50	66.20	62.20	58.90	53.70	58.00	56.30	58.50
9.1	Audio-visual, photographic and information processing equipment	6.70	5.60	7.20	6.40	6.70	7.20	7.10	8.60	8.00	7.10	7.90	6.70	7.80	7.10
9.1.1	Audio equipment and accessories, CD players	1.00	1.20	1.60	1.10	1.20	2.00	1.40	1.50	1.70	1.40	1.70	1.30	1.00	1.40
9.1.2	TV, video and computers	5.40	3.90	4.80	4.60	4.00	4.50	4.70	6.20	5.90	4.90	4.90	5.10	6.30	4.90
9.1.3	Photographic, cine and optical equipment	[0.40]	0.50	0.80	0.70	1.50	0.70	1.10	1.00	0.40	0.80	[1.30]	0.30	0.50	0.80
9.2	Other major durables for recreation and culture	[0.50]	2.90	3.10	3.30	4.10	2.00	[5.80]	2.60	2.80	3.20	5.50	2.40	5.50	3.30
9.3	Other recreational items and equipment, gardens and pets	8.70	10.30	9.40	12.10	10.60	12.60	8.70	12.90	12.80	11.00	9.10	11.00	10.10	10.90
9.3.1	Games, toys and hobbies	2.00	2.10	1.50	1.90	1.80	2.40	1.80	2.10	2.50	2.00	1.70	1.70	2.40	2.00
9.3.2	Computer software and games	1.50	1.50	1.20	1.80	1.70	1.90	1.30	2.20	1.60	1.60	0.80	2.20	1.60	1.70
9.3.3	Equipment for sport, camping and open-air recreation	0.70	1.00	1.80	0.70	0.60	1.10	1.00	0.90	0.90	1.00	0.70	0.90	1.00	0.90
9.3.4	Horticultural goods, garden equipment and plants	1.80	2.60	1.90	2.70	2.50	2.80	2.30	3.10	2.90	2.60	2.20	2.70	2.30	2.60
9.3.5	Pets and pet food	2.70	3.20	3.00	5.00	4.00	4.40	2.50	4.50	4.90	3.80	3.60	3.50	2.90	3.70
9.4	Recreational and cultural services	15.60	16.90	15.10	16.40	16.60	18.20	20.30	21.80	17.70	17.90	14.00	17.70	18.30	17.70
9.4.1	Sports admissions, subscriptions, leisure class fees and equipment hire	3.40	4.60	3.90	4.60	4.10	5.90	7.80	5.90	5.60	5.30	2.70	4.30	4.60	5.10
9.4.2	Cinema, theatre and museums etc.	1.60	2.00	1.40	2.20	1.80	2.10	3.30	2.30	2.40	2.20	2.00	2.00	2.50	2.20
9.4.3	TV, video, satellite rental, cable subscriptions, TV licences and the Internet	5.50	6.00	5.80	5.60	5.40	6.00	5.60	5.90	5.30	5.70	5.50	6.00	5.80	5.70
9.4.4	Miscellaneous entertainments	0.80	1.20	0.80	1.40	1.10	1.10	1.10	1.10	1.50	1.10	0.70	1.20	1.70	1.10
9.4.5	Development of film, deposit for film development, passport photos, holiday and school photos	0.20	0.20	0.30	0.30	0.40	0.20	0.30	0.30	0.50	0.30	[0.50]	0.30	0.30	0.30
9.4.6	Gambling payments	4.10	2.90	3.00	2.40	3.70	2.80	2.10	6.30	2.40	3.30	2.70	3.90	3.40	3.30

Table A35   Detailed household expenditure by UK countries and regions, 2008-2010 (cont.)

Average weekly household expenditure (£)

Commodity or service		North East	North West	Yorks & the Humber	East Midlands	West Midlands	East	London	South East	South West	England	Wales	Scotland	Northern Ireland	United Kingdom
9.4	Recreational and cultural services	15.60	16.90	15.10	16.40	16.60	18.20	20.30	21.80	17.70	17.90	14.00	17.70	18.30	17.70
9.4.1	Sports admissions, subscriptions, leisure class fees and equipment hire	3.40	4.60	3.90	4.60	4.10	5.90	7.80	5.90	5.60	5.30	2.70	4.30	4.60	5.10
9.4.2	Cinema, theatre and museums etc.	1.60	2.00	1.40	2.20	1.80	2.10	3.30	2.30	2.40	2.20	2.00	2.00	2.50	2.20
9.4.3	TV, video, satellite rental, cable subscriptions, TV licences and the Internet	5.50	6.00	5.80	5.60	5.40	6.00	5.60	5.90	5.30	5.70	5.50	6.00	5.80	5.70
9.4.4	Miscellaneous entertainments	0.80	1.20	0.80	1.40	1.10	1.10	1.10	1.10	1.50	1.10	0.70	1.20	1.70	1.10
9.4.5	Development of film, deposit for film development, passport photos, holiday and school photos	0.20	0.20	0.30	0.30	0.40	0.20	0.30	0.30	0.50	0.30	[0.50]	0.30	0.30	0.30
9.4.6	Gambling payments	4.10	2.90	3.00	2.40	3.70	2.80	2.10	6.30	2.40	3.30	2.70	3.90	3.40	3.30
9.5	Newspapers, books and stationery	5.10	5.40	5.10	5.90	5.60	6.20	5.80	6.60	6.90	5.90	5.70	6.20	6.50	5.90
9.5.1	Books	1.00	1.10	1.10	1.40	1.10	1.40	1.60	1.70	1.80	1.40	1.20	1.20	1.30	1.40
9.5.2	Diaries, address books, cards etc.	1.60	1.80	1.60	1.90	1.90	2.00	1.80	2.10	2.10	1.90	1.50	1.80	1.70	1.90
9.5.3	Newspapers	1.70	1.60	1.50	1.70	1.80	1.80	1.50	1.70	2.00	1.70	2.10	2.30	2.50	1.80
9.5.4	Magazines and periodicals	0.80	0.80	0.80	1.00	0.80	0.90	0.80	1.00	1.10	0.90	0.90	0.90	1.00	0.90
9.6	Package holidays	10.90	13.20	15.70	13.40	12.70	15.30	13.90	13.80	14.00	13.80	11.50	14.00	8.00	13.50
9.6.1	Package holidays - UK	0.60	1.10	1.20	1.10	1.10	1.50	0.60	1.10	1.30	1.10	0.80	0.70	[0.50]	1.00
9.6.2	Package holidays - abroad	10.30	12.10	14.50	12.20	11.60	13.80	13.30	12.70	12.60	12.70	10.70	13.30	7.50	12.50
10	Education	2.30	5.40	3.80	9.10	4.90	5.80	16.60	9.80	8.30	7.90	9.70	5.20	8.10	7.70
10.1	Education fees	1.90	5.10	3.30	8.90	4.70	5.50	16.30	9.40	8.00	7.60	9.50	5.00	7.60	7.40
10.2	Payments for school trips, other ad-hoc expenditure	[0.40]	0.30	0.50	[0.20]	0.20	0.30	0.40	0.40	0.30	0.30	[0.10]	0.20	0.50	0.30
11	Restaurants & hotels	30.50	36.90	34.10	36.60	36.30	38.60	51.40	41.00	36.90	38.90	30.30	36.50	46.30	38.40
11.1	Catering services	26.10	30.50	28.20	29.30	29.70	30.60	42.20	31.50	29.00	31.40	27.40	30.30	42.40	31.40
11.1.1	Restaurant and café meals	9.50	12.40	11.90	13.00	11.70	14.50	17.80	14.90	13.70	13.60	11.70	12.90	16.30	13.50
11.1.2	Alcoholic drinks (away from home)	7.30	7.90	6.70	6.90	7.20	6.80	8.40	6.70	6.50	7.20	6.70	7.10	9.80	7.20
11.1.3	Take away meals eaten at home	3.60	4.00	3.80	3.30	4.00	3.60	4.40	3.60	3.40	3.80	3.70	4.20	7.10	3.90
11.1.4	Other take-away and snack food	3.70	4.00	3.90	3.60	3.60	3.90	6.70	4.00	3.40	4.20	3.40	4.20	6.10	4.20
11.1.5	Contract catering (food) and canteens	2.00	2.20	1.90	2.30	3.30	1.70	4.80	2.30	2.10	2.60	1.90	2.00	3.10	2.50
11.2	Accommodation services	4.40	6.40	5.90	7.30	6.50	8.00	9.30	9.50	7.90	7.50	2.90	6.20	4.00	7.10
11.2.1	Holiday in the UK	1.60	2.70	3.40	3.30	3.10	3.40	2.40	4.00	3.50	3.10	1.70	2.70	1.50	2.90
11.2.2	Holiday abroad	2.80	3.60	2.40	4.00	3.40	4.70	6.80	5.50	4.30	4.30	1.20	3.40	2.50	4.10
11.2.3	Room hire	[0.00]	[0.10]	[0.00]	[0.00]	[0.10]	[0.00]	[0.10]	[0.00]	[0.00]	0.10	[0.00]	[0.10]	[0.00]	0.10
12	Miscellaneous goods & services	27.70	33.70	29.30	35.30	33.20	39.10	42.90	41.40	36.70	36.30	26.30	33.10	37.00	35.50
12.1	Personal care	8.20	9.90	9.50	9.80	9.60	11.10	11.20	12.00	10.20	10.30	8.70	10.40	12.80	10.30
12.1.1	Hairdressing, beauty treatment	2.30	3.30	3.30	2.80	2.90	3.60	3.30	4.10	3.20	3.30	2.40	3.00	4.10	3.20
12.1.2	Toilet paper	0.60	0.70	0.70	0.80	0.80	0.80	0.80	0.80	0.80	0.80	0.80	0.70	1.00	0.80
12.1.3	Toiletries and soap	1.70	2.00	1.80	2.10	1.90	2.20	2.50	2.40	2.20	2.10	2.00	2.20	2.60	2.10
12.1.4	Baby toiletries and accessories (disposable)	0.50	0.60	0.60	0.60	0.70	0.60	0.70	0.70	0.60	0.60	0.70	0.50	0.80	0.60
12.1.5	Hair products, cosmetics and electrical personal appliances	3.00	3.30	3.00	3.50	3.30	3.80	3.80	3.90	3.40	3.50	2.80	3.90	4.30	3.50

Table A35   **Detailed household expenditure by UK countries and regions, 2008-2010 (cont.)**

Commodity or service		North East	North West	Yorks & the Humber	East Midlands	West Midlands	East	London	South East	South West	England	Wales	Scotland	Northern Ireland	United Kingdom
		Average weekly household expenditure (£)													
12.2	Personal effects	2.90	3.20	2.10	2.90	2.90	3.40	3.80	3.40	4.30	3.30	1.70	3.00	2.60	3.20
12.3	Social protection	2.60	2.90	2.00	3.20	3.00	3.50	5.70	3.30	3.40	3.40	1.90	3.70	3.80	3.40
12.4	Insurance	11.80	14.30	13.60	14.90	14.60	16.00	17.40	16.70	15.30	15.20	12.30	12.20	14.30	14.80
12.4.1	Household insurances - structural, contents and appliances	4.50	4.80	4.80	5.00	4.90	5.20	5.60	5.40	5.10	5.10	4.90	4.70	4.00	5.00
12.4.2	Medical insurance premiums	0.60	1.10	1.00	1.40	1.30	1.90	2.90	2.60	1.60	1.70	0.50	0.90	0.80	1.60
12.4.3	Vehicle insurance including boat insurance	6.70	8.40	7.70	8.20	8.40	8.80	8.60	8.60	8.50	8.30	6.70	6.30	9.30	8.10
12.4.4	Non-package holiday, other travel insurance	[0.00]	[0.00]	[0.10]	[0.40]	[0.00]	[0.00]	[0.30]	[0.10]	[0.20]	0.10	[0.10]	[0.30]	[0.10]	0.20
12.5	Other services	2.30	3.40	2.10	4.40	3.10	5.20	4.70	5.90	3.50	4.00	1.70	3.90	3.60	3.90
12.5.1	Moving house	1.00	1.60	0.70	2.30	1.60	2.60	2.70	3.20	1.60	2.00	0.60	1.80	0.70	1.90
12.5.2	Bank, building society, post office, credit card charges	0.40	0.50	0.40	0.40	0.40	0.40	0.60	0.50	0.60	0.50	0.30	0.50	0.30	0.50
12.5.3	Other services and professional fees	1.00	1.30	1.00	1.70	1.10	2.10	1.40	2.20	1.30	1.50	0.70	1.60	2.60	1.50
1-12	**All expenditure groups**	317.90	361.70	344.50	383.20	367.80	408.00	475.60	432.60	407.30	396.00	343.10	374.60	416.70	391.90
13	**Other expenditure items**	54.70	68.80	61.00	66.20	62.20	85.30	102.20	90.50	75.30	76.50	50.90	72.60	66.10	74.60
13.1	Housing: mortgage interest payments council tax etc.	40.40	45.70	41.50	47.30	44.90	58.80	65.80	63.80	52.90	52.60	37.20	50.30	36.70	51.20
13.2	Licences, fines and transfers	2.40	3.00	3.10	3.30	3.30	3.60	3.40	3.70	4.10	3.40	2.90	3.00	3.60	3.30
13.3	Holiday spending	4.20	10.10	8.00	6.50	5.00	8.60	18.20	10.70	7.50	9.50	4.00	7.20	10.70	9.00
13.4	Money transfers and credit	7.80	10.00	8.50	9.10	9.00	14.40	14.80	12.40	10.80	11.10	6.80	12.00	15.00	11.00
13.4.1	Money, cash gifts given to children	[0.10]	0.10	0.10	0.10	0.20	0.10	[0.20]	0.10	0.10	0.10	0.20	0.10	0.30	0.10
13.4.2	Cash gifts and donations	6.30	8.80	7.20	7.40	7.10	12.30	12.10	9.90	8.60	9.10	5.40	10.50	13.00	9.20
13.4.3	Club instalment payments (child) and interest on credit cards	1.30	1.10	1.20	1.50	1.80	2.00	2.50	2.30	2.20	1.80	1.20	1.40	1.70	1.70
	Total expenditure	372.70	430.50	405.50	449.40	430.10	493.40	577.80	523.20	482.60	472.40	394.00	447.20	482.80	466.50
14	**Other items recorded**														
14.1	Life assurance, contributions to pension funds	15.40	17.60	16.20	20.20	17.20	20.80	24.40	23.70	21.10	20.10	17.40	21.70	15.30	20.00
14.2	Other insurance inc. friendly societies	0.80	1.20	1.30	1.40	1.10	1.80	1.20	1.70	1.60	1.40	0.60	1.20	0.80	1.30
14.3	Income tax, payments less refunds	59.00	76.70	63.80	84.50	74.90	98.40	163.90	118.70	81.20	95.80	70.80	89.60	67.70	93.20
14.4	National insurance contributions	22.00	25.70	23.90	27.10	25.80	30.10	35.30	32.10	25.20	28.00	23.30	29.00	24.70	27.80
14.5	Purchase or alteration of dwellings, mortgages	31.70	35.60	34.80	43.30	45.70	55.30	54.90	63.90	47.20	46.90	35.50	37.30	29.70	45.00
14.6	Savings and investments	3.90	4.70	4.90	5.00	4.40	7.00	7.70	8.10	8.90	6.30	2.60	6.60	3.60	6.10
14.7	Pay off loan to clear other debt	2.40	2.10	1.90	2.20	2.90	2.40	2.10	3.30	2.40	2.40	1.20	1.90	[0.50]	2.30
14.8	Windfall receipts from gambling etc[3]	2.20	2.00	1.30	1.10	1.80	1.60	1.60	3.80	1.20	1.90	1.20	2.60	2.10	1.90

Table A36   **Household expenditure by urban/rural areas (GB)[1], 2008-2010**

	Urban	Rural
Average number of weighted households (thousands)	19,680	5,620
Total number of households in sample (over 3 years)	11,920	3,690
Total number of persons in sample (over 3 years)	27,770	8,590
Total number of adults in sample (over 3 years)	21,440	6,760
Weighted average number of persons per household	2.3	2.3

Commodity or service	Average weekly household expenditure (£)	
1 Food & non-alcoholic drinks	50.80	55.80
2 Alcoholic drinks, tobacco & narcotics	10.80	12.20
3 Clothing & footwear	21.50	21.30
4 Housing (net)[2], fuel & power	58.00	54.50
5 Household goods & services	28.70	34.00
6 Health	5.00	5.60
7 Transport	58.30	76.00
8 Communication	12.40	12.30
9 Recreation & culture	55.80	68.00
10 Education	7.50	8.40
11 Restaurants & hotels	37.90	39.30
12 Miscellaneous goods & services	34.60	38.60
1-12 All expenditure groups	381.40	426.10
13 Other expenditure items	73.20	80.20
Total expenditure	**454.60**	**506.30**
Average weekly expenditure per person (£)		
Total expenditure	**194.20**	**218.00**

Note: The commodity and service categories are not comparable to those in publications before 2001-02.
      Please see page ix for symbols and conventions used in this report.
      This table is based on a three year average.

1 Combined urban/rural classification for England & Wales and Scotland - see definitions in Appendix B.
2 Excludes mortgage interest payments, council tax.

Table A37   Income and source of income by household composition, 2010
United Kingdom

Composition of household	Weighted number of households	Number of households in the sample	Weekly household income		Source of income					
			Disposable	Gross	Wages and salaries	Self employment	Invest-ments	Annuities and pensions[1]	Social security benefits[2]	Other sources
	(000s)	Number	£	£	Percentage of gross weekly household income					
All households	**26,320**	**5,260**	**578**	**700**	**65**	**10**	**2**	**8**	**14**	**1**
Composition of household										
One adult	7,800	1,510	293	340	45	6	4	16	28	1
Retired households mainly dependent on state pensions[3]	790	150	176	177	-	-	1	4	95	[0]
Other retired households	2,980	570	256	276	-	-	6	40	54	[0]
Non-retired households	4,030	800	343	419	71	10	3	5	10	1
One adult, one child	690	160	239	260	44	[1]	1	[1]	47	6
One adult, two or more children	720	170	343	368	29	[11]	0	[0]	53	6
One man and one woman	7,770	1,680	622	759	61	7	3	14	14	1
Retired households mainly dependent on state pensions[3]	440	110	278	281	[0]	-	1	8	90	[0]
Other retired households	2,140	500	474	522	4	[1]	7	45	42	[0]
Non-retired households	5,180	1,070	712	898	76	8	3	7	5	1
Two men or two women	750	130	602	738	75	[5]	2	6	9	2
Two men or two women with children	150	30	462	516	61	[3]	[0]	[1]	35	[1]
One man one woman, one child	2,190	430	694	867	79	11	1	1	8	1
One man one woman, two children	2,130	450	884	1,119	78	14	1	[0]	7	1
One man one woman, three children	530	110	909	1,164	73	14	2	[0]	11	0
Two adults, four or more children	170	40	605	688	58	[5]	[1]	-	36	[1]
Three adults	1,560	270	951	1,118	64	19	2	6	7	2
Three adults, one or more children	910	160	857	1,039	72	10	1	3	12	1
Four or more adults	640	90	978	1,169	72	10	1	5	8	5
Four or more adults, one or more children	320	40	976	1,163	79	[2]	[1]	[3]	13	[2]

Note: Please see page ix for symbols and conventions used in this report.

1 Other than social security benefits.

2 Excluding housing benefit and council tax benefit (rates rebate in Northern Ireland) - see definitions in Appendix B.

3 Mainly dependent on state pension and not economically active - see definitions in Appendix B.

Table A38   **Income and source of income by age of household reference person, 2010**
United Kingdom

Age of head of household	Weighted number of house-holds	Number of house-holds in the sample	Weekly household income Dispo-sable	Gross	Wages and salaries	Self employ-ment	Invest-ments	Annuities and pensions[1]	Social security benefits[2]	Other sources
	(000s)	Number	£	£	Percentage of gross weekly household income					
Less than 30	2,810	450	473	569	79	5	1	[0]	10	4
30 to 49	9,540	1,890	714	894	80	10	1	0	8	1
50 to 64	7,020	1,490	641	783	64	13	3	10	9	1
65 to 74	3,420	800	420	459	15	8	4	31	41	0
75 or over	3,530	630	325	350	3	[2]	6	36	52	[0]

Note: Please see page xiii for symbols and conventions used in this report.

1 Other than social security benefits.
2 Excluding housing benefit and council tax benefit (rates rebate in Northern Ireland) - see defintions in Appendix B.

Table A39   **Income and source of income by gross income quintile group, 2010**
United Kingdom

Gross income quintile group	Weighted number of house-holds	Number of house-holds in the sample	Weekly household income Dispo-sable	Gross	Wages and salaries	Self employ-ment	Invest-ments	Annuities and pensions[1]	Social security benefits[2]	Other sources
	(000s)	Number	£	£	Percentage of gross weekly household income					
Lowest twenty per cent	5,270	1,040	151	157	8	3	2	10	76	2
Second quintile group	5,260	1,090	297	319	27	4	2	16	48	2
Third quintile group	5,270	1,080	456	525	53	7	2	15	22	1
Fourth quintile group	5,270	1,040	673	815	73	7	2	8	8	1
Highest twenty per cent	5,260	1,020	1,317	1,688	76	13	3	5	2	1

Note: Please see page ix for symbols and conventions used in this report.

1 Other than social security benefits.
2 Excluding housing benefit and council tax benefit (rates rebate in Northern Ireland) - see defintions in Appendix B.

Table A40 **Income and source of income by household tenure, 2010**
United Kingdom

Tenure of dwelling	Weighted number of house-holds	Number of house-holds in the sample	Weekly household income		Source of income					
			Dispo-sable	Gross	Wages and salaries	Self employ-ment	Invest-ments	Annuities and pensions[1]	Social security benefits[2]	Other sources
	(000s)	Number	£	£	Percentage of gross weekly household income					
Owners										
Owned outright	8,340	1,770	544	635	39	12	5	22	21	0
Buying with a mortgage[3]	9,100	1,830	793	1,001	80	10	1	3	5	1
All	17,430	3,600	674	826	65	11	3	10	11	1
Social rented from										
Council[4]	2,590	500	285	313	47	[2]	0	4	45	1
Registered social landlord[5]	2,150	430	309	337	41	4	0	5	48	3
All	4,750	920	296	324	44	3	0	4	46	2
Private rented[6]										
Rent free	370	70	392	463	62	[8]	2	7	20	[1]
Rent paid, unfurnished[7]	2,890	540	495	598	75	7	2	2	13	2
Rent paid, furnished	880	130	561	679	79	[3]	1	[1]	5	10
All	4,150	740	500	603	75	6	2	2	12	4

Note: Please see page ix for symbols and conventions used in this report.
1 Other than social security benefits.
2 Excluding housing benefit and council tax benefit (rates rebate in Northern Ireland) - see defintions in Appendix B.
3 Including shared owners (who own part of the equity and pay mortgage, part rent).
4 "Council" includes local authorities, new towns, and scottish homes, but see note 5 below.
5 Formerly housing association.
6 All tenants whose accomodation goes with the job of someone in the household are allocated to "rented privately", even if the landlord is a local authority, housing association, or housing action trust, or if the accomodation is rent free. Squatters are also included in this category.
7 Unfurnished' includes the answers: 'partly furnished'.

Table A41   Income and source of income by UK countries and regions, 2008-2010

| | Weighted number of house-holds | Total number of house-holds | Weekly household income | | Source of income | | | | | |
			Dispo-sable	Gross	Wages and salaries	Self employ-ment	Invest-ments	Annuities and pensions[1]	Social security benefits[2]	Other sources
Government Office Regions	(000s)	Number	£	£	Percentage of gross weekly household income					
United Kingdom	26,000	16,930	573	699	66	9	3	8	13	1
North East	1,280	730	459	543	64	7	2	7	19	1
North West	3,130	1,770	512	619	66	7	2	8	17	1
Yorkshire and the Humber	2,170	1,460	477	570	66	7	2	8	17	1
East Midlands	1,980	1,210	542	661	68	7	3	7	14	1
West Midlands	2,200	1,470	520	624	64	9	2	8	16	1
East	2,270	1,550	616	749	66	12	2	7	11	1
London	3,030	1,410	776	982	68	12	5	6	9	1
South East	3,080	2,190	651	810	67	10	3	8	11	1
South West	2,510	1,520	555	666	61	10	4	10	15	1
England	21,640	13,300	582	711	66	9	3	8	13	1
Wales	1,280	800	507	604	61	10	2	9	18	1
Scotland	2,380	1,510	547	669	70	7	2	7	13	1
Northern Ireland	690	1,320	511	609	63	10	1	7	18	1

Note: Please see page ix for symbols and conventions used in this report.
This table is based on a three year average.
1 Other than social security benefits.
2 Excluding housing benefit and council tax benefit (rates rebate in Northern Ireland) - see defintions in Appendix B.

Table A42   Income and source of income by GB urban/rural area, 2008-2010

| | Weighted number of house-holds | Total number of house-holds | Weekly household income | | Source of income | | | | | |
			Dispo-sable	Gross	Wages and salaries	Self employ-ment	Invest-ments	Annuities and pensions[1]	Social security benefits[2]	Other sources
GB urban rural areas	(000s)	Number	£	£	Percentage of gross weekly household income					
Urban	19,680	11,920	561	685	67	8	3	7	14	1
Rural	5,620	3,690	621	758	62	11	4	10	13	1

Note: Please see page ix for symbols and conventions used in this report.
This table is based on a three year average.
1 Other than social security benefits.
2 Excluding housing benefit and council tax benefit (rates rebate in Northern Ireland) - see definitions in Appendix B.

Table A43   Income and source of income by socio-economic classification, 2010
United Kingdom

| | Weighted number of house-holds | Number of house-holds in the sample | Weekly household income | | Source of income | | | | | |
			Dispo-sable	Gross	Wages and salaries	Self employ-ment	Invest-ments	Annuities and pensions[1]	Social security benefits[2]	Other sources
NS-SEC Group[3]	(000s)	Number	£	£	Percentage of gross weekly household income					
Large employers/higher managerial	1,260	260	1,209	1,653	92	2	1	2	2	0
Higher professional	1,770	350	1,131	1,410	67	26	2	3	2	0
Lower managerial and professional	4,620	910	794	1,014	86	5	2	3	4	1
Intermediate	1,350	270	582	708	82	[3]	2	4	8	2
Small employers	1,580	330	683	744	20	60	4	4	10	1
Lower supervisory	1,690	320	603	749	89	2	0	2	6	1
Semi-routine	1,770	350	460	542	80	[2]	0	3	14	1
Routine	1,520	290	513	612	84	[1]	0	3	11	1
Long-term unemployed[4]	490	100	250	258	[15]	[0]	[5]	[3]	75	[2]
Students	590	100	477	554	64	[5]	1	[1]	9	20
Occupation not stated[5]	9,680	1,990	325	354	10	1	5	33	50	1

Note: Please see page ix for symbols and conventions used in this report.
1 Other than social security benefits.
2 Excluding housing benefit and council tax benefit (rates rebate in Northern Ireland) - see definitions in Appendix B.
3 National Statistics Socio-economic classification (NS-SEC) - see defintions in Appendix B.
4 Includes those who have never worked.
5 Includes those who are economically inactive - see defintions in Appendix B.

Table A44   Income and source of income, 1970 to 2010

United Kingdom

	Weighted number of households	Number of households in the sample	Weekly household income[1]				Source of income					
			Current prices		Constant prices		Percentage of gross weekly household income					
			Dispo-sable	Gross	Dispo-sable	Gross	Wages and salaries	Self employ-ment	Invest-ments	Annuities and pensions[2]	Social security benefits[3]	Other sources
	(000s)	Number	£	£	£	£						
1970		6,390	28	34	344	414	77	7	4	3	9	1
1980		6,940	115	140	384	470	75	6	3	3	13	1
1990		7,050	258	317	457	562	67	10	6	5	11	1
1995-96		6,800	307	381	457	567	64	9	5	7	14	2
1996-97		6,420	325	397	472	577	65	9	4	7	14	1
1997-98		6,410	343	421	483	593	67	8	4	7	13	1
1998-99[4]	24,660	6,630	371	457	507	624	68	8	4	7	12	1
1999-2000	25,340	7,100	391	480	526	645	66	10	5	7	12	1
2000-01	25,030	6,640	409	503	534	656	67	9	4	7	12	1
2001-02[5]	24,450	7,470	442	541	569	695	68	8	4	7	12	1
2002-03	24,350	6,930	453	552	571	696	68	8	3	7	12	1
2003-04	24,670	7,050	464	570	569	699	67	9	3	7	13	1
2004-05	24,430	6,800	489	601	581	714	68	8	3	7	13	1
2005-06	24,800	6,790	500	616	579	713	67	8	3	7	13	1
2006[6]	24,790	6,650	521	642	588	724	67	9	3	7	12	1
2006[7]	25,440	6,650	515	635	582	716	67	9	3	7	13	1
2007	25,350	6,140	534	659	578	714	67	8	4	7	13	1
2008	25,690	5,850	582	713	605	742	67	9	4	7	12	1
2009	25,980	5,830	558	683	584	714	66	8	3	8	14	1
2010	26,320	5,260	578	700	578	700	65	10	2	8	14	1

Note: Please see page ix for symbols and conventions used in this report.

1 Does not include imputed income from owner-occupied and rent-free households.
2 Other than social security benefits.
3 Excluding housing benefit and council tax benefit (rates rebate in Northern Ireland) and their predecessors in earlier years - see Appendix B.
4 Based on weighted data from 1998-99.
5 From 2001-02 onwards, weighting is based on the population estimates from the 2001 census.
6 From 1998-99 to this version of 2006, figures shown are based on weighted data using non-reponsive weights based on the 1991 Census and population figures from the 1991 and 2001 Censuses.
7 From this version of 2006, figures shown are based on weighted data using updated weights, with non-response weights and poulation figures based on the 2001 Census.

Table A45 **Percentage of households with durable goods 1970 to 2010**

United Kingdom

	Car/ van	Central heating[1]	Washing machine	Tumble dryer	Dish- washer	Micro-wave	Tele-phone	Mobile phone	DVD Player	Satellite receiver[2]	Cd player	Home computer	Internet connection
1970	52	30	65	--	--	--	35	--	--	--	--	--	--
1975	57	47	72	--	--	--	52	--	--	--	--	--	--
1980	60	59	79	--	--	--	72	--	--	--	--	--	--
1985	63	69	83	--	--	--	81	--	--	--	--	13	--
1990	67	79	86	--	--	--	87	--	--	--	--	17	--
1994-95	69	84	89	50	18	67	91	--	--	--	46	--	--
1995-96	70	85	91	50	20	70	92	--	--	--	51	--	--
1996-97	69	87	91	51	20	75	93	16	--	19	59	27	--
1997-98	70	89	91	51	22	77	94	20	--	26	63	29	--
1998-99	72	89	92	51	24	80	95	26	--	27	68	32	9
1998-99[3]	72	89	92	51	23	79	95	27	--	28	68	33	10
1999-2000	71	90	91	52	23	80	95	44	--	32	72	38	19
2000-01	72	91	93	53	25	84	93	47	--	40	77	44	32
2001-02[4]	74	92	93	54	27	86	94	64	--	43	80	49	39
2002-03	74	93	94	56	29	87	94	70	31	45	83	55	45
2003-04	75	94	94	57	31	89	92	76	50	49	86	58	49
2004-05	75	95	95	58	33	90	93	78	67	58	87	62	53
2005-06	74	94	95	58	35	91	92	79	79	65	88	65	55
2006[5]	76	95	96	59	38	91	91	80	83	71	88	67	59
2006[6]	74	95	96	59	37	91	91	79	83	70	87	67	58
2007	75	95	96	57	37	91	89	78	86	77	86	70	61
2008	74	95	96	59	37	92	90	79	88	82	86	72	66
2009	76	95	96	58	39	93	88	81	90	86	84	75	71
2010	75	96	96	57	40	92	87	80	88	88	83	77	73

Note: Please see page ix for symbols and conventions used in this report.

-- Data not available.

1 Full or partial.

2 Includes digital and cable receivers.

3 From this version of 1998-99, figures shown are based on weighted data and including children's expenditure.

4 From 2001-02 onwards, weighting is based on the population figures from the 2001 census.

5 From 1998-99 to this version of 2006, figures shown are based on weighted data using non-response weights based on the 1991 Census and population figures from the 1991 and 2001 Census.

6 From this version of 2006, figures shown are based on weighted data using updated weights, with non response weights and population figures based on the 2001 Census.

Table A46 **Percentage1 of households with durable goods by income group and household composition, 2010**

United Kingdom

	Central heating[2]	Washing machine	Tumble dryer	Micro- wave	Dish- washer	CD player
All households	*96*	*96*	*57*	*92*	*40*	*83*
Gross income decile group						
Lowest ten per cent	92	87	39	86	12	68
Second decile group	96	92	43	92	17	73
Third decile group	95	95	45	91	23	77
Fourth decile group	96	97	56	92	26	82
Fifth decile group	96	98	61	92	37	87
Sixth decile group	97	97	60	93	39	89
Seventh decile group	97	99	65	92	50	86
Eighth decile group	97	99	60	95	51	85
Ninth decile group	98	100	66	94	65	89
Highest ten per cent	97	100	74	92	81	92
Household composition						
One adult, retired households [3]	93	88	37	90	18	68
One adult, non-retired households	93	92	40	88	23	79
One adult, one child	99	99	49	91	13	74
One adult, two or more children	99	98	59	89	30	79
Two adults, retired households[3]	96	98	57	93	26	77
Two adults, non-retired households	96	99	60	92	49	87
Two adults, one child	96	99	65	93	45	82
Two adults, two or more children	98	99	73	95	59	88
All other households without children	97	99	65	92	50	87
All other households with children	97	100	73	97	58	90

	Home computer	Internet connection	Tele- phone	Mobile phone	Satellite receiver[4]	DVD Player
All households	*77*	*73*	*87*	*80*	*88*	*88*
Gross income decile group						
Lowest ten per cent	46	39	68	67	76	76
Second decile group	48	43	81	65	82	78
Third decile group	55	49	86	70	84	81
Fourth decile group	72	67	86	81	90	87
Fifth decile group	80	74	88	81	87	90
Sixth decile group	88	84	89	84	92	94
Seventh decile group	92	91	89	86	90	93
Eighth decile group	96	94	93	87	90	93
Ninth decile group	97	96	95	89	94	95
Highest ten per cent	98	97	96	90	93	96
Household composition						
One adult, retired households[3]	18	16	98	44	78	66
One adult, non-retired households	71	63	70	86	82	86
One adult, one child	78	70	65	88	91	91
One adult, two or more children	88	85	71	82	88	100
Two adults, retired households [3]	39	33	97	62	87	82
Two adults, non-retired households	92	89	89	88	90	93
Two adults, one child	92	91	86	87	96	95
Two adults, two or more children	97	95	91	88	93	97
All other households without children	92	88	91	84	90	91
All other households with children	97	96	92	86	95	94

Note: Please see page ix for symbols and conventions used in this report.

1 See table A47 for number of recording households.
2 Full or partial.
3 Mainly dependent on state pensions and not economically active - see Appendix B.
4 Includes digital and cable receivers.

Table A47   **Percentage of households with cars by income group, tenure and household composition, 2010**
United Kingdom

	One car/van	Two cars/vans	Three or more cars/vans	All with cars/vans	Weighted number of house-holds (000s)	House-holds in the sample (number)
All households	**45**	**24**	**6**	**75**	**26,320**	**5,260**
Gross income decile group						
Lowest ten per cent	29	[3]	[0]	32	2,630	510
Second decile group	42	[3]	[1]	46	2,640	530
Third decile group	52	7	[1]	60	2,620	540
Fourth decile group	58	11	[2]	70	2,630	550
Fifth decile group	63	15	[3]	80	2,630	550
Sixth decile group	60	25	[3]	88	2,640	530
Seventh decile group	46	36	7	89	2,630	530
Eighth decile group	41	45	8	93	2,640	520
Ninth decile group	32	50	14	96	2,630	510
Highest ten per cent	25	50	21	95	2,630	500
Tenure of dwelling[1]						
Owners						
Owned outright	53	22	7	81	8,340	1,770
Buying with a mortgage	42	41	9	92	9,100	1,830
All	47	32	8	87	17,430	3,600
Social rented from						
Council	34	5	[0]	40	2,590	500
Registered social landlord [2]	37	5	-	42	2,150	430
All	35	5	[0]	41	4,750	920
Private rented						
Rent free	41	[18]	[3]	62	370	70
Rent paid, unfurnished	47	18	[3]	68	2,890	540
Rent paid, furnished	39	[7]	[2]	48	880	130
All	45	16	3	63	4,150	740
Household composition						
One adult, retired mainly dependent on state pensions[3]	33	-	-	33	790	150
One adult, other retired	44	[1]	[0]	46	2,980	570
One adult, non-retired	58	2	[1]	61	4,030	800
One adult, one child	45	-	-	45	690	160
One adult, two or more children	48	[2]	[1]	51	720	170
Two adults, retired mainly dependent on state pensions[3]	80	[7]	[2]	89	450	110
Two adults, other retired	59	21	[2]	82	2,300	530
Two adults, non-retired	42	40	5	87	5,760	1,170
Two adults, one child	45	38	[3]	86	2,300	450
Two adults, two children	37	49	7	93	2,150	450
Two adults, three children	41	44	[4]	89	540	120
Two adults, four or more children	[40]	[37]	-	77	170	40
Three adults	28	35	25	89	1,560	270
Three adults, one or more children	30	44	17	91	910	160
All other households without children	[18]	28	39	85	640	90
All other households with children	[41]	[19]	[30]	90	320	40

Note: Please see page ix for symbols and conventions used in this report.

1 See footnotes in Table A32.
2 Formerly housing association.
3 Mainly dependent on state pensions and not economically active - see Appendix B.

Table A48   Percentage of households with durable goods by UK countries and regions, 2008-2010

	North East	North West	Yorks and the Humber	East Midlands	West Midlands	East	London	South East	South West	England	Wales	Scotland	Northern Ireland	United Kingdom
Average weighted number of households (thousands)	1,280	3,130	2,170	1,980	2,200	2,270	3,030	3,080	2,510	21,640	1,280	2,380	690	26,000
Total number of households in sample (over 3 years)	730	1,770	1,460	1,210	1,470	1,550	1,410	2,190	1,520	13,300	800	1,510	1,320	16,930
Percentage of households														
by region and Country														
Car/van	68	73	73	78	76	82	65	81	83	76	74	70	77	75
One	43	44	44	43	44	45	46	42	44	44	43	42	44	44
Two	21	24	24	28	25	29	16	30	31	25	26	23	27	25
Three or more	4	6	4	7	8	9	3	9	7	6	5	5	6	6
Central heating full or partial	98	95	94	95	95	97	93	96	95	95	94	97	99	95
Fridge-freezer or deep freezer	97	97	97	97	97	96	98	97	98	97	97	95	96	97
Washing machine	96	96	96	96	95	96	95	96	96	96	96	97	97	96
Tumble dryer	59	58	59	60	63	61	45	59	60	58	62	59	56	58
Dishwasher	27	34	34	40	35	44	39	47	42	39	35	39	48	39
Microwave	94	94	94	93	94	92	88	90	91	92	95	92	94	92
Telephone	84	87	85	87	87	91	90	91	92	89	88	88	84	88
Mobile phone	80	81	79	86	82	82	81	84	83	82	84	84	49	80
DVD player	91	90	89	90	90	90	85	89	88	89	86	89	78	88
Satellite receiver[1]	86	90	86	85	86	83	79	86	86	85	83	88	86	85
CD player	84	83	82	88	85	87	81	86	88	85	86	82	77	84
Home computer	70	73	71	74	74	80	81	78	76	76	70	74	66	75
Internet connection	64	67	66	68	69	75	77	75	71	71	65	69	61	70

Note: Please see page ix for symbols and conventions used in this report.
This table is based on a three year average.
1 Includes digital and cable receivers.

Table A49 **Percentage of households by size, composition and age in each gross income decile group, 2010**

United Kingdom

	Lowest ten per cent	Second decile group	Third decile group	Fourth decile group	Fifth decile group	Sixth decile group	Seventh decile group	Eighth decile group	Ninth decile group	Highest ten per cent	All house-holds
Lower boundary of group (£ per week)		160	238	315	413	522	651	801	1,015	1,368	
Weighted number of households (thousands)	2,630	2,640	2,620	2,630	2,630	2,640	2,630	2,640	2,630	2,630	26,320
Number of households in the sample	510	530	540	550	550	530	530	520	510	500	5,260
Size of household											
One person	77	61	49	34	24	21	12	8	5	4	30
Two persons	18	20	37	41	44	40	40	40	37	33	35
Three persons	[3]	13	6	15	16	20	20	25	23	24	16
Four persons	[1]	4	6	7	9	13	20	20	24	26	13
Five persons	[1]	[1]	[1]	[2]	4	[3]	5	6	9	10	4
Six or more persons	[0]	[0]	[1]	[1]	[3]	[2]	[2]	[2]	[2]	[3]	2
All sizes	100	100	100	100	100	100	100	100	100	100	100
Household composition											
One adult, retired mainly dependent on state pensions[1]	15	9	5	[1]	-	-	-	-	-	-	3
One adult, other retired	23	36	26	13	7	5	[1]	[1]	[0]	[0]	11
One adult, non-retired	40	16	17	20	17	17	11	7	5	4	15
One adult, one child	10	4	4	4	[3]	[1]	[0]	[0]	-	[0]	3
One adult, two or more children	[2]	7	5	6	4	[2]	[1]	[0]	[1]	[0]	3
Two adults, retired mainly dependent on state pensions[1]	[0]	5	7	3	[1]	[0]	-	-	-	-	2
Two adults, other retired	[1]	5	13	20	17	15	7	5	[3]	[2]	9
Two adults, non-retired	6	6	12	14	23	24	33	35	34	31	22
Two adults, one child	[2]	7	[3]	8	7	12	11	14	12	11	9
Two adults, two children	[0]	[3]	[3]	[3]	7	10	13	12	14	17	8
Two adults, three children	[1]	[0]	[1]	[1]	[3]	[1]	[2]	[3]	4	4	2
Two adults, four or more children	[0]	[0]	[0]	[1]	[2]	[1]	[1]	[1]	[0]	[1]	1
Three adults	[0]	[1]	[1]	[3]	6	6	9	10	10	13	6
Three adults, one or more children	-	[1]	[1]	[1]	[3]	[3]	8	6	7	7	3
All other households without children	[0]	-	-	[1]	[1]	[2]	[3]	[4]	7	7	2
All other households with children	[0]	-	[0]	[0]	[1]	[2]	[0]	[1]	[3]	[4]	1
All compositions	100	100	100	100	100	100	100	100	100	100	100
Age of household reference person											
15 and under 20 years	[1]	[0]	[1]	[1]	-	-	[0]	[0]	-	-	[0]
20 and under 25 years	6	5	[3]	[3]	[4]	[5]	[5]	[1]	[2]	[0]	3
25 and under 30 years	5	6	[4]	8	7	10	9	10	8	[3]	7
30 and under 35 years	5	6	[4]	8	7	7	11	14	10	11	8
35 and under 40 years	7	[3]	6	5	8	9	11	14	13	10	9
40 and under 45 years	6	5	6	9	10	9	12	11	16	19	10
45 and under 50 years	8	4	5	6	7	9	9	12	14	16	9
50 and under 55 years	8	4	5	6	10	7	12	13	12	20	10
55 and under 60 years	8	5	4	6	9	8	9	9	9	9	8
60 and under 65 years	11	8	9	11	10	11	9	6	9	7	9
65 and under 70 years	7	12	10	11	8	9	6	4	4	[2]	7
70 and under 75 years	7	10	11	8	8	6	[2]	[1]	[2]	[1]	6
75 and under 80 years	9	11	12	8	6	5	[2]	[2]	[1]	[1]	6
80 and under 85 years	5	9	9	6	[3]	[3]	[1]	[1]	[0]	[1]	4
85 and under 90 years	[5]	7	7	[4]	[3]	[1]	[1]	[0]	[0]	-	3
90 years or more	[2]	[3]	[3]	[1]	[0]	[1]	[1]	-	-	-	1
All ages	100	100	100	100	100	100	100	100	100	100	100

Note: Please see page ix for symbols and conventions used in this report.

1 Mainly dependent on state pensions and not economically active - see Appendix B.

Table A50 **Percentage of households by economic activity, tenure and socio-economic classification in each gross income decile group, 2010**

United Kingdom

	Lowest ten per cent	Second decile group	Third decile group	Fourth decile group	Fifth decile group	Sixth decile group	Seventh decile group	Eighth decile group	Ninth decile group	Highest ten per cent	All house-holds
Lower boundary of group (£ per week)		160	238	315	413	522	651	801	1015	1,368	
Weighted number of households (thousands)	2,630	2,640	2,620	2,630	2,630	2,640	2,630	2,640	2,630	2,630	26,320
Number of households in the sample	510	530	540	550	550	530	530	520	510	500	5,260
Number of economically active persons in household											
No person	71	71	62	45	29	19	11	6	5	[2]	32
One person	27	24	30	41	44	40	26	20	15	15	28
Two persons	[2]	5	8	13	24	36	52	60	59	59	32
Three persons	-	[1]	-	[1]	[3]	5	10	13	17	15	6
Four or more persons	-	-	-	[0]	[0]	[1]	[1]	[1]	[5]	9	2
All economically active persons	100	100	100	100	100	100	100	100	100	100	100
Tenure of dwelling[1]											
Owners											
Owned outright	26	38	46	39	38	36	26	22	22	23	32
Buying with a mortgage	9	8	11	20	27	36	48	56	64	66	35
All	35	46	58	60	65	72	74	78	86	89	66
Social rented from											
Council	27	24	13	10	9	6	5	[3]	[1]	[0]	10
Registered social landlord [2]	21	13	13	11	9	7	[3]	[4]	[1]	[0]	8
All	48	38	27	21	18	13	8	7	[2]	[1]	18
Private rented											
Rent free	[2]	[3]	[1]	[2]	[1]	[1]	[1]	[1]	[1]	[0]	1
Rent paid, unfurnished	12	10	11	14	13	11	13	9	8	7	11
Rent paid, furnished	[4]	[3]	[3]	[3]	[3]	[3]	[4]	[5]	[3]	[3]	3
All	17	17	16	20	17	15	18	15	13	10	16
All tenures	100	100	100	100	100	100	100	100	100	100	100
Socio-economic classification											
Higher managerial and professional											
Large employers/higher managerial	-	[0]	[0]	[0]	[1]	[2]	[3]	7	12	23	5
Higher professional	[1]	[1]	[0]	[1]	[3]	5	8	15	13	22	7
Lower managerial and professional	[2]	[1]	5	8	16	18	27	28	35	35	18
Intermediate	[1]	[1]	4	6	8	7	7	8	8	[2]	5
Small employers	4	4	5	7	9	7	10	4	5	5	6
Lower supervisory	-	[1]	4	6	8	11	10	11	9	[4]	6
Semi-routine	[3]	5	7	12	11	10	9	6	[3]	[1]	7
Routine	[3]	4	5	7	7	10	8	7	[4]	[2]	6
Long-term unemployed[3]	7	[3]	[4]	[2]	[2]	[1]	[0]	[1]	-	-	2
Students	[3]	[3]	[2]	[2]	[2]	[2]	[2]	[2]	[2]	[1]	2
Occupation not stated[4]	77	77	63	50	34	27	15	10	9	4	37
All occupational groups	100	100	100	100	100	100	100	100	100	100	100

Note: Please see page ix for symbols and conventions used in this report.

1 See footnotes in Table A32.
2 Formerly housing association.
3 Includes those who have never worked.
4 Includes those who are economically inactive - see definitions in Appendix B.

Table A51   **Average weekly household expenditure by OAC supergroup, 2010**

United Kingdom

	Blue collar Communities	City living	Countryside	Prospering suburbs	Constrained by circumstances	Typical traits	Multicultural	
	OAC Super-group 1	OAC Super-group 2	OAC Super-group 3	OAC Super-group 4	OAC Super-group 5	OAC Super-group 6	OAC Super-group 7	All house-holds
Weighted number of households (thousands)	4,380	1,810	3,380	5,900	3,280	5,320	2,260	26,320
Total number of households in sample	890	310	740	1,230	640	1,060	390	5,260
Total number of persons in sample	2,240	610	1,750	2,910	1,250	2,370	1,060	12,180
Total number of adults in sample	1,610	520	1,380	2,340	970	1,870	730	9,430
Weighted average number of persons per household	2.5	2.0	2.3	2.4	1.9	2.3	2.8	2.3
Commodity or service	Average weekly household expenditure (£)							
1 Food & non-alcoholic drinks	49.70	48.00	59.90	61.00	40.50	51.60	55.90	53.20
2 Alcoholic drinks, tobacco & narcotics	13.50	10.80	13.80	11.00	11.60	11.40	9.70	11.80
3 Clothing & footwear	20.80	26.60	23.70	26.50	16.00	24.50	25.10	23.40
4 Housing (net)[1], fuel & power	53.50	115.10	60.10	48.90	51.30	56.30	83.00	60.40
5 Household goods & services	22.40	39.00	39.30	41.00	19.40	30.50	25.60	31.40
6 Health	3.40	6.00	5.30	6.40	2.50	5.40	6.20	5.00
7 Transport	50.80	61.70	83.00	84.80	33.00	64.40	63.00	64.90
8 Communication	12.40	13.80	13.50	13.40	9.70	13.30	16.00	13.00
9 Recreation & culture	49.50	48.30	81.10	73.20	36.90	53.70	49.80	58.10
10 Education	1.50	37.50	15.30	8.00	1.60	10.70	12.50	10.00
11 Restaurants & hotels	30.80	52.10	40.70	48.10	24.40	38.70	42.70	39.20
12 Miscellaneous goods & services	25.70	41.90	45.50	44.60	20.60	37.20	32.80	35.90
1-12 All expenditure groups	**333.90**	**500.90**	**481.20**	**467.00**	**267.40**	**397.80**	**422.50**	**406.30**

Note: Please see page ix for Symbols and conventions used in this report.
1 Excluding mortgage interest payments, council tax and Northern Ireland rates

Table A52 **Average weekly household expenditure by OAC group, 2010**
United Kingdom

	Terraced blue collar	Younger blue collar	Older blue collar	Transient communities	Settled in the city	Village life	Agricultural	Accessible countryside	Prospering younger families	Prospering older families	Prospering semis
	OAC group 1A	OAC group 1B	OAC group 1C	OAC group 2A	OAC group 2B	OAC group 3A	OAC group 3B	OAC group 3C	OAC group 4A	OAC group 4B	OAC group 4C
Weighted number of households (thousar	980	1,650	1,750	620	1,200	1,330	950	1,090	1,100	1,680	1,830
Total number of households in sample	200	330	370	100	210	300	220	230	230	360	380
Total number of persons in sample	500	850	880	170	430	690	520	540	610	840	860
Total number of adults in sample	350	600	670	160	370	550	410	430	460	680	700
Weighted average number of persons per household	2.5	2.6	2.5	1.9	2.0	2.3	2.4	2.3	2.7	2.3	2.3
Commodity or service					Average weekly household expenditure (£)						
1 Food & non-alcoholic drinks	49.20	49.30	50.40	44.50	49.80	58.20	61.20	60.70	65.30	62.30	56.60
2 Alcoholic drinks, tobacco & narcoti	13.00	16.10	11.40	8.70	11.80	13.90	15.00	12.80	11.40	10.70	10.10
3 Clothing & footwear	23.80	18.20	21.50	26.10	26.90	21.50	26.70	23.80	31.60	28.80	22.80
4 Housing (net)[1] fuel & power	52.00	58.40	49.70	148.70	97.90	57.50	61.50	62.10	47.60	50.70	44.90
5 Household goods & services	28.50	16.60	24.40	18.30	49.70	36.70	38.10	43.60	40.50	41.10	33.00
6 Health	2.20	3.60	3.90	5.50	6.20	4.50	5.10	6.50	4.90	7.10	5.30
7 Transport	47.20	39.20	63.90	57.70	63.70	82.10	81.80	85.10	88.80	84.90	69.70
8 Communication	12.80	13.00	11.60	14.90	13.20	14.20	11.60	14.20	14.80	12.90	12.90
9 Recreation & culture	43.20	42.60	59.50	39.00	53.10	84.60	72.80	84.20	78.10	80.40	60.70
10 Education	2.50	0.90	1.50	36.00	38.30	9.50	18.70	19.30	5.20	4.10	6.80
11 Restaurants & hotels	29.80	27.00	34.90	51.40	52.50	37.20	41.70	43.90	48.90	50.70	40.00
12 Miscellaneous goods & services	21.30	22.60	31.10	40.20	42.80	45.50	42.00	48.70	50.00	47.00	34.10
1-12 **All expenditure groups**	**325.40**	**307.40**	**363.70**	**491.20**	**506.00**	**465.40**	**476.30**	**504.90**	**487.10**	**480.70**	**397.00**

Note: Please see page xiii for symbols and conventions used in this report.
1 Excluding mortgage interest payments, council tax and Northern Ireland rates

Table A52 **Average weekly household expenditure by OAC group, 2010 (cont.)**
United Kingdom

	Thriving suburbs	Senior communities	Older workers	Public housing	Settled households	Least divergent	Young families in terraced homes	Aspiring households	Asian communities	Afro-Caribbean communities	All house-holds
	OAC group 4D	OAC group 5A	OAC group 5B	OAC group 5C	OAC group 6A	OAC group 6B	OAC group 6C	OAC group 6D	OAC group 7A	OAC group 7B	
Weighted number of households (thousands)	1,300	490	2,040	750	1,470	1,490	1,360	1,000	1,400	860	26,320
Total number of households in sample	260	90	410	150	300	300	270	200	250	140	5,260
Total number of persons in sample	600	130	780	330	680	670	590	430	720	340	12,180
Total number of adults in sample	500	120	620	240	550	520	460	340	490	250	9,430
Weighted average number of persons per household	2.3	1.5	1.9	2.2	2.3	2.3	2.3	2.2	3.0	2.5	2.3
Commodity or service					Average weekly household expenditure (£)						
1 Food & non-alcoholic drinks	62.00	32.40	42.50	40.20	52.90	54.50	44.90	54.50	60.80	47.90	53.20
2 Alcoholic drinks, tobacco & narcotics	12.20	11.30	10.60	14.40	10.70	13.10	10.90	10.90	10.30	8.90	11.80
3 Clothing & footwear	24.50	11.00	18.20	13.30	25.50	26.40	19.50	27.10	26.30	23.20	23.40
4 Housing (net)[1], fuel & power	53.30	55.30	51.60	47.80	46.20	56.40	59.90	65.90	78.00	91.10	60.40
5 Household goods & services	52.60	16.20	20.90	17.30	39.30	24.20	32.80	23.80	24.90	26.60	31.40
6 Health	8.60	2.70	2.90	1.20	6.40	7.00	3.30	4.20	6.10	6.50	5.00
7 Transport	102.40	19.30	37.10	30.90	63.30	74.90	48.00	72.70	69.90	51.80	64.90
8 Communication	13.80	8.20	9.50	11.40	13.10	13.00	12.70	14.90	15.00	17.50	13.00
9 Recreation & culture	77.20	20.80	40.00	39.00	63.20	55.30	39.40	56.70	48.50	52.00	58.10
10 Education	17.20	0.90	2.20	0.40	6.50	7.00	1.30	35.20	11.20	14.70	10.00
11 Restaurants & hotels	55.60	17.30	26.90	22.20	39.80	37.80	30.60	49.70	38.20	50.10	39.20
12 Miscellaneous goods & services	51.60	12.50	21.30	24.20	47.30	35.20	26.80	39.60	33.70	31.30	35.90
1-12 **All expenditure groups**	**530.80**	**207.70**	**283.70**	**262.40**	**414.40**	**404.80**	**330.00**	**455.20**	**423.00**	**421.70**	**406.30**

Table A53   **Average gross normal weekly household income by OAC supergroup, 2010**
United Kingdom

	Blue collar Communities	City living	Countryside	Prospering suburbs	Contstrained by circumstances	Typical traits	Multicultural	All house-holds
	OAC Super-group 01	OAC Super-group 02	OAC Super-group 03	OAC Super-group 04	OAC Super-group 05	OAC Super-group 06	OAC Super-group 07	All house-holds
Percentage of households by Government Office Region and Country								
Weighted number of households (thousands)	4380	1810	3380	5900	3280	5320	2260	26320
Total number of households in sample	890	310	740	1230	640	1060	390	5260
Total number of persons in sample	2240	610	1750	2910	1250	2370	1060	12180
Total number of adults in sample	1610	520	1380	2340	970	1870	730	9430
Weighted average number of persons per household	3	2	2	2	2	2	3	2
Gross normal weekly household income (£)	**537.6**	**861.5**	**776.3**	**851.9**	**416**	**714.7**	**757.5**	**700.5**

Note: Please see page ix for Symbols and conventions used in this report.

Methodology

Description and response rate of the survey

The survey

A household expenditure survey has been conducted each year in the UK since 1957. From 1957 to March 2001 the Family Expenditure Survey (FES) and National Food Survey (NFS) provided information on household expenditure patterns and food consumption. In April 2001 these surveys were combined to form the Expenditure and Food Survey (EFS).

In 2008 selected Government household surveys, on which the Office for National Statistics (ONS) leads, were combined into one Integrated Household Survey (IHS). In anticipation of this, the EFS moved to a calendar-year basis in January 2006. The EFS questionnaire became known as the Living Costs and Food (LCF) module of the IHS in 2008, to accommodate the insertion of a core set of IHS questions.

More information about the IHS can be found on the National Statistics website: http://www.ons.gov.uk/ons/about-ons/surveys/a-z-of-surveys/integrated-household-survey/index.html. In summary, the survey design allows for the collection of common core data across the pooled samples of the constituent surveys, achieving the biggest pool of UK social data after the Census. The large sample allows a detailed level of analysis to be conducted, and allows results to be reported for smaller geographic areas. The IHS has become the key vehicle for high-profile national data collection initiatives including questions on subjective well-being, and on sexual identity.

The LCF is a voluntary sample survey of private households. The basic unit of the survey is the household. A household comprises one person or a group of people who have the accommodation as their only or main residence and (for a group) either share at least one meal a day or share the living accommodation, that is, a living room or sitting room. (See 'Definitions'.)

Each individual aged 16 and over in the household visited is asked to keep diary records of daily expenditure for two weeks. Information about regular expenditure, such as rent and mortgage payments, is obtained from a household interview along with retrospective information on certain large, infrequent expenditures such as those on vehicles. Children aged 7 to 15 are asked to keep a simplified version of the diary.

Detailed questions are asked about the income of each adult member of the household. In addition, personal information such as age, sex and marital status is recorded for each household member. A copy of the LCF questionnaire is available from the Economic and Social Data Service.

The survey is continuous, interviews being spread evenly over the year to ensure that seasonal effects are covered. The questionnaire content is reviewed thoroughly to ensure that it is up-to-date and captures information efficiently. Some changes reflect new forms of expenditure or new sources of income, especially benefits. Others are the result of new requirements by the survey's users. (See the section on 'Improvements' for more information.)

The sample design

The LCF sample for Great Britain is a multi-stage stratified random sample with clustering. It is drawn from the Small Users file of the Postcode Address File (PAF) – the Post Office's list of addresses. All Scottish offshore islands and the Isles of Scilly are excluded from the sample because of excessive interview travel costs. Postal sectors are the primary sample unit. 638 postal sectors are randomly selected after being arranged in strata defined by Government Office Regions (sub-divided into metropolitan and non-metropolitan areas) and two 2001 Census variables: socio-economic group of the head of household and ownership of cars. These census variables were new stratifiers originally introduced for the 1996/97 survey, and updated following the results of the 2001 Census. The results of the 2011 Census will be used in due course. The Northern Ireland sample is drawn as a random sample of addresses from the Land and Property Services Agency list.

Response to the survey

Great Britain

A total of 11,484 households were selected in 2010 for the LCF in Great Britain. However, it is not possible to get full response. A small number of households cannot be contacted at all, and in other households one or more members decline to co-operate. 5,116 households in Great Britain co-operated fully in the survey in 2010; that is, they answered the household questionnaire and all adults in the household answered the full income questionnaire and kept the expenditure diary. A further 204 households provided sufficient information to be included as valid responses. The overall response rate for the 2010 LCF was 50 per cent in Great Britain, the same as in 2009.

Details of response are shown in the following table.

Response in 2010 – Great Britain

	No of households or addresses	Percentage of effective sample
i. Sam pled addresses	11,484	-
ii. Ineligible addresses: businesses, institutions, empty, demolished/derelict	1,208	-
iii. Extra households (multi-household addresses)	38	-
iv. Total eligible (that is i less ii, plus iii)	10,314	100
v. Co-operating households (which includes 191 partials)	5,116	49.6
vi. R efusals	3,810	36.9
vii. Households at which no contact could be obtained	1,388	13.5

Northern Ireland

In the Northern Ireland survey, the eligible sample was 250 households. The number of co-operating households who provided usable data was 147, giving a response rate of 59 per cent. This represents an increase of 3 percentage points from the 2009 survey year.

Northern Ireland was over-sampled in the years 1997/98 to 2009 in order to provide a large enough sample for some separate analysis. This boost to the Northern Ireland sample was discontinued in 2010.

Partial response

Three types of partial response are accepted on the LCF:

- all adults complete the full income section of the interview, but one or more adults in the household refuse to keep the diary

- all adults in the household keep the diary, but one or more adults provides only partial income information

- one or more adults refuse to keep the diary and one or more adults provide only partial income information

All partial responses must contain a diary from the Main Diary Keeper (MDK), who is the person who does most of the shopping in the household. If the MDK refuses to complete the diary the household is classified as a refusal.

In 2010 partial responses accounted for 3.7 per cent (187 households) of all co-operating households. Of these partials, the majority (89 per cent) occurred because one or more adults in the household refused to keep the diary. Partial income information was provided in 3 per cent of cases. The remaining 7 per cent of households contained adults who refused to keep the diary and adults who provided only partial income information.

Type of partial response in 2010 – Great Britain

Type of partial response	Number of households	Percentage of partials
1. One or more adults refuse to keep the diary[1]	167	89
2. One or more adults provide only partial income information	6	3
3. One or more adults i) refuse to keep the diary and ii) provide only partial income information [1]	14	7
All	**187**	**100**

[1] Diary is present for the main diary keeper.

LCF response rates over time

Response rates to household surveys have been declining in recent years. In 2010 the LCF's response rate for Great Britain was 50 per cent (see 'Response to the survey' for a detailed breakdown), compared with around 60 per cent in 2000/01. It should be noted that the LCF requires satisfactory completion of both the household questionnaire and diary (see 'Eligible response' for more information).

Response rates over time – Great Britain

Year	Response rate
	Percentage
2000/01	59
2001/02	62
2002/03	58
2003/04	58
2004/05	57
2005/06	57
2006	55
2007	53
2008	51
2009	50
2010	50

N.B. in 2006, the survey moved to a calendar year. In 2007 the
sample size w as reduced by 5%

Response rates are sometimes used as an indicator of a survey's quality and how representative a sample is of the target population. It is generally assumed that the lower the response rates the greater the likelihood of bias in the results. This is because the characteristics of non-responding households may differ from those of responding households causing certain types of households to be under-represented in the sample. Analysis of the UK Labour Force Survey (LFS) tentatively reported an effect for some variables; a separate study[1] for the LCF found that weekly variation in the number of LCF cases achieved did not result in different expenditure patterns being reported. Overall, there is little evidence to suggest that a limited drop in response will affect bias to a large degree. Nevertheless, to maintain quality of the data, methods such as proxy and partial interviews are set in place to minimise non-response for the LCF. Increasing incentive payments is another possible way to improve response rates. This is being investigated by ONS, despite the additional costs involved.

Currently, non-responders are accounted for in the weighting process for LCF data, which compensates for non-responders recognised from analysis of the 2001 Census (see 'Weighting' for more information). At present, the LCF is contributing towards the 2011 Census non-response linkage project, which will enable non-response weights to be updated.

Lower response also potentially impacts on precision, with fewer cases being completed. However, considering the years from 2007 (when the current set sample size was adopted for Great Britain), percentage standard errors at COICOP level have not changed measurably (see Table A1 for 2010 standard errors).

For users to fully utilise LCF data in an informed way, sample sizes and response rates must be considered. Therefore, sample sizes are provided within each table of *Family Spending*, and small sample sizes are highlighted for users' attention. Where necessary, tables with detailed breakdowns are averaged across three years to overcome issues of accuracy and disclosure.

The fieldwork

The fieldwork is conducted by the Office for National Statistics (ONS) in Great Britain and by the Northern Ireland Statistics and Research Agency (NISRA) of the Department of Finance and Personnel in Northern Ireland using almost identical questionnaires. Households at the selected addresses are visited and asked to co-operate in the survey. In order to maximise response, interviewers make at least four separate calls, and sometimes many more, at different times of day on households which are difficult to contact. Interviews are conducted by Computer Assisted Personal Interviewing (CAPI) using laptop computers. During the interview information is collected about the household; certain regular payments such as rent, gas, electricity and telephone accounts; expenditure on certain large items (for example vehicle purchases over the previous 12 months); and income. Each individual aged 16 and over in the household is asked to keep a detailed record of expenditure every day for two weeks. Children aged between 7 and 15 are also asked to keep a simplified diary of daily expenditure. In 2010 a total of 1,516 children aged between 7 and 15 in responding households in the UK were asked to complete expenditure diaries; 216, or about 14 per cent, did not do so. This number includes both refusals and children who had no expenditure during the two weeks. Information provided by all members of the household is kept strictly confidential. Each person aged 16 and over in the household who keeps a diary (and whose income information is collected) is subsequently paid £10 as a token of appreciation. Children who keep a diary are given a £5 payment.

In the last two months of the 1998/99 survey, as an experiment, a small book of postage stamps was enclosed with the introductory letter sent to every address. Response seemed to increase as a result of this experiment and it has become a permanent feature of the survey. It is difficult to quantify the exact effect on response but the cognitive work that was carried out as part of the EFS development indicated that it was having a positive effect.

Some survey cases are reissued if a response is not obtained, that is, the cases are reallocated to field interviewers at a later date to attempt to achieve response. Criteria are applied to determine which cases should be reissued. Until 2010 there was a strict reissue criterion that restricted the number of cases that could be reissued. In 2010 this was changed to include cases where the interviewer was 'not sure' whether it was worth reissuing the case. It is then left to field staff at headquarters to asses whether or not to reissue. Previously the only cases reissued were those where the interviewer reported that the household had expressed an interest in participating or would be likely to participate.

In 2010 some 322 addresses were reissued, of which 48 were converted into responding households. This increased the overall response rate by 0.1 percentage points.

Eligible response

Under LCF rules, a refusal by just one person to respond to the income section of the questionnaire invalidates the response of the whole household. Similarly, a refusal by the household's main shopper to complete the two-week expenditure diary also results in an invalid response.

Proxy response

Questions about general household affairs are put to all household members or to the household representative person (HRP), and questions about work and income are put to the individual members of the household. Where a member of the household is not present during the household interview, another member of the household (for example a spouse) may be able to provide information about the absent person. The individual's interview is then identified as a proxy interview. Under LCF rules, the expenditure diary cannot be completed by proxy; if a household member is not present during the diary period they are classified as an absent spender.

In 2001/02 the EFS began including households that contained a proxy interview. In that year, 12 per cent of all responding households contained at least one proxy interview. In 2010 the percentage of responding households with a proxy interview was 22 per cent. The rise in the percentage of proxy interviews over time reflects general response trends for social surveys; interviewers are finding it increasingly difficult to make initial contact with households and to secure interviews with each household member.

Households containing proxy interviews, 2003/04 to 2010 – Great Britain

	2003/04	2004/05	2005/06	2006	2007	2008	2009	2010
	Percentage of households							
Proxy interviews	12	15	13	14	17	21	23	22
Total number of households	**6,432**	**6,265**	**6,258**	**6,059**	**5,545**	**5,271**	**5,222**	**5,116**

Analysis of the 2010 data revealed that the inclusion of proxy interviews increased response from above average income households. For the 2010 survey, the average gross normal weekly household income was 16 per cent higher than it would have been if proxy interviews had not been accepted. Similar findings were obtained with respect to expenditure: total spending was 11 per cent higher than if proxy interviews had not been included. Use of proxies enhances the sample size and hence the precision of the figures obtained. It also enables the survey to capture the income and expenditure from (on average) higher-earning households and hence ensures that these households are represented fully in the survey. This must be weighed against the risk that the proxy interviews may not provide exactly the same information as direct interviews, but the available evidence suggestions that including proxies provides higher data quality overall.

Reliability

Great care is taken in collecting information from households, and comprehensive checks are applied during processing so that errors in recording and processing are minimised. The main factors that affect the reliability of the survey results are sampling variability, non-response bias and some incorrect reporting of certain items of expenditure and income. Measures of sampling variability are given alongside some results in this report and are discussed in detail in 'Standard errors and estimates of precision'.

The households which decline to respond to the survey tend to differ in some respects from those that co-operate. It is therefore possible that their patterns of expenditure and income also differ. A comparison was made of the households responding in the 1991 FES with those not responding, based on information from the 1991 Census (A comparison of the Census characteristics of respondents and non-respondents to the 1991 FES by K Foster[2], ONS Survey Methodology Bulletin No. 38, Jan 1996). Results from the study indicate that response was lower than average in Greater London, higher in non-metropolitan areas, and that non-response tended to increase with increasing age of the head of the household – up to age 65. Households that contained three or more adults, or where the head was born outside the United Kingdom or was classified to an ethnic minority group, were also more likely than others to be non-responding. Non-response was also above average where the head of the household had no post-school qualifications, was self-employed, or was in a manual social class group. The data were re-weighted to compensate for the main non-response biases identified from the 1991 Census comparison, as described in 'Weighting'. ONS has completed a similar comparative exercise with the 2001 Census data, which resulted in an update of the non-response weights for the estimates for 2007 onwards. Further analysis will be conducted using the results of the 2011 Census.

Checks are included in the computer assisted personal interviewing (CAPI) program, which are applied to the responses given during the interview. Other procedures are also in place to ensure that users are provided with high quality data. For example, quality control is carried out to ensure that any unusual values (outliers) are genuine, and checks are made on any unusual changes in average spending compared with the previous year.

When aspects of the survey change, rigorous tests are used to ensure the proposed changes are sensible and work both in the field and on the processing system. During 2010 a set of questions were developed to capture expenditure on combined communications packages, for example, where television, internet and telephone services are purchased from a single provider. These questions were developed following an expert review and cogitative testing conducted by the ONS Methodology directorate. These questions have been included in the 2011 questionnaire and will feed into the next edition of *Family Spending*.

Income and expenditure balancing

The LCF is designed primarily as a survey of household expenditure on goods and services. It also gathers information about the income of household members, and is an important and detailed source of income data. However, it is not possible to draw up a balance sheet of income and expenditure either for individual households or groups of households.

The majority of expenditure information collected relates to the two-week period immediately following the interview, whereas income components can refer to a much longer period (the most recent 12 months). LCF income does not include withdrawal of savings; loans and money received in payment of loans; receipts from maturing insurance policies; proceeds from the sale of assets (such as a car); and winnings from betting or windfalls, such as legacies. Despite this, recorded expenditure might reflect these items, as well as the effects of living off savings, using capital, borrowing money or income – either recent or from a previous period.

Hence, there is no reason why income and expenditure should balance. In fact measured expenditure exceeds measured income at the bottom end of the income distribution. However, this difference cannot be regarded as a reliable measure of savings or dis-saving.

For further information of what is included in income on the LCF see 'Income headings'.

Imputation of missing information

Although LCF response is generally based on complete households responding, there are areas in the survey for which missing information is imputed (inferred, sometimes in conjunction with other sources). This falls into two broad categories: item imputation and diary imputation. Using a combination of reliable imputation procedures ensures that the LCF data provide a comprehensive picture of the spending patterns and income sources for each household.

Questionnaire item imputation

Although LCF interviewers are trained to obtain full answers or best estimates for all questions, and encourage respondents to refer to up-to-date bills and statements, missing values in the questionnaire can sometimes occur when respondents are unable to provide an answer.

Missing data for questions in the expenditure and income parts of the questionnaire are imputed using the following procedures:

- By reference to tables based on external (non-LCF) data published elsewhere. For example, rates and eligibility rules for state benefits; council tax rates for properties in different council tax bands; and the Annual Survey of Hours and Earnings (ASHE).

- By reference to tables based on LCF data from previous years showing average amounts according to gross household income quintile. Expenditures imputed using this method include insurances; utilities; rent; personal pensions; and child income.

- Using information collected elsewhere in the questionnaire. For example, missing tax and National Insurance payments can be imputed if a gross salary figure has been provided.

Respondents who have failed to answer important questions, such that the main streams of income and expenditure are not captured, are converted to refusals to ensure that a sufficiently high level of data accuracy is maintained.

Markers are included in the LCF datasets to record whether values at key expenditure and state benefit questions were imputed or amended; this gives a broad indication of the amount of imputation carried out for each individual case.

Diary imputation

The LCF accepts households with missing diaries, as long as the diary of the Main Diary Keeper (MDK) is present. Diaries which are missing are imputed, that is they will receive the diary data from a person in another responding household with matching characteristics. Missing diaries for households in Great Britain and Northern Ireland are imputed from the pool of responding cases for the appropriate country.

The first step in the diary imputation process involves scoring each person in the pool of potential donors for suitability as a match for the person with a missing diary.

The scoring system is:
- matching age = 8 points
- matching relationship to the HRP = 4 points
- matching employment status = 2 points
- matching survey month = 1 point

In the next stage of the process the potential donor with the highest score is selected and the diary data from the donor is copied to the receiving person. To be used as a donor a diary must achieve a minimum score of 8 points.

In 2010, 181 households had imputed diaries, accounting for 3.5 per cent of responding households.

Variables that indicate whether or not a diary has been imputed, and the number of diaries imputed per household, are included in the LCF datasets.

Uses of the survey

LCF expenditure data

Retail Prices Index – The main reason, historically, for instituting a regular survey on expenditure by households has been to provide information on spending patterns for the Retail Prices Index (RPI) and the Consumer Prices Index (CPI). From April 2011 the CPI rather than the RPI is used as basis for indexation of benefits, tax credit and state and public service pensions. The RPI and CPI measure the change in the cost of a selection of goods and services (the 'basket of goods') representative of the expenditure of the vast majority of households. The pattern of expenditure gradually changes from one year to the next, and the composition of the basket of goods needs to be kept up-to-date. Accordingly, regular information is required on spending patterns and much of this is supplied by the LCF. The expenditure weights for the general RPI and CPI need to relate to people within given income limits, for which the LCF is the only source of information.

Household expenditure and Gross Domestic Product (GDP) – LCF data on spending are an important source used in compiling national estimates of household final consumption expenditure which are published regularly in *United Kingdom National Accounts* (ONS Blue Book). Household final consumption expenditure estimates feed into the National Accounts and estimates of GDP. They will also provide the weights for Purchasing Power Parities (PPPs) for international price comparisons. LCF data are also used in the estimation of taxes on expenditure, in particular VAT.

Regional accounts – LCF expenditure information is one of the sources used by ONS to derive regional estimates of consumption expenditure. It is also used in compiling some of the other estimates for the regional accounts.

The statistical office of the European Union (Eurostat) collates information from family budget surveys conducted by the member states. The LCF is the UK's contribution to this important EU initiative to collect data on household expenditure from member countries.

Other government uses – The Department of Energy and Climate Change and the Department for Transport both use LCF expenditure data in their own fields relating to, for example, energy, housing, cars and transport.

Non-government uses – There are also numerous users outside central government, including academic researchers and business and market researchers. One example is an academic study that has used LCF data, as part of a wider study, to obtain a clear picture of utility expenditure patterns across the European Union.

LCF income data

Redistribution of income – LCF information on income and expenditure is used to study how government taxes and benefits affect household income. The Government's interdepartmental tax benefit model is based on the LCF and enables the economic effects of policy measures to be analysed across households. This model is used by HM Treasury and HM Revenue and Customs to estimate the impact on different households of possible changes in taxes and benefits.

Non-government users – As with the expenditure data, LCF income data are also studied extensively outside government. In particular, academic researchers in the economic and social science areas of many universities use the LCF. For example the Institute for Fiscal Studies uses LCF data in research it carries out both for government and on its own account to inform public debate.

Comparability with other sources

The main comparator for LCF estimates of expenditure data are the figures on final household consumption expenditure (HHFCE) published in *Consumer Trends* and used in *UK National Accounts*. These can be found via the following link:

www.ons.gov.uk/ons/publications/all-releases.html?definition=tcm%3A77-23619

LCF data feed into some of the estimates published in *Consumer Trends*, but other sources are also used. While differences occur in the estimates published, the differences are credible. Research is ongoing into the different estimates produced, and their causes. This is due to be published in late 2012 or early 2013.

Other LCF data

The Department for Environment, Food and Rural Affairs (Defra) publishes separate reports using LCF data on food expenditure to estimate consumption and nutrient intake. The Department for Transport uses LCF data to monitor and forecast levels of car ownership and use, and in studies on the effects of motoring taxes.

Note: **Great care is taken to ensure complete confidentiality of information and to protect the identity of LCF households. Only anonymised data are supplied to users.**

Standard errors and estimates of precision

The Living Costs and Food Survey (LCF) is a sample of households and not a census of the whole population. Therefore, the results are liable to differ to some degree from those that would have been obtained if every single household had been covered. Some of the differences will be systematic, in that lower proportions of certain types of household respond than of others. That aspect is discussed in 'Description and response rate of the survey' and 'Weighting'. This section discusses the effect of sampling variability; in other words, the effect of differences in expenditure and income between the households in the sample and in the whole population that arise from random chance.

The degree of variability will depend on the sample size and how widely particular categories of expenditure (or income) vary between households. The sampling variability is smallest for the average expenditure of large groups of households on items purchased frequently and when the level of spending does not vary greatly between households. Conversely, it is largest for small groups of households, and for items purchased infrequently or for which expenditure varies considerably between households. A numerical measure of the likely magnitude of such differences (between the sample estimate and the value of the entire population) is provided by the quantity known as the standard error.

The calculation of standard errors takes into account the fact that the LCF sample is drawn in two stages: first a sample of areas (primary sampling units), then a sample of addresses within each of these areas. The main features of the sample design are described in 'Description and response rate of the survey'. The calculation also takes account of the effect of weighting. The two-stage sample increases sampling variability slightly, but the weighting reduces it for some items.

Standard errors for detailed expenditure items are presented in relative terms in Table A1 (standard error as a percentage of the average to which it refers). As the calculation of full standard errors is complex, this is the only table where they are shown. Tables B1and B2 in this section show the design factor (DEFT), a measure of the efficiency of the survey's sample design. The DEFT is calculated by dividing the 'full' standard error by the standard error that would have applied if the survey had used a simple random sample ('simple method').

Table B1   Percentage standard errors of expenditure of households and number of recording households, 2010

United Kingdom

Commodity or service	Weighted average weekly household expenditure (£)	Percentage standard error — Simple method	Design factor (DEFT)	Percentage standard error — Full method	Households recording expenditure — Recording households in sample	Percentage of all households
All expenditure groups	**406.30**	**1.1**	**1.1**	*1.2*	**5,263**	*100*
Food and non-alcoholic drinks	53.20	0.9	1.0	*0.9*	5,232	*99*
Alcoholic drink, tobacco & narcotics	11.80	2.2	1.0	*2.3*	3,330	*63*
Clothing and footwear	23.40	2.5	1.1	*2.7*	3,507	*67*
Housing, fuel and power	60.40	1.3	1.1	*1.5*	5,245	*100*
Household goods and services	31.40	4.3	1.0	*4.2*	4,798	*91*
Health	5.00	5.6	1.1	*6.0*	2,634	*50*
Transport	64.90	2.1	1.1	*2.3*	4,507	*86*
Communication	13.00	1.4	1.0	*1.4*	5,004	*95*
Recreation and culture	58.10	2.4	0.9	*2.2*	5,219	*99*
Education	10.00	9.5	1.3	*12.4*	402	*8*
Restaurants and hotels	39.20	1.8	1.1	*2.1*	4,592	*87*
Miscellaneous goods and services	35.90	2.4	1.0	*2.3*	5,132	*98*

Table B2   **Percentage standard errors of income of households and numbers of recording households, 2010**
United Kingdom

Source of income	Weighted average weekly household income (£)	Percentage standard error		Percentage standard error	Households recording income	
		Simple method	Design factor (DEFT)	Full method	Recording households in sample	Percentage of all households
Gross household income	**700**	*1.4*	**1.1**	*1.6*	**5,252**	*100*
Wages and salaries	452	*1.9*	1.0	*1.9*	3,013	*57*
Self-employment	67	*8.0*	1.2	*9.4*	655	*12*
Investments	16	*7.2*	0.9	*6.3*	2,370	*45*
Annuities and pensions (other than social security benefits)	58	*3.7*	0.9	*3.2*	1,707	*33*
Social security benefits	99	*1.4*	0.8	*1.1*	3,926	*75*
Other sources	8	*8.0*	1.3	*10.1*	652	*12*

Using the standard errors – confidence intervals

A common use of standard errors is in calculating 95 per cent confidence intervals. Simplifying a little, these can be taken to mean that there is only a 5 per cent chance that the true population value lies outside the 95 per cent confidence interval, which is calculated as 1.96 times the standard error on either side of the mean. For example, the average expenditure on food and non-alcoholic drinks is £53.20 and the corresponding percentage standard error (full method) is 0.9 per cent. The amount either side of the mean for 95 per cent confidence is then:

1.96 x (0.9 ÷100) x £53.20 = £0.90 (rounded to nearest 10p)
Lower limit is 53.20 – 0.90 = £52.30 (rounded to nearest 10p)
Upper limit is 53.20 + 0.90 = £54.10 (rounded to nearest 10p)

Similar calculations can be carried out for other estimates of expenditure and income. The 95 per cent confidence intervals for main expenditure categories are given in Table B3.

Table B3 **95 per cent confidence intervals for average household expenditure, 2010**

United Kingdom

Commodity or service	Weighted average weekly household expenditure (£)	95% confidence interval	
		Lower limit	Upper limit
All expenditure groups	**406.30**	**396.70**	**415.90**
Food and non-alcoholic drinks	53.20	52.30	54.10
Alcoholic drink, tobacco & narcotics	11.80	11.30	12.30
Clothing and footwear	23.40	22.20	24.60
Housing, fuel and power	60.40	58.60	62.20
Household goods and services	31.40	28.80	34.00
Health	5.00	4.40	5.60
Transport	64.90	62.00	67.80
Communication	13.00	12.60	13.40
Recreation and culture	58.10	55.60	60.60
Education	10.00	7.60	12.40
Restaurants and hotels	39.20	37.60	40.80
Miscellaneous goods and services	35.90	34.30	37.50

Calculation of standard errors – confidence intervals
Simple method

This formula treats the LCF sample as though it had arisen from a much simpler design with no multi-stage sampling, stratification, differential sampling or non-response weights. The weights are used but only to estimate the true population standard deviation in what is, in fact, a weighted design. The method of calculation is as follows: Let n be the total number of responding households in the survey, x_r the expenditure on a particular item of the r-th household, w_r be the weight attached to household r, and $\bar{x}$ the average expenditure per household on that item (averaged over the n households). Then the standard error $\bar{x}$, sesrs, is given by:

$$sesrs = \sqrt{\frac{\sum_{r=1}^{n} w_r (x_r - \bar{x})^2}{(n-1)\sum_{r=1}^{n} w_r}}$$

Full method

In fact, the sample in Great Britain is a multi-stage, stratified, random sample described further in 'Description and response rate of the survey'. First a sample of areas, the Primary Sampling Units (PSUs), is drawn from an ordered list. Then within each PSU a random sample of households is drawn. In Northern Ireland, however, the sample is drawn in a single stage and there is no clustering. The results are also weighted for non-response and calibrated to match the population separately by sex, by 5-year age ranges, and by region, as described in 'Weighting'.

The method for calculating complex standard errors for the weighted estimates used on this survey is quite complex. First, methods that take account of the clustering, stratification and differential sampling (and initial non-response weights) used in the design are applied. These are then modified to allow for the calibration weighting used on the survey. The exact formulae also depend on whether standard errors are being estimated for an estimated total or a mean or proportion. Here the method for a total is outlined.

Consecutive PSUs in the ordered list are first grouped into pairs or triples, at the end of a regional stratum. The standard error of a weighted total is estimated by:

$$sedes = \sqrt{\sum_h \frac{k_h}{k_h - 1} \sum_i (x_{hi} - \bar{x}_h)^2}$$

where h denotes the stratum (PSU pairs or triples), k_h is the number of PSUs in the stratum h (either 2 or 3), the x_{hi} is the weighted total in PSU i and the $\bar{x}_h$ is the mean of these totals in stratum h. Further details of this method of estimating sampling errors are described in *A Sampling Errors Manual* (B Butcher and D Elliot, ONS 1987).

The effect of the calibration weighting is calculated using a jackknife linearisation estimator. It uses the formula given above but with each household's expenditure, x_r, replaced by a residual from a linear regression of expenditure on the number of people in each household in each of the region and age by sex categories used in the weighting.

The formulae have been expressed in terms of expenditures on a particular item, but of course they can also be applied to expenditures on groups of items, commodity groups and incomes from particular sources.

Definitions

Major changes in definitions since 1991 are described in 'Changes to definitions, 1991 to 2010'. Changes made between 1980 and 1990 are summarised in Appendix E of *Family Spending 1994–95*. For earlier changes see Annex 5 of Family Expenditure Survey 1980.

Household

A household comprises one person or a group of people who have the accommodation as their only or main residence and (for a group):

> either share at least one meal a day
> or   share the living accommodation, that is, a living room or sitting room

Resident domestic servants are included. The members of a household are not necessarily related by blood or marriage. As the survey covers only private households, people living in hostels, hotels, boarding houses or institutions are excluded. Households are included if some or all members are not British subjects, however, information is not collected from households containing members of the diplomatic service of another country or members of the United States armed forces.

Retired households

Retired households are those where the household reference person is retired. The household reference person is defined as retired if they have reached state pension age and are economically inactive. From May 2010 the state pension age for women is increasing gradually to be in line with the male pension age of 65 by 2018. Therefore, if for example a male household reference person is aged over 65 years of age, but working part-time or waiting to take up a part-time job, this household would not be classified as a retired household. For analysis purposes two categories are used in this report:

- 'A retired household mainly dependent upon state pensions' is one in which at least three-quarters of the total income of the household is derived from national insurance retirement and similar pensions, including housing and other benefits paid in supplement to or instead of such pensions. The term 'national insurance retirement and similar pensions' includes national insurance disablement and war disability pensions, and income support in conjunction with these disability payments
- 'Other retired households' are retired households which do not fulfil the income conditions of 'retired household mainly dependent upon state pensions' because more than a quarter of the household's income derives from other sources. For example, occupational retirement pensions and/or income from investments, or annuities.

Household reference person (HRP)

From 2001/02 the concept of household reference person (HRP) was adopted on all government-sponsored surveys in place of head of household. The household reference person is the householder who:

- owns the household accommodation, or
- is legally responsible for the rent of the accommodation, or
- has the household accommodation as an emolument or perquisite, or
- has the household accommodation by virtue of some relationship to the owner who is not a member of the household.

If there are joint householders the household reference person will be the one with the higher income. If the income is the same, then the eldest householder is taken.

Members of household

In most cases the members of co-operating households are easily identified as the people who satisfy the conditions in the definition of a household, see above, and are present during the record-keeping period. However, difficulties of definition arise where people are temporarily away from the household or else spend their time between two residences. The following rules apply in deciding whether or not such persons are members of the household:

- Married people living and working away from home for any period are included as members, provided they consider the sampled address to be their main residence. In general, other people (such as relatives, friends and boarders) who are either temporarily absent or who spend their time between the sampled address and another address, are included as members if they consider the sampled address to be their main residence. However, there are exceptions which override the subjective main residence rule:

    i.   children under 16 years of age away at school are included as members;

    ii. older people receiving education away from home, including children aged 16 and 17, are excluded unless they are at home for all or most of the record-keeping period;

    iii. visitors staying temporarily with the household, and others who have been in the household for only a short time are treated as members, provided they will be staying with the household for at least one month from the start of record-keeping.

Household composition

A consequence of these definitions is that household compositions quoted in this report include some households where certain members are temporarily absent, for example, a 'two-adult and children' household where one parent is temporarily away from home.

Adult

In the report, people who have reached the age of 18 are classed as adults. In addition, those aged 16 to 18 who are not in full-time education, or who are married, are classed as adults.

Children

In the report, people who are under 18 years of age, in full-time education and have never been married are classed as children.

However, in the definition of clothing, clothing for people aged 16 years and over is classified as clothing for men and women; clothing for those aged 5 to 15 as clothing for boys and girls; and clothing for those under five as babies clothing.

Main Diary Keeper (MDK) (or main shopper)

The MDK is the person in the household who is normally responsible for most of the food shopping. This includes people who organise and pay for the shopping although they do not physically do the shopping themselves.

Spenders

Household members aged 16 and over, excluding those who, for special reasons, are not capable of keeping diary record-books, are described as spenders.

Absent spenders

If a spender is absent for longer than 7 days they are defined as an 'absent spender'. Absent spenders do not keep a diary and consequently are not eligible for the monetary gift that is paid to diary keepers.

Non-spenders

If a household member is completely incapable of contributing to the survey by answering questions or keeping a diary, then they are defined as a 'non-spender'. However, incapable people living on their own cannot be designated as non-spenders as they comprise the whole expenditure unit. If this is the case, the interviewer should enlist the help of the person outside of the household who looks after their interests. If there is no-one able or willing to help, the address is coded as incapable.

Economically active

These are people aged 16 and over who fall into the following categories:

- **Employees at work** – those who at the time of interview were working full-time or part-time as employees or were away from work on holiday. Part-time work is defined as normally working 30 hours a week or less (excluding meal breaks) including regularly worked overtime.

- **Employees temporarily away from work** – those who, at the time of interview, had a job but were temporarily absent due to, for example, illness, temporary lay-off, or strike.

- **Government supported training schemes** – those participating in government programmes and schemes who, in the course of their participation, receive training such as Employment Training, and including those who are also employees in employment.

- **Self-employed** – those who, at the time of interview, said they were self-employed.

- **Unemployed** – those who, at the time of interview, were out of employment and have sought work within the last four weeks and were available to start work within two weeks, or were waiting to start a job already obtained.

- **Unpaid family workers** – those working unpaid for their own or a relative's business. In this report, unpaid family workers are included under economically inactive in analyses by economic status (Tables A17 and B5) because insufficient information is available to assign them to an economic status group.

Economically inactive

- **Retired** – people who have reached national insurance retirement age and are not economically active. From May 2010 the female state pension age is gradually increasing to align with the male pension age of 65 by 2018.

- **Unoccupied** – people under national insurance retirement age who are not working, nor actively seeking work. This category includes certain self-employed people such as mail order agents and baby-sitters who are not classified as economically active.

National Statistics Socio-economic Classification (NS-SEC)

From 2001 the National Statistics Socio-economic Classification (NS-SEC) was adopted for all official surveys, in place of Social Class based on Occupation (SC) and Socio-economic Groups (SEG). NS-SEC is itself based on the Standard Occupational Classification 2000 (SOC2000) and details of employment status. Although NS-SEC is an occupation-based classification, there are procedures for classifying those not in work.

The main categories used for analysis in *Family Spending* are:

1		Higher managerial and professional occupations, sub-divided into:
	1.1	Large employers and higher managerial occupations
1.2		Higher professional occupations
2		Lower managerial and professional occupations
3		Intermediate occupations
4		Small employers and own account workers
5		Lower supervisory and technical occupations
6		Semi-routine occupations
7		Routine occupations
8		Never worked and long-term unemployed
9		Students
10		Occupation not stated
11		Not classifiable for other reasons

The long-term unemployed are defined as those unemployed and seeking work for 12 months or more. Members of the armed forces, who were assigned to a separate category in social class, are included within the NS-SEC classification. Individuals that have retired within the last 12 months are classified according to their employment. Other retired individuals are assigned to the 'Not classifiable for other reasons' category.

Socio-economic Classification (SE-SEC) regions

These are the same areas as UK regions and countries, see region map on page 158 for more details.

Urban and rural areas

This classification introduced in 2005/06 replaces the previous Department for Transport, Local Government and the Regions (DTLR) 1991 Census-based urban and rural classification, which was used in previous editions of *Family Spending*. The new classification is applied across Great Britain and is an amalgamation of the Rural and Urban Classification 2004 for England and Wales and the Scottish Executive Urban Rural Classification. These classifications are based on 2001 Census data and have been endorsed as the standard National Statistics Classifications for identifying urban and rural areas across GB. In broad terms, an area is defined as urban or rural depending on whether the population falls inside a settlement of 10,000 or more. For further details concerning these classifications please refer to the Rural/Urban Definition and LA Classification on the Office for National Statistics (ONS) website.

Expenditure

Any definition of expenditure is to some extent arbitrary, and the inclusion of certain types of payment is a matter of convenience or convention depending on the purpose for which the information is to be used. In the tables in this report, total expenditure represents current expenditure on goods and services. Total expenditure, defined in this way, excludes those recorded payments that are really savings or investments: for example, purchases of national savings certificates, life assurance premiums, and contributions to pension funds. Similarly, income tax payments, national insurance contributions, mortgage capital repayments and other payments for major additions to dwellings are excluded. Expenditure data are collected in the diary record-book and in the household schedule. Informants are asked to record in the diary any payments made during the 14 days of record-keeping, whether or not the goods or services paid for have been received. Certain types of expenditure which are usually regular though infrequent, such as insurance, licences and season tickets, and the periods to which they relate, are recorded in the household schedule as well as regular payments such as utility bills.

The cash purchase of motor vehicles is also entered in the household schedule. In addition, expenditure on some items purchased infrequently (thereby being subject to high sampling errors) has been recorded in the household schedule using a retrospective recall period of either 3 or 12 months. These items include carpets, furniture, holidays and some housing costs. In order to avoid duplication, all payments shown in the diary record-book which relate to items listed in the household or income schedules are omitted in the analysis of the data, irrespective of whether there is a corresponding entry on the latter schedules. Amounts paid in respect of periods longer than a week are converted to weekly values.

Expenditure tables in this report show the 12 main commodity groups of spending and these are broken down into items which are numbered hierarchically (see 'Changes to definitions, 1991 to 2010' which details a major change to the coding frame used from 2001/02). Table A1 shows a further breakdown in the items themselves into components which can be separately identified. The items are numbered as in the main expenditure tables, and the average weekly household expenditure and percentage standard error is shown against each item or component.

Qualifications which apply to this concept of expenditure are described in the following paragraphs:

- **Goods supplied from a household's own shop or farm**
  Spenders are asked to record and give the value of goods obtained from their own shop or farm, even if the goods are withdrawn from stock for personal use without payment. The value is included as expenditure

- **Hire purchase and credit sales agreements, and transactions financed by loans repaid by instalments**
  Expenditure on transactions under hire purchase or credit sales agreements, or financed by loans repaid by instalments, consists of all instalments that are still being paid at the date of interview, together with down payments on commodities acquired within the preceding three months. These two components (divided by the periods covered) provide the weekly averages which are included in the expenditure on the separate items given in the tables in this report

- **Club payments and budget account payments, instalments through mail order firms and similar forms of credit transaction**
  When goods are purchased by forms of credit other than hire purchase and credit sales

agreement, the expenditure on them may be estimated either from the amount of the instalment which is paid or from the value of the goods which are acquired. Since the particular commodities to which the instalment relates may not be known, details of goods ordered through, for example, clubs or mail order firms, during the month prior to the date of interview, are recorded in the household schedule. The weekly equivalent of the value of the goods is included in the expenditure on the separate items given in the tables in this report. This procedure has the advantage of enabling club transactions to be related to specific articles. Although payments into clubs, etc. are shown in the diary record-book, these entries are excluded from expenditure estimates

- **Credit card transactions**
  From 1988 purchases made by credit card or charge card have been recorded in the survey on an acquisition basis rather than the formerly used payment basis. Thus, if a spender acquired an item (by use of credit/charge card) during the two week survey period, the value of the item would be included as part of expenditure in that period whether or not any payment was made in this period to the credit card account. Payments made to the card account are ignored. However any payment of credit/charge card interest is included in expenditure if made in the two week period

- **Income tax**
  Amounts of income tax deducted under the Pay As You Earn (PAYE) scheme or paid directly by those who are employers or self-employed are recorded (together with information about tax refunds). For employers and the self-employed the amounts comprise the actual payments made in the previous 12 months and may not correspond to the tax due on the income arising in that period, for example if no tax has been paid but is due or if tax payments cover more than one financial year. However, the amounts of tax deducted at source from some of the items which appear in the Income Schedule are not directly available. Estimates of the tax paid on bank and building society interest and amounts deducted from dividends on stocks and shares are therefore made by applying the appropriate rates of tax. In the case of income tax paid at source on pensions and annuities, similar adjustments are made. These estimates mainly affect the relatively few households with high incomes from interest and dividends, and households including someone receiving a pension from previous employment

- **Rented dwellings**
  Expenditure on rented dwellings is taken as the sum of expenditure on a number of items such as rent, council tax, and water rates. For local authority tenants the expenditure is gross rent less any rebate (including rebate received in the form of housing benefit), and for other tenants it is gross rent less any rent allowance received under statutory schemes including the Housing Benefit Scheme. Rebate on council tax or rates (Northern Ireland) is deducted from expenditure on council tax or rates. Receipts from sub-letting part of the dwelling are not deducted from housing costs but appear (net of the expenses of the sub-letting) as investment income

- **Rent-free dwellings**
  Rent-free dwellings are those owned by someone outside the household and where either no rent is charged or the rent is paid by someone outside the household. Households whose rent is paid directly to the landlord by the DWP do not live rent-free. Payments for council tax for example are regarded as the cost of housing. Rebate on rates (Northern Ireland)/Council tax/water rates (Scotland) (including rebate received in the form of housing benefit), is deducted from expenditure on rates/council tax/water rates. Receipts from sub-letting part of

the dwelling are not deducted from housing costs but appear (net of the expenses of the sub-letting) as investment income

- **Owner-occupied dwellings**

    In the LCF, payments for water rates, ground rent, fuel, maintenance and repair of the dwelling, among other items, are regarded as the cost of housing. Receipts from letting part of the dwelling are not deducted from housing costs but appear (net of the expenses of the letting) as investment income. Mortgage capital repayments and amounts paid for the outright purchase of the dwelling or for major structural alterations are not included as housing expenditure, but are entered under 'other items recorded', as are council tax, rates (Northern Ireland) and mortgage interest payments. Structural insurance is included in 'miscellaneous goods and services'

- **Second-hand goods and part-exchange transactions**

    The survey expenditure data are based on information about actual payments and therefore include payments for second-hand goods and part-exchange transactions. New payments only are included for part-exchange transactions, that is the costs of the goods obtained less the amounts allowed for the goods which are traded in. Receipts for goods sold or traded in are not included in income

- **Business expenses**

    The survey covers only private households and is concerned with payments made by members of households as private individuals. Spenders are asked to state whether expenditure that has been recorded on the schedules includes amounts that will be refunded as expenses from a business or organisation or that will be entered as business expenses for income tax purposes, for example rent, telephone charges, travelling expenses and meals out. Any such amounts are deducted from the recorded expenditure

Income

The standard concept of income in the survey is, as far as possible, that of gross weekly cash income current at the time of interview, that is before the deduction of income tax actually paid, national insurance contributions and other deductions at source. However, for a few tables a concept of disposable income is used, defined as gross weekly cash income less the statutory deductions and payments of income tax (taking refunds into account) and national insurance contributions. Analysis in Chapter 3 of this report and some other analyses of LCF data use 'equivalisation' of incomes: in other words adjustment of household income to allow for the different size and composition of each household. For more information see Chapter 3. The cash levels of certain items of income (and expenditure) recorded in the survey by households receiving supplementary benefit were affected by the Housing Benefit Scheme introduced in stages from November 1982. From 1984 housing expenditure is given on a strictly net basis and all rent/council tax rebates and allowances and housing benefit are excluded from gross income.

Although information about most types of income is obtained on a current basis, some data, principally income from investment and from self-employment, are estimated over a 12-month period.

The following are excluded from the assessment of income:

- money received by one member of the household from another (for example housekeeping money, dress allowance, children's pocket money) other than wages paid to resident domestic servants;
- withdrawals of savings, receipts from maturing insurance policies, proceeds from sale of financial and other assets (such as houses, cars, and furniture), winnings from betting, lump-sum gratuities and windfalls such as legacies;
- the value of educational grants and scholarships not paid in cash;
- the value of income in kind, including the value of goods received free and the abatement in cost of goods received at reduced prices, and of bills paid by someone who is not a member of the household;
- loans and money received in repayment of loans.

Details are obtained of the income of each member of the household. The income of the household is taken to be the sum of the incomes of all its members. The information does not relate to a common or a fixed time period. Items recorded for periods greater than a week are converted to a weekly value.

Particular points relating to some components of income are as follows:

- **Wages and salaries of employees**

  The normal gross wages or salaries of employees are taken to be their earnings. These are calculated by adding to the normal 'take home' pay amounts deducted at source, such as income tax payments, national insurance contributions and other deductions (for example payments into firm social clubs, superannuation schemes, works transport, and benevolent funds). Employees are asked to give the earnings actually received including bonuses and commission the last time payment was made and, if different, the amount usually received. It is the amount usually received that is regarded as the normal take-home pay. Additions are made so as to include in normal earnings the value of occasional payments, such as bonuses or commissions received quarterly or annually. One of the principal objects in obtaining data on income is to enable expenditure to be classified in ranges of normal income. Average household expenditure is likely to be based on the long-term expectations of the various members of the household as to their incomes rather than be altered by short-term changes affecting individuals. Hence, if employees have been away from work without pay for 13 weeks or less, they are regarded as continuing to receive their normal earnings instead of social security benefits, such as unemployment or sickness benefit, that they may be receiving. Otherwise, normal earnings are disregarded and current short-term social security benefits taken instead. Wages and salaries include any earnings from subsidiary employment as an employee and the earnings of HM Forces

- **Income from self-employment**

  Income from self-employment covers any personal income from employment other than as an employee: for example, as a sole trader, professional or other person working on his own account or in partnership, including subsidiary work on his own account by a person whose main job is as an employee. It is measured from estimates of income or trading profits, after deduction of business expenses but before deduction of tax, over the most recent 12-month period for which figures can be given. Should either a loss have been made or no profit, income would be taken as the amounts drawn from the business for own use or as any other income received from the job or business. People working as mail order agents or baby-sitters, with no other employment, have been classified as unoccupied rather than as self-employed, and the earnings involved have been classified as earnings from 'other sources' rather than self-employment income

- **Income from investment**

  Income from investments or from property, other than that in which the household is residing, is the amount received during the 12 months immediately prior to the date of the initial interview. It includes receipts from sub-letting part of the dwelling (net of the expenses of the sub-letting). If income tax has been deducted at source the gross amount is estimated by applying a conversion factor during processing

- **Social security benefits**

  Income from social security benefits does not include the short-term payments such as unemployment or sickness benefit, received by an employee who has been away from work for 13 weeks or less, and who is therefore regarded as continuing to receive his normal earnings as described on page 156.

Quantiles

The quantiles of a distribution divide it into a number of equal parts; each of which contains the same number of households. In *Family Spending*, quantiles are applied to both household expenditure and income distributions.

For example, the median of a distribution divides it into two equal parts, so that half the households in a distribution of household income will have income more than the median, and the other half will have income less than the median. Similarly, quartiles, quintiles and deciles divide the distribution into four, five and ten equal parts respectively.

Most of the analysis in *Family Spending* is done in terms of quintile groups and decile groups.

In the calculation of quantiles for this report, zero values are counted as part of the distribution.

Income headings

Headings used for identifying 2010 income information

Source of Income

References in tables	Components separately identified	Explanatory notes
a. Wages and salaries	Normal 'take-home' pay from main employment	(i) In the calculation of household income in this report, where an employee has been away from work without pay for 13 weeks or less his normal wage or salary has been used in estimating his total income instead of social security benefits, such as unemployment or sickness benefits that he may have received. Otherwise such benefits are used in estimating total income (see notes at reference e).
	'Take-home' pay from subsidiary employment	
	Employees' income tax deduction	
	Employees' National Insurance contribution	
	Superannuation contributions deducted from pay	
	Other deductions	
		(ii) Normal income from wages and salaries is estimated by adding to the normal 'take-home' pay deductions made at source last time paid, together with the weekly value of occasional additions to wages and salaries (see page 153).
		(iii) The components of wages and salaries, for which figures are separately available, amount in total to the normal earnings of employees, regardless of the operation of the 13 week rule in note (i) above. Thus the sum of the components listed here does not in general equal the wages and salaries figure in tables of this report.

Income headings

Headings used for identifying 2010 income information

Source of income

b. Self-employment	Income from business or profession, including subsidiary self-employment	The earnings or profits of a trade or profession, after deduction of business expenses but before deduction of tax.
c. Investments	Interest on building society shares and deposits Interest on bank deposits and savings accounts, including National Savings Bank Interest on ISAs Interest on Gilt-edged stock and War Loans Interest and dividends from stocks, shares, bonds, trusts, debentures and other securities Rent or income from property, after deducting expenses but inclusive of income tax (including receipts from letting or sub-letting part of own residence, net of the expenses of the letting or sub-letting). Other unearned Income	
d. Annuities and pensions, other than social security	Annuities and income from trust or covenant Pensions from previous employers Personal pensions	
e. Benefits	Child benefit Guardian's allowance Carer's allowance (formerly Invalid care allowance) Retirement pension (National Insurance) or old person's pension credit Widow's pension/bereavement allowance or widowed parent's allowance War disablement pension or war widow/widower's pension Severe disablement allowance Care component of disability living allowance Mobility component of disability living allowance Attendance allowance Job seekers allowance Winter fuel allowance Cold Weather Payment Income support Working tax credit Child tax credit Incapacity benefit	(i) The calculation of household income in this report takes account of the 13 week rule described at reference a, note (i). (ii) The components of social security benefits, for which figures are separately available, amount in total to the benefits received in the week before interview. That is to say, they include amounts that are discounted from the total by the operation of the 13 week rule in note (i). Thus the sum of the components listed here differs from the total of social security benefits used in the income tables of this report. (iii) Housing Benefit is treated as a reduction in housing costs and not as income.

Income headings

Headings used for identifying 2010 income information

Source of income

	Statutory sick pay (from employer)
	Industrial injury disablement benefit
	Maternity allowance
	Statutory maternity pay
	Statutory paternity pay
	Statutory adoption pay
	Health in pregnancy grant
	Any other benefit including lump sums and grants
	Social security benefits excluded from Income calculation by 13 week rule.

f. Other sources

Married person's allowance from husband/wife temporarily away from home
Alimony or separation allowances; allowances for foster children, allowances from members of the Armed Forces or Merchant Navy, or any other money from friends or relatives, other than husbands outside the household Benefits from trade unions, friendly societies etc. other than pensions
Value of meal vouchers
Earnings from intermittent or casual work over 12 months, not included in a or b above
Student loans and money scholarships received by persons aged 16 and over and aged under 16
Other income for children under 16 e.g. from spare time j income from Trusts or investments.

Regions of the United Kingdom[1]

Orkney
Islands

Shetland
Islands

Scotland

Northern
Ireland

North
East

North
West

Yorkshire
and The
Humber

—— Country and Regional
boundaries

East
Midlands

West
Midlands

Wales

East of
England

London

South East

South West

1 Government Offices for the Regions closed at the end of March 2011. The former Government Office
Regions (England) are now referred to as 'regions' for statistical purposes. Scotland, Northern Ireland
and Wales are not regions, but are often used as equivalents for the purpose of representing statistics
that cover the whole of the UK.

Contains Ordnance Survey data © Crown copyright and database right 2011

Change in definitions, 1991 to 2010

1991

No significant changes.

1992

Housing – Imputed rent for owner occupiers and households in rent-free accommodation was discontinued. For owner occupiers this had been the rent they would have had to pay themselves to live in the property they own, and for households in rent-free accommodation it was the rent they would normally have had to pay. Until 1990 these amounts were counted both as income and as a housing cost. Mortgage interest payments were counted as a housing cost for the first time in 1991.

1993

Council Tax – Council Tax was introduced to replace the Community Charge in Great Britain from April 1993.

1994/95

New expenditure items – The definition of expenditure was extended to include two items previously shown under 'other payments recorded'. These were:
• gambling payments, and
• mortgage protection premiums

Expenditure classifications – A new classification system for expenditures was introduced in April 1994. The system is hierarchical and allows more detail to be preserved than the previous system. New categories of expenditure were introduced and are shown in detail in Table 7.1. The 14 main groups of expenditure were retained, but there were some changes in the content of these groups.

Gambling payments – data on gambling expenditure and winnings are collected in the expenditure diary. Previously these were excluded from the definition of household expenditure used in the FES. The data are shown as memoranda items under the heading 'Other payments recorded' on both gross and net bases. The net basis corresponds approximately to the treatment of gambling in the National Accounts. The introduction of the National Lottery stimulated a reconsideration of this treatment. From April 1994, (gross) gambling payments have been included as expenditure in 'Leisure Services'. Gambling winnings continued to be noted as a memorandum item under 'Other items recorded'. They are treated as windfall income. They do not form a part of normal household income, nor are they subtracted from gross gambling payments. This treatment is in line with the PRODCOM classification of the statistical office of the European Union (Eurostat) for expenditure in household budget surveys.

1995/96

Geographical coverage – The FES geographical coverage was extended to mainland Scotland north of the Caledonian Canal.

Under 16s diaries – Two-week expenditure diaries for 7 to 15-year-olds were introduced following three feasibility pilot studies which found that children of that age group were able to cope with the task of keeping a two-week expenditure record. Children are asked to record everything they buy with their own money but to exclude items bought with other people's money. Purchases are coded according to the same coding categories as adult diaries except for meals and snacks away from home which are coded as school meals, hot meals and snacks, and cold meals and snacks. Children who keep a diary are given a £5 incentive payment. A refusal to keep an under 16's diary does not invalidate the household from inclusion in the survey.

Pocket money given to children is still recorded separately in adult diaries, and money paid by adults for school meals and school travel is recorded in the Household Questionnaire. Double counting is eliminated at the processing stage.

Tables in *Family Spending* reports did not include the information from the children's diaries until the 1998/99 report. Appendix F in the 1998/99 and 1999/2000 reports show what difference the inclusion made.

1996/97

Self-employment –- The way in which information about income from self-employment is collected was substantially revised in 1996/97 following various tests and pilot studies. The quality of such data was increased but this may have lead to a discontinuity. Full details are shown in the Income Questionnaire, available from the address in the introduction.

Cable/satellite television – Information on cable and satellite subscriptions is now collected from the household questionnaire rather than from the diary, leading to more respondents reporting this expenditure.

Mobile phones – Expenditure on mobile phones was previously collected through the diary. From 1996/97 this has been included in the questionnaire.

Job Seekers Allowance (JSA) – Introduced in October 1996 as a replacement for Unemployment Benefit and any Income Support associated with the payment of Unemployment Benefit. Receipt of JSA is collected with NI Unemployment Benefit and with Income Support. In both cases the number of weeks a respondent has been in receipt of these benefits is taken as the number of weeks receiving JSA in the last 12 months and before that period the number of weeks receiving Unemployment Benefit/Income Support.

Retrospective recall – The period over which information is requested has been extended from 3 to 12 months for vehicle purchase and sale. Information on the purchase of car and motorcycle spare parts is no longer collected by retrospective recall. Instead expenditure on these items is collected through the diary.

State benefits – The lists of benefits specifically asked about was reviewed in 1996/97. See the Income Questionnaire for more information.

Sample stratifiers – New stratifiers were introduced in 1996/97 based on standard regions, socio-economic group and car ownership.

Government Office Regions – Regional analyses presented using the Government Office Regions (GORs) formed in 1994. Previously all regional analyses used Standard Statistical Regions (SSRs). For more information see Appendix F in the 1996/97 report.

1997/98

Bank/building society service charges – Collection of information on service charges levied by banks has been extended to include building societies.

Payments from unemployment/redundancy insurances – Information is now collected on payments received from private unemployment and redundancy insurance policies. This information is then incorporated into the calculation of income from other sources.

Retired households – The definition of retired households has been amended to exclude households where the head of the household is economically active.

Rent-free tenure – The definition of rent-free tenure has been amended to include those households for which someone outside the household, except an employer or an organisation, is paying a rent or mortgage on behalf of the household.

National Lottery – From February 1997 expenditure on National Lottery tickets was collected as three separate items: tickets for the Wednesday draw only, tickets for the Saturday draw only and tickets for both draws.

Northern Ireland sample boost – From 1997/89 Northern Ireland was over sampled in order to provide enough households to conduct separate regional analysis.

1998/99

Children's income – Three new expenditure codes were introduced: pocket money to children; money given to children for specific purposes and cash gifts to children. These replaced a single code covering all three categories.

Main job and last paid job – Harmonised questions were adopted.

1999/2000

Disabled Persons Tax Credit replaced Disability Working Allowance and **Working Families Tax Credit** replaced Family Credit from October 1999.

2000/01

Household definition – The definition was changed to the harmonised definition which has been in use in the Census and nearly all other government household surveys since 1981. The effect is to group together into a single household some people who would have been allocated to separate households on the previous definition. The effect is fairly small but not negligible.

Up to 1999/2000 the FES definition was based on the pre-1981 Census definition and required members to share eating and budgeting arrangements as well as share living accommodation.

The definition of a household was:

- One person or a group of people who have the accommodation as their only or main residence, and (for a group) share the living accommodation, that is a living or sitting room, **and** share meals together (or have common housekeeping)

The harmonised definition is less restrictive:

- One person or a group of people who have the accommodation as their only or main residence and (for a group) share the living accommodation, that is a living or sitting room **or** share meals together or have common housekeeping

The effect of the change is probably to increase average household size by 0.6 per cent.

Question reductions – A thorough review of the questionnaire showed that a number of questions were no longer needed by government users. These were cut from the 2000/01 survey to reduce the burden on respondents. The reduction was fairly small but it did make the interview flow better. All the questions needed for a complete record of expenditure and income were retained.

Redesigned diary – The diary was redesigned to be easier for respondents to keep and to look cleaner. The main change of substance was to delete the column for recording whether each item was purchased by credit, charge or shop card.

Ending of MIRAS – Tax relief on interest on loans for house purchase was abolished from April 2000. Questions related to MIRAS (Mortgage Interest Relief at Source) were therefore dropped. They included some that were needed to estimate the amount if the respondent did not know it. A number were retained for other purposes, however, such as the amount of the loan still outstanding which is still asked for households paying a reduced rate of interest because one of them works for the lender.

2001/02

Expenditure and Food Survey (EFS) introduced, replacing the Family Expenditure Survey (FES) and National Food Survey (NFS)

Household reference person – this replaced the previous concept of head of household. The household reference person is the householder, that is the person who:

- owns the household accommodation, or
- is legally responsible for the rent of the accommodation, or
- has the household accommodation as an emolument or perquisite, or
- has the household accommodation by virtue of some relationship to the owner who is not a member of the household.

If there are joint householders the household reference person is the one with the higher income. If the income is the same, then the eldest householder is taken.

A key difference between household reference person and head of household is that the household reference person must always be a householder, whereas the head of household was always the husband, who might not even be a householder himself.

National Statistics Socio-economic Classification (NS-SEC) – the National Statistics Socio-economic Classification (NS-SEC) was adopted for all official surveys, in place of Social Class based on Occupation (SC) and Socio-economic Group (SEG). NS-SEC is itself based on the Standard Occupational Classification 2000 (SOC2000) and details of employment status. The long-term unemployed, which fall into a separate category, are defined as those unemployed and seeking work for 12 months or more. Members of the armed forces, who were assigned to a separate category in Social Class, are included within the NS-SEC classification. Residual groups that remain unclassified include students and those with inadequately described occupations.

COICOP – From 2001/02, the Classification Of Individual COnsumption by Purpose (COICOP/HBS, referred to as COICOP in this report) was introduced as a new coding frame for expenditure items. COICOP has been adapted to the needs of household budget surveys (HBS) across the EU and, as a consequence, is compatible with similar classifications used in national accounts and consumer price indices. This allows the production of indicators which are comparable Europe-wide, such as the Harmonised Indices of Consumer Prices (computed for all goods as well as sub-categories such as food and transport). The main categorisation of spending used in this report (namely 12 categories relating to food and non-alcoholic beverages; alcoholic beverages, tobacco and narcotics; clothing and footwear; housing, fuel and power; household goods and services; health; transport; communication; recreation and culture; education; restaurants and hotels; and miscellaneous goods and services) is only comparable between the two frames at a broad level. Table 4.1 has been produced by mapping COICOP to the FES 14 main categories. However the two frames are not comparable for any smaller categories, leading to a break in trends between 2000/01 and 2001/02 for any level of detail below the main 12-fold categorisation. A complete listing of COICOP and COICOP plus (an extra level of detail added by individual countries for their own needs) is available on request from the address in the introduction.

Proxy interviews – While questions about general household affairs are put to all household members or to a main household informant, questions about work and income are put to the individual members of the household. Where a member of the household is not present during the household interview, another member of the household (for example, a spouse) may be able to provide information about the absent person. The individual's interview is then identified as a proxy interview. From 2001/02 the EFS began accepting responses that contained a proxy interview.

Short income – From 2001/02 the EFS accepted responses from households that answered the short income section. This was designed for respondents who were reluctant to provide more detailed income information.

2002/03

Main shopper – At the launch of the EFS in April 2001, the respondent responsible for buying the household's main shopping was identified as the 'main diary keeper' also known as 'main shopper'.

The importance of the main shopper is to ensure that we have obtained information on the bulk of the shopping in the household. Without this person's co-operation we have insufficient information to use the other diaries kept by members of the household in a meaningful way. The main shopper must therefore complete a diary for the interview to qualify as a full or partial interview. Without their participation, the outcome will be a refusal no matter who else is willing to complete a diary.

2003/04

Working Tax Credit replaced Disabled Persons Tax Credit and Working Families Tax Credit from April 2003.

Pension Credit replaced Minimum Income Guarantee from October 2003.

Child Tax Credit replaced Children's Tax Credit and Childcare Tax Credit from April 2003.

2004/05

No significant changes.

2005/06

Urban and rural definition – A new urban and rural area classification based on 2001 Census data has been introduced onto the EFS dataset and is presented in Tables A38, A45 and A48 of this publication. The classification replaces the Department for Transport, Local Government and the Regions (DTLR) 1991 Census-based urban and rural classification that was used in previous editions of *Family Spending*. The new classification is the standard National Statistics classification for identifying urban and rural areas in England and Wales, and Scotland. Please refer to 'Definitions' for further details.

Motor vehicle road taxation refunds – Questions on road tax refunds were inadvertently omitted from the 2005/06 questionnaire. Within the Appendix A tables of the 2005/06 report, the heading for category 13.2.3 'Motor vehicle road taxation payments less refunds' has been changed to reflect this omission.

Purchase of vehicles – During April to December 2005 respondents who had sold a vehicle were not asked whether they had bought that same vehicle in the previous year. This was corrected from January 2006, but means that some expenditure on vehicles may have been missed.

2006

Reporting period –The LCF started reporting on a calendar-year basis, rather than for financial years.

2007

Mortgage interest payments – An improvement to the imputation of mortgage interest payments has been implemented and applied to 2006 and 2007 data in this publication, which should lead to more accurate figures. This will also lead to a slight discontinuity.

Mortgage capital repayments – An error was discovered in the derivation of mortgage capital repayments which was leading to double counting. This has been amended for the 2006 and 2007 data in this publication, which will cause a minor discontinuity.

2008

IHS – The LCF joined the Integrated Household Survey (IHS)

NS-SEC – The LCF question used to derive the student category for NS-SEC B was changed in 2008 due to the introduction of the Integrated Household Survey (IHS). Prior to the IHS, respondents were asked if they were currently in full-time education and those who responded yes to this question were classified as students. Since 2008, respondents have been asked if they are enrolled on any full-time or part-time education course and those who respond 'yes' have then been asked to select the course they are attending from a set of options. Respondents who select any of the full-time course options have been classified as students under NS-SEC. This more stringent definition of full-time student has resulted in a decrease in the number of people classified as students.

2009

Gas & electricity payment methods – Following consultation with the Department for Energy and Climate Change (DECC), the payment methods have been updated for the gas and electricity questions. This has brought the LCF questions in line with those on the EHS. This may cause a slight discontinuity in the data.

A question was added to capture the take up of the Health in Pregnancy grant, a benefit introduced in April 2009.

A question capturing Cold Weather Payments was included from July 2009 onwards.

2010

Multiple household interviews – In 2010 the LCF in line with other ONS social surveys changed the way interviews were conducted at addresses with more than one private household. Until July 2010 interviewers were instructed to interview all households (up to a maximum of three at any one address) when a sampled address contained more than one private household. From July onwards interviewers were no longer required to interview multiple households at these addresses, instead a random selection method is used to select one household to interview.

Northern Ireland sample – Until 2010 the LCF ran a boosted Northern Ireland sample which allowed for separate Northern Ireland analysis to be conducted, at the end of 2009 this boost came to an end.

Female pension age – Due to changes in the female pension age, the way in which pensioner households are identified has changed. Until May 2010 any female over 60 and male over 65 was defined as a pensioner. From May 2010 onwards the female pension age increases gradually, which is reflected in the LCF definition.

Internet subscription fees – For 2010 the internet subscription fees sub-category has been moved from recreation and culture to the communications category of COICOP.

Survey improvements

Quality project

During 2010 work was done to improve the quality of the LCF outputs. A review of the quality assurance and validation processes was conducted and a number of areas of improvement identified. An executive summary outlining the key aspects of the project will be published shortly.

Questionnaire review

In order to ensure the LCF questionnaire is up-to-date it is important that questions are regularly reviewed and relevant changes made. During 2010 a number of sections of the questionnaire were updated following a review in 2009.

The questionnaire review project is ongoing with further topics being review during 2010 and 2011. This includes the new section deigned to capture expenditure on combined communication packages which was tested in 2010 and included in the 2011 questionnaire.

Weighting

Since 1998/99 the survey has been weighted to reduce the effect of non-response bias and produce population totals and means. The weights are produced in two stages. First, the data are weighted to compensate for non-response (sample-based weighting). Second, the sample distribution is weighted so that it matches the population distribution in terms of region, age group and sex (population-based weighting).

Sample based weighting using the Census

Weighting for non-response involves giving each respondent a weight so that they represent the non-respondents that are similar to them in terms of the survey characteristics. From 1998/99 the EFS used results from the 1991 Census-linked study of non-respondents to carry out non-response weighting[2]. From 2007 onwards the EFS/LCF non-response classes and weights have been annually updated using 2001 Census-linked data.

The Census-linked studies matched census addresses with the sampled addresses of some of the large continuous surveys, including FES for the1991 link study and EFS for the 2001 link study. In this way it was possible to match the address details of the respondents as well as the non-respondents with corresponding information gathered from the Census for the same address. The information collected during the 1991 and then the 2001 Census/FES/EFS matching work was then used to identify the types of households that were being under-represented in the survey.

For the 1991 Census-based non-response weights, a combination of household variables were analysed with the software package AnswerTree (using the chi-squared statistics CHAID)[3] to identify which characteristics were most significant in distinguishing between responding and non-responding households. These characteristics were sorted by the program to produce 10 weighting classes with different response rates. For the updated 2001 Census-based non-response weights, a combination of household variables were analysed using a mixed model approach. The mixed model is a combined approach to modelling, designed to benefit from the underlying statistical model of logistic regression as well as utilising AnswerTree. Updated weighting classes were produced, using this analysis, to further improve non-response weighting from 2007. The results of the 2011 Census-linked studies will be used to further update non-response weighting in due course.

Population-based weighting

The second stage of the weighting adjusts the non-response weights so that weighted totals match population totals. As the LCF sample is based on private households, the population totals used in the weighting need to relate to people living in private households. The population totals used are the most up-to-date official figures available; from 2006 onwards, these totals have been population projections based on estimates rolled forward from the 2001 Census. These estimates used exclude residents of institutions not covered by the EFS/LCF, such as those living in bed-and-breakfast accommodation, hostels, residential homes and other institutions.

The non-response weights were calibrated[4] so that weighted totals matched population totals for males and females in different age groups and for regions. An important feature of the population-based weighting is that it is done by adjusting the factors for households not individuals.

The weighting is carried out separately for each quarter of the survey. The main reason is that sample sizes vary more from quarter to quarter than in the past. This is due to reissuing addresses after an interval of a few months where there had previously been no contact or a refusal to a new interviewer. This results in more interviews in the later quarters of the year than in the first quarter. Quarterly weighting, therefore, counteracts any potential bias from the uneven spread of interviews through the year. Quarterly weighting also results in small sample numbers in some of the age/sex categories that were used in previous years. The categories have therefore been widened slightly to avoid this.

Effects of weighting on the data

Table B4 shows the effects of the weighting by comparing unweighted and weighted data from 2010.

Table B4 The effect of weighting on expenditure, 2010
United Kingdom

Commodity or service	Average weekly household expenditure		Absolute difference	*Percentage difference*
	Unweighted	Weighted as published		
All expenditure groups	**403.30**	**406.30**	**2.99**	*0.7*
Food and non-alcoholic drinks	53.60	53.20	-0.43	*-0.8*
Alcoholic drink, tobacco & narcotics	11.90	11.80	-0.04	*-0.3*
Clothing and footwear	22.80	23.40	0.55	*2.4*
Housing, fuel and power	57.70	60.40	2.63	*4.6*
Household goods and services	32.00	31.40	-0.59	*-1.9*
Health	5.10	5.00	-0.04	*-0.8*
Transport	64.70	64.90	0.12	*0.2*
Communication	12.70	13.00	0.36	*2.9*
Recreation and culture	59.50	58.10	-1.38	*-2.3*
Education	8.90	10.00	1.12	*12.5*
Restaurants and hotels	38.30	39.20	0.91	*2.4*
Miscellaneous	36.10	35.90	-0.22	*-0.6*
Weekly household income:				
Disposable	570	578	9	*1.5*
Gross	689	700	11	*1.7*

The weighting increased the estimate of total average expenditure by £2.99 a week. It had the largest impact on average weekly expenditure on education, increasing the estimate by 12.5 per cent. It also increased the estimate of spending on clothing and footwear by 2.4 per cent, and the estimate for housing, fuel and power by 4.6 per cent. It reduced the estimate of spending on household goods and services by 1.9 per cent and the estimate for recreation and culture by 2.3 per cent. Weighting also increased the estimates of average income, by £9 a week (1.5 per cent) for disposable household income and by £11 a week (1.7 per cent) for gross household income, which is the income used in most tables in the report.

Re-weighting also has an effect on the variance of estimates. In an analysis on the 1999/2000 data, weighting increased variance slightly for some items and reduced it for others. Overall the effect was to reduce variance slightly.

Further information

Further information on the method used to produce the weights is available from the contacts given on page ii of this publication.

1 www.ccsr.ac.uk/esds/events/2008-12-02/bright.ppt
2 See Foster, K. (1994) *Weighting the FES to compensate for non-response, Part 1: An investigation into Census-based weighting schemes*, London: OPCS.
3 CHAID is an acronym that stands for Chi-squared Automatic Interaction Detection. As is suggested by its name, CHAID uses chi-squared statistics to identify optimal splits or groupings of independent variables in terms of predicting the outcome of a dependent variable, in this case response.
4 Implemented by the CALMAR software package before 2007 and GES for 2006–08 (updated weights).

Table B5 Characteristics of households, 2010
United Kingdom

	Percentage [1] of all house- holds	Weighted number of house- holds (000s)	House- holds in sample number		Percentage [1] of all house- holds	Weighted number of house- holds (000s)	House- holds in sample number
Total number of households	100	26,320	5,260	**Composition of household (cont)**			
				Four adults	2	490	80
Size of household							
One person	30	7,800	1,510	Four adults, one child	1	160	20
Two persons	35	9,210	1,960	Four adults, two or more children	[0]	70	10
Three persons	16	4,330	830				
Four persons	13	3,420	680	Five adults	0	120	20
Five persons	4	1,110	210				
Six persons	1	300	50	Five adults, one or more children	[0]	40	..
Seven persons	[0]	90	20				
Eight persons	[0]	20	..	All other households without children	[0]	30	..
Nine or more persons	[0]	40	..	All other households with children	[0]	50	..
Composition of household							
One adult	30	7,800	1,510	**Number of economically active persons in household**			
Retired households mainly							
dependent on state pensions[2]	3	790	150	No person	32	8,430	1,760
Other retired households	11	2,980	570	One person	28	7,420	1,520
Non-retired households	15	4,030	800	More than one person	40	10,470	1,980
One man	13	3,510	670	Two persons	32	8,320	1,630
Aged under 65	9	2,320	410	Three persons	6	1,660	280
Aged 65 and over	5	1,190	250	Four persons	1	390	60
One woman	16	4,290	850	Five persons	[0]	70	..
Aged under 60	5	1,360	310	Six or more persons	[0]	20	..
Aged 60 and over	11	2,940	540				
				Households with married women	46	12,020	2,540
One adult, one child	3	690	160	Households with married women			
One man, one child	[0]	40	..	economically active	26	6,790	1,380
One woman, one child	2	650	150	With no dependent children	14	3,580	730
One adult, two or more children	3	720	170	With dependent children	12	3,210	650
One man, two or more children	0	110	20	One child	6	1,450	280
One woman, two or more children	2	610	150	Two children	5	1,450	300
				Three children	1	260	60
One man, one woman	29	7,770	1,680	Four or more children	[0]	50	10
Retired households mainly							
dependent on state pensions[2]	2	440	110	Households with married women			
Other retired households	8	2,140	500	not economically active	20	5,240	1,160
Non-retired households	20	5,180	1,070	With no dependent children	15	3,980	910
Two men or two women	3	750	130	With dependent children	5	1,260	250
				One child	2	470	80
Two adults with children	20	5,160	1,060	Two children	2	520	110
One man one woman, one child	8	2,190	430	Three children	1	150	30
Two men or two women, one child	0	110	20	Four or more children	0	120	30
One man one woman, two children	8	2,130	450				
Two men or two women, two children	[0]	20	..				
One man one woman, three children	2	530	110	**Economic status of household reference person**			
Two men or two women, three children	[0]	20	..	Economically active	63	16,490	3,250
Two adults, four children	0	110	30	Employee at work	50	13,290	2,610
Two adults, five children	[0]	50	..	Full-time	42	10,960	2,130
Two adults, six or more children	[0]	10	..	Part-time	9	2,320	480
Three adults	6	1,560	270	Government-supported training	[0]	20	..
Three adults with children	3	910	160	Unemployed	4	940	180
Three adults, one child	2	600	100	Self-employed	9	2,250	460
Three adults, two children	1	250	40				
Three adults, three children	[0]	50	..	Economically inactive	37	9,830	2,020
Three adults, four or more children	[0]	20	..				

Note : Please see page xiii for symbols and conventions used in this report.

1 Based on weighted number of households
2 Mainly dependent on state pensions and not economically active - see definitions

Table B5   Characteristics of households, 2010 (cont.)

United Kingdom

	Percentage[1] of all house- holds	Weighted number of house- holds (000s)	House- holds in sample (number)		Percentage[1] of all house- holds	Weighted number of house- holds (000s)	House- holds in sample (number)
Age of household reference person				**GB urban/rural areas 2008-2010 (over 3 years)**			
15 and under 20 years	[0]	90	20				
20 and under 25 years	3	880	130	GB Urban	78	19,680	3,970
25 and under 30 years	7	1,840	300	GB rural	22	5,620	1,230
30 and under 35 years	8	2,140	410				
35 and under 40 years	9	2,270	450	**Tenure of dwelling[5]**			
40 and under 45 years	10	2,740	540	Owners			
				Owned outright	32	8,340	1,770
45 and under 50 years	9	2,400	490	Buying with a mortgage	35	9,100	1,830
50 and under 55 years	10	2,560	500	All	66	17,430	3,600
55 and under 60 years	8	2,040	460	Social rented from			
				Council	10	2,590	500
60 and under 65 years	9	2,420	530	Registered social landlord	8	2,150	430
65 and under 70 years	7	1,920	450	All	18	4,750	920
70 and under 75 years	6	1,500	360	Private rented			
				Rent free	1	370	70
75 and under 80 years	6	1,500	300	Rent paid, unfurnished	11	2,890	540
80 and under 85 years	4	1,050	190	Rent paid, furnished	3	880	130
85 and under 90 years	3	710	100	All	16	4,150	740
90 years or more	1	280	40				
				Households with durable goods			
Government Office Regions and Countries				Car/van	75	19,730	4,060
2008-2010 (3-year average)				One	45	11,780	2,440
United Kingdom	100	26,000	5,640	Two	24	6,390	1,330
				Three or more	6	1,560	290
North East	5	1,280	240				
North West	12	3,130	590	Central heating, full or partial	96	25,300	5,070
Yorkshire and the Humber	8	2,170	490	Fridge-freezer or deep freezer	97	25,510	5,110
				Washing machine	96	25,390	5,100
East Midlands	8	1,980	400	Tumble dryer	57	15,010	3,070
West Midlands	8	2,200	490	Dishwasher	40	10,570	2,170
East of England	9	2,270	520	Microwave oven	92	24,190	4,860
London	12	3,030	470	Telephone	87	22,890	4,630
South East	12	3,080	730	Mobile phone	80	21,100	4,250
South West	10	2,510	510	DVD player	88	23,230	4,700
				Satellite receiver[6]	88	23,110	4,680
England	83	21,640	4,430	Compact disc player	83	21,800	4,420
Wales	5	1,280	270	Home computer	77	20,370	4,090
Scotland	9	2,380	500	Internet connection	73	19,330	3,890
Northern Ireland	3	690	440				
Socio-economic classification							
of household reference person							
Higher managerial and professional	12	3,030	610				
Large employers/higher managerial	5	1,260	260				
Higher professional	7	1,770	350				
Lower managerial and professional	18	4,620	910				
Intermediate	5	1,350	270				
Small employers	6	1,580	330				
Lower supervisory	6	1,690	320				
Semi-routine	7	1,770	350				
Routine	6	1,520	290				
Long-term unemployed[3]	2	490	100				
Students	2	590	100				
Occupation not stated[4]	37	9,680	1,990				

Note : Please see page xiii for symbols and conventions used in this report.
1 Based on weighted number of households
2 Mainly dependent on state pensions and not economically active - see definitions

Table B6 Characteristics of persons, 2010

United Kingdom

	Males				Females				All persons		
	Percentage[1] of		Weighted number of persons (000s)	Persons in the sample (number)	Percentage[1] of		Weighted number of persons (000s)	Persons in the sample (number)	%[1] of	Weighted number of persons (000s)	Persons in the sample (number)
	all males	all persons			all females	all persons			all persons		
All persons	**100**	**49**	**30,240**	**5,860**	**100**	**51**	**31,130**	**6,320**	**100**	**61,370**	**12,180**
Adults	**78**	**38**	**23,460**	**4,460**	**79**	**40**	**24,710**	**4,970**	**78**	**48,170**	**9,430**
Persons aged under 60	57	28	17,140	3,030	56	28	17,380	3,440	56	34,510	6,470
Persons aged 60 or under 65	6	3	1,920	420	6	3	1,970	440	6	3,890	860
Persons aged 65 or under 70	5	2	1,460	350	5	2	1,440	340	5	2,900	690
Persons aged 70 or over	10	5	2,940	650	13	6	3,920	760	11	6,860	1,410
Children	**22**	**11**	**6,780**	**1,400**	**21**	**10**	**6,430**	**1,350**	**22**	**13,210**	**2,750**
Children under 2 years of age	3	1	850	170	2	1	760	160	3	1,610	330
Children aged 2 or under 5	4	2	1,090	240	4	2	1,130	240	4	2,230	480
Children aged 5 or under 16	13	7	3,990	850	12	6	3,740	810	13	7,730	1,660
Children aged 16 or under 18	3	1	850	140	3	1	800	140	3	1,650	280
Economic activity											
Persons active (aged 16 or over)	55	27	16,680	3,040	46	24	14,430	2,870	51	31,110	5,910
Persons not active	45	22	13,560	2,820	54	27	16,700	3,450	49	30,260	6,270
Men 65 or over and Women 60 or over	13	6	3,940	900	20	10	6,340	1,310	17	10,280	2,210
Others (Including children under 16)	32	16	9,610	1,920	33	17	10,360	2,140	33	19,980	4,060

Note : Please see page xiii for symbols and conventions used in this report

1   based on weighted number of households

Table B7   Index to tables in reports on the Family Expenditure Survey in 1999/00 to 2000/01 and the Living Costs and Food Survey 2001/02 to 2010

2010 tables		2009	2008	2007	2006	2005/ 06	2004/ 05	2003/ 04	2002/ 03	2001/ 02[1]	2000/ 01	1999/ 2000
Detailed expenditure and place of purchase												
A1	Detailed expenditure with full-method standard errors	A1	A1	A1	A1	A1	A1	A1	7.1	7.1	7.1	7.1
..	Expenditure on alcoholic drink by type of premises	A2	A2	A2	A2	A2	A2	A2	7.2	7.2	7.2	7.2
A2	Expenditure on food by place of purchase	A3	A3	A3	A3	A3	A3	A3	7.3	7.3	7.3	7.3
..	Expenditure on alcoholic drink by place of purchase	..	..	..	..	..	..	..	-	-	-	7.4
..	Expenditure on selected items by place of purchase	A4	A4	A4	A4	A4	A4	A4	7.4	7.4	7.4	-
..	Expenditure on petrol, diesel and other motor oils by place of purchase	..	..	..	..	..	..	..	-	-	-	7.5
..	Selected household goods and personal goods and services by place of purchase	..	..	..	..	..	..	..	-	-	-	7.6
..	Selected regular purchases by place of purchase	..	..	..	..	..	..	..	-	-	-	7.7
A3	Expenditure on clothing and footwear by place of purchase	A5	A5	A5	A5	A5	A5	A5	7.5	7.5	7.5	7.8
Expenditure by income												
A4	Main items by gross income decile	A6	A6	A6	A6	A6	A6	A6	1.1	1.1	1.1	1.1
A5	Percentage on main items by gross income decile	A7	A7	A7	A7	A7	A7	A7	1.2	1.2	1.2	1.2
A6	Detailed expenditure by gross income decile	A8	A8	A8	A8	A8	A8	A8	1.3	1.3	1.3	1.3
..	(Housing expenditure in each tenure group)	..	..	..	..	..	..	..	-	-	-	-
A7	Main items by disposable income decile	A9	A9	A9	A9	A9	A9	A9	1.4	1.4	1.4	1.4
A8	Percentage on main items by disposable income decile	A10	A10	A10	A10	A10	A10	A10	1.5	1.5	1.5	1.5
Expenditure by age and income												
A9	Main items by age of HRP	A11	A11	A11	A11	A11	A11	A11	2.1	2.1	2.9	-
..	Main items by age of head of household	..	..	..	..	..	..	..	-	-	2.1	2.1
A10	Main items as a percentage by age of HRP	A12	A12	A12	A12	A12	A12	A12	2.2	2.2	2.2	2.2
A11	Detailed expenditure by age of HRP	A13	A13	A13	A13	A13	A13	A13	2.3	2.3	2.3	2.3
A12	Aged under 30 by income	A14	A14	A14	A14	A14	A14	A14	2.4	2.4	2.4	2.4
A13	Aged 30 and under 50 by income	A15	A15	A15	A15	A15	A15	A15	2.5	2.5	2.5	2.5
A14	Aged 50 and under 65 by income	A16	A16	A16	A16	A16	A16	A16	2.6	2.6	2.6	2.6
A15	Aged 65 and under 75 by income	A17	A17	A17	A17	A17	A17	A17	2.7	2.7	2.7	2.7
A16	Aged 75 or over by income	A18	A18	A18	A18	A18	A18	A18	2.8	2.8	2.8	2.8
Expenditure by socio-economic characteristics												
A17	By economic activity status of HRP	A19	A19	A19	A19	A19	A19	A19	3.1	3.1	3.9	-
..	By economic activity status of HoH	..	..	..	..	..	..	..	-	-	3.1	3.1
..	By occupation	..	..	..	..	..	..	..	-	-	3.2	3.2
A18	HRP is a full-time employee by income	A20	A20	A20	A20	A20	A20	A20	3.2	3.2	3.3	3.3
A19	HRP is self-employed by income	A21	A21	A21	A21	A21	A21	A21	3.3	3.3	3.4	3.4
..	By social class c	..	..	..	..	..	..	..	-	-	3.5	3.5
A20	By number of persons working	A22	A22	A22	A22	A22	A22	A22	3.4	3.4	3.6	3.6
A21	By age HRP completed continuous full-time education	A23	A23	A23	A23	A23	A23	A23	3.5	3.5	3.7	3.7
..	By occupation of HRP	..	..	..	..	..	..	..	-	-	3.8	-
A22	By socio-economic class of HRP	A24	A24	A24	A24	A24	A24	A24	3.6	3.6	-	-
Expenditure by composition, income and tenure												
A23	Expenditure by household composition	A25	A25	A25	A25	A25	A25	A25	4.1	4.1	4.1	4.1
A24	One adult retired households mainly dependent on state pensions	A26	A26	A26	A26	A26	A26	A26	4.2	4.2	4.2	4.2
A25	One adult retired households not mainly dependent on state pensions	A27	A27	A27	A27	A27	A27	A27	4.3	4.3	4.3	4.3
A26	One adult non-retired	A28	A28	A28	A28	A28	A28	A28	4.4	4.4	4.4	4.4
A27	One adult with children	A29	A29	A29	A29	A29	A29	A29	4.5	4.5	4.5	4.5
A28	Two adults with children	A30	A30	A30	A30	A30	A30	A30	4.6	4.6	4.6	4.6

Notes

..   Tables do not appear in these publications

1   Household Reference Person (HRP) replaced Head Of Household (HOH) in 2001/02

Table B7 Index to tables in reports on the Family Expenditure Survey in 1999/00 to 2000/01 and the Living Costs and Food Survey 2001/02 to 2010 (cont.)

2010 tables	2009	2008	2007	2006	2005/ 06	2004/ 05	2003/ 04	2002/ 03	2001/ 02[1]	2000/ 01	1999/ 2000
Expenditure by composition, income and tenure (cont.)											
A29 Two adults non-retired	A31	A31	A31	A31	A31	A31	A31	4.7	4.7	4.7	4.7
A30 Two adults retired mainly dependent on state pensions	A32	A32	A32	A32	A32	A32	A32	4.8	4.8	4.8	4.8
A31 Two adults retired not mainly dependent on state pensions	A33	A33	A33	A33	A33	A33	A33	4.9	4.9	4.9	4.9
A32 Household expenditure by tenure	A34	A34	A34	A34	A34	A34	A34	4.10	4.10	4.10	4.10
.. Household expenditure by type of dwelling	..	..	..	..	..	..	..	-	-	-	-
Expenditure by region[2]											
A33 Main items of expenditure by region	A35	A35	A35	A35	A35	A35	A35	5.1	5.1	5.1	5.1
A34 Main items as a percentage of expenditure by region	A36	A36	A36	A36	A36	A36	A36	5.2	5.2	5.2	5.2
A35 Detailed expenditure by region	A37	A37	A37	A37	A37	A37	A37	5.3	5.3	5.3	5.3
.. (Housing expenditure in each tenure group)	..	..	..	..	..	..	..	-	-	-	-
.. Expenditure by type of administrative area	..	..	..	..	..	..	..	-	-	5.4	5.4
A36 Expenditure by urban/rural areas (GB only)	A38	A38	A38	A38	A38	A38	A38	5.4	5.4	5.5	-
Household income											
A37 Income by household composition	A40	A40	A40	A40	A40	A40	A40	8.1	8.1	8.1	8.1
A38 Income by age of HRP	A41	A41	A41	A41	A41	A41	A41	8.2	8.2	8.10	-
.. By age of head of household	..	..	..	..	..	..	..	-	-	8.2	8.2
A39 Income by income group	A42	A42	A42	A42	A42	A42	A42	8.3	8.3	8.3	8.3
A40 Income by household tenure	A43	A43	A43	A43	A43	A43	A43	8.4	8.4	8.4	8.4
.. Income by economic status of oH H	..	..	..	..	..	..	..	-	-	8.5	8.5
.. Income by occupational grouping of HoH	..	..	..	..	..	..	..	-	-	8.6	8.6
A41 Income by regions	A44	A44	A44	A44	A44	A44	A44	8.5	8.5	8.7	8.7
A42 Income by GB urban/rural areas	A45	A45	A45	A45	A45	A45	A45	8.6	8.6	8.8	-
A43 Income by socio-economic class	A46	A46	A46	A46	A46	A46	A46	8.7	-	-	-
A44 Income 1970 to 2010	A47	A47	A47	A47	A47	A47	A47	8.8	8.7	8.9	8.8
.. Income by economic activity status of HRP	..	..	..	..	..	..	..	-	-	8.11	-
.. Income by occupation of HRP	..	..	..	..	..	..	..	-	-	8.12	-
Households characteristics and ownership of durable goods											
A45 Percentage with durable goods 1970 to 2010	A50	A50	A50	A50	A50	A50	A50	9.3	9.3	9.3	9.3
A46 Percentage with durable goods by income group & household composition	A51	A51	A51	A51	A51	A51	A51	9.4	9.4	9.4	9.4
A47 Percentage with cars	A52	A52	A52	A52	A52	A52	A52	9.5	9.5	9.5	9.5
A48 Percentage with durable goods by UK Countries and regions	A53	A53	A53	A53	A53	A53	A53	9.6	9.6	9.6	9.6
A49 Percentage by size, composition, age, in each income group	A54	A54	A54	A54	A54	A54	A54	9.7	9.7	9.7	9.7
.. Percentage by occupation, economic activity, tenure in each income group	..	..	..	..	..	..	..	-	-	9.8	9.8
A50 Percentage by economic activity, tenure and socio-economic class in each income group	A55	A55	A55	A55	A55	A55	A55	9.8	9.8	-	-
Output Area Classification											
A51 Expenditure by OAC supergroup	5.4	A56	5.3	..	..	..	..	..	..	..	..
A52 Expenditure by OAC group	5.5	A57	5.4	..	..	..	..	..	..	..	..
A53 Average gross weekly household income by OAC supergroup	5.6	A58	5.4	..	..	..	..	..	..	..	..
Methodology											
B5 Household characteristics	A48	A48	A48	A48	A48	A48	A48	9.1	9.1	9.1	9.1
B6 Person characteristics	A49	A49	A49	A49	A49	A49	A49	9.2	9.2	9.2	9.2
Trends in household expenditure (moved to Chapter 4)											
4.1 FES main items 1995/96 - 2010	4.1	4.1	4.1	4.1	4.1	4.1	4.1	6.1	6.1	6.1	6.1
4.2 FES as a percentage of total expenditure 1995/96 - 2010	4.2	4.2	4.2	4.2	4.2	4.2	4.2	6.2	6.2	6.2	6.2
by region[3]	..	..	..	..	..	..	..	-	-	6.3	6.3
4.3 COICOP main items 2004/05 to 2010	4.3	4.3	4.3	4.3	4.3	..	..	..	..	..	..
4.4 COICOP as a percentage of total expenditure 2004/05 to 2010	4.4	4.4	4.4	4.4	4.4	..	..	..	..	..	..
4.5 Household expenditure 2004/05 to 2010 COICOP based current prices	4.5	4.5	..	..	..	..	..	..	..	..	..

Notes
.. Tables do not appear in these publications
1 Household Reference Person (HRP) replaced Head Of Household (HOH) in 2001/02